DISEASES OF FIELD CROPS: DIAGNOSIS AND MANAGEMENT

VOLUME 1

Cereals, Small Millets, and Fiber Crops

DISEASES OF FIELD CROPS: DIAGNOSIS AND MANAGEMENT

VOLUME 1

Cereals, Small Millets, and Fiber Crops

Edited by

J. N. Srivastava

Department of Plant Pathology
Bihar Agricultural University, Sabour–813210,
Bhagalpur, Bihar, India

A. K. Singh

Division of Plant Pathology
Sher-e-Kashmir University of Agricultural Sciences and Technology,
Jammu, (J & K), India

APPLE
ACADEMIC
PRESS

Apple Academic Press Inc.
4164 Lakeshore Road
Burlington ON L7L 1A4, Canada

Apple Academic Press Inc.
1265 Goldenrod Circle NE
Palm Bay, Florida 32905, USA

First issued in paperback 2021

Exclusive worldwide distribution by CRC Press, a member of Taylor & Francis Group

Diseases of Field Crops: Diagnosis and Management, Volume 1: Cereals, Small Millets, and Fiber Crops
ISBN-13: 978-1-77188-839-4 (hbk)
ISBN-13: 978-1-77463-961-0 (pbk)
ISBN-13: 978-0-42932-184-9 (eBook)

Diseases of Field Crops: Diagnosis and Management, Two Volumes set
ISBN-13: 978-1-77188-841-7 (Hardcover)
ISBN-13: 978-0-42932-177-1 (eBook)

Library and Archives Canada Cataloguing in Publication

Title: Diseases of field crops : diagnosis and management / edited by J. N. Srivastava (Department of Plant Pathology, Bihar Agricultural University, Sabour-813210, Bhagalpur, Bihar, India), A. K. Singh (Division of Plant Pathology, Sher-e-Kashmir University of Agricultural Sciences and Technology, Jammu, (J & K), India).

Other titles: Diseases of field crops (Burlington, Ont.)

Names: Srivastava, J. N. (Plant pathologist), editor. | Singh, A. K. (Plant pathologist), editor.

Description: Includes bibliographical references and indexes. | Content: Volume 1. Cereals, small millets, and fiber crops.

Identifiers: Canadiana (print) 2020020274X | Canadiana (ebook) 20200202855 | ISBN 9781771888417 (set ; hardcover) | ISBN 9781771888394 (v. 1 ; hardcover) | ISBN 9780429321771 (set ; eBook) | ISBN 9780429321849 (v. 1 ; electronic bk.)

Subjects: LCSH: Plant diseases—Molecular aspects. | LCSH: Phytopathogenic microorganisms—Control.

Classification: LCC SB732.65 .D57 2020 | DDC 632/.3—dc23

Library of Congress Cataloging-in-Publication Data

Names: Srivastava, J. N. (Plant pathologist), editor. | Singh, A. K. (Plant pathologist), editor.

Title: Diseases of field crops : diagnosis and management / edited by J. N. Srivastava, A. K. Singh.

Description: Palm Bay, Florida, USA : Apple Academic Press, 2020. | Includes bibliographical references and index. | Contents: v. 1. Cereals, small millets, and fiber crops -- v. 2. Pulses, oil seeds, narcotics, and sugar. | Summary: "Plant diseases cause yield loss in crop production, poor quality of produce, and great economic losses as well. Knowledge of the perpetuation and spread of the pathogens and various factors affecting disease development is an important need. Disease diagnosis is the prime requirement for determining preventive or curative measures for effective disease management. This new 2-volume set, Diseases of Field Crops: Diagnosis and Management, helps to fill the need for research on plant diseases, their effects, how they spread, and effective management measures to mitigate their harmful effects. The volumes in this set showcase recent advances in molecular plant pathology and discuss appropriate diagnostic techniques for identification of causal agents and diseases, providing the information necessary to establish management strategies. The chapters in these two volumes include detailed description of symptoms, causal organisms, disease cycles, epidemiology, and management techniques of economically important diseases. The volumes explore existing strategies and offer new methods that can be used in an integrated manner and with a comprehensive approach for the management of major diseases of the field crops. Also taken into consideration is the impact of global climate change on the spread and severity of plant diseases. The first volume, Volume 1: Cereals, Small Millets, and Fiber Crops, covers a selection of cereal crops or grains for fodder and human food and the diseases that affect them. The crops include rice, maize, wheat, millet, sorghum, jute, and more. Volume 2: Volume 2: Pulses, Oil Seeds, Narcotics, and Sugar Crops is comprised of eighteen chapters with each chapter focusing on only one crop, with detailed account of symptoms, causal organisms, disease cycles, epidemiology, and management of the diseases caused by fungi, bacteria and viruses. Some crops discussed include green gram, chickpeas and peas, lentils, soybeans, groundnuts, sunflowers, sugarcane, tobacco, and others. With chapters from authors actively engaged in teaching, research, and extension, the volumes are vividly enhanced with an abundance of photos and illustrations of typical symptoms, portrayed with graphs, tables, and line drawings. These volumes will provide a wealth of valuable information on plant diseases and effective management strategies. They will be a useful resource and guide for growers, students and faculty, industry professionals, and researchers and scientists engaged in agriculture and in resolving plant disease diagnosis, disease etiology, and management. Key features: Provides appropriate diagnostic techniques for identification of causal agents and diseases Explores existing strategies and offers new methods for the management of major diseases of the field crops Considers the impact of global climate change on the spread and severity of plant diseases"-- Provided by publisher.

Identifiers: LCCN 2020012233 (print) | LCCN 2020012234 (ebook) | ISBN 9781771888394 (v. 1 ; hardcover) | ISBN 9781771888400 (v. 2 ; hardcover) | ISBN 9780429321849 (v. 1 ; ebook) | ISBN 9780429321962 (v. 2 ; ebook)

Subjects: LCSH: Plant diseases--Molecular aspects. | Phytopathogenic microorganisms--Control.

Classification: LCC SB732.65 .D54 2020 (print) | LCC SB732.65 (ebook) | DDC 632--dc23

LC record available at https://lccn.loc.gov/2020012233

LC ebook record available at https://lccn.loc.gov/2020012234

Apple Academic Press also publishes its books in a variety of electronic formats. Some content that appears in print may not be available in electronic format. For information about Apple Academic Press products, visit our website at **www.appleacademicpress.com** and the CRC Press website at **www.crcpress.com**

About the Editors

J. N. Srivastava, PhD

*Associate Professor cum Sr. Scientist (Plant Pathology),
Bihar Agricultural University, Sabour, Bhagalpur,
Bihar, India*

J. N. Srivastava, PhD, is currently an Associate Professor
cum Senior Scientist (Plant Pathology) at Bihar Agricultural
University, Sabour, Bhagalpur, Bihar, India. Formerly, he
was Assistant Professor / Jr. Scientist (Plant pathology) at
Sher-e-Kashmir University of Agriculture Sciences and Technology, Jammu
and Kashmir, India. He has acted as Principal Investigator and Co-Principal
Investigator for several projects on biological control / integrated disease
management. In addition to teaching and research, he is also engaged in various
agriculture extension activities, such as farm advisory services, including video
conferencing, radio and TV talks, vocational trainings for the rural youth skill,
entrepreneurship development programs, farmer training, in-service training,
etc. Dr. Srivastava has published many research papers in both Indian and
international journals, book chapters, extension articles, practical and technical
bulletins, and leaflets. He was the recipient of many awards for achievement
and teaching, including an Excellence in Teaching Award (2017), Excellence in
Science Communication Award (2017), Distinguished Faculty Award (2017),
Eminent Scientist Award (2018), etc. He is a member of many academic and
scientific organizations and is associated with many international, national,
and provincial scientific, cultural, and academic / educational bodies, and
also serving as an editorial board member of the International Journal of Plant
Protection. He is also a reviewer for various scientific journals.

A. K. Singh, PhD

*Assistant Professor, Division of Plant Pathology,
Sher-e-Kashmir University of Agricultural Sciences and
Technology, Jammu (J & K), India*

A. K. Singh, PhD, is Assistant Professor in the Division of Plant Pathology at Sher-e-Kashmir University
of Agricultural Sciences and Technology, Jammu and

Kashmir, India. He has been engaged for more than 12 years in teaching (both undergraduate and postgraduate levels) and research and also involved in the transfer of technology through different extension activities. He has published more than 30 research papers in national and international journals of repute, one practical manual, and several book chapters and popular articles and has presented many research papers at international, national and regional symposiums and seminars. He is a recipient of several prestigious awards for his research contributions. Also, he served as a member of the editorial board of *Krishi Vikas Patrika*, published by the Directorate of Extension, SKUAST of Jammu. He is a member of several societies of plant pathologists in India.

Contents

Contributors

Kottramma C. Addangadi
Regional Research Station, S. D. Agricultural University,
Bhachau–370140, Gujarat, India

Durga Prasad Awasthi
Department of Plant Pathology, College of Agriculture, Tripura, India

Ashwani K. Basandrai
CSKHPKV, Rice, and Wheat Research Center, Malan District Kangra, Himachal Pradesh–176047, India

Daisy Basandrai
CSKHPKV, Rice, and Wheat Research Center, Malan District Kangra,
Himachal Pradesh–176047, India

V. Bhuvaneswari
Department of Plant Pathology, Andhra Pradesh Rice Research Station and Regional Agricultural Research Station, Maruteru–534 122, West Godavari District, Andhra Pradesh, India

R. N. Bunker
Department of Plant Pathology, Rajasthan College of Agriculture, Maharana Pratap University of Agriculture and Technology, Udaipur–313 001, Rajasthan, India

Niranjan Chinara
Department of Plant Pathology, College of Agriculture, Odisha University of Agriculture and Technology, Bhubaneswar–751 003, Odisha, India

Rajib Kumar De
ICAR - Central Research Institute for Jute and Allied Fibers, (ICAR), Nilganj, Barrackpore, Kolkata–700120, India

Devanshu Dev
Department of Plant Pathology, College of Agriculture, G.B. Pant University of Agriculture and Technology, Pantnagar, Uttarakhand–263145, India

Anil Gupta
Division of Plant Pathology, Sher-E-Kashmir University of Agricultural Sciences and Technology (SKUAST-J), Chatha, Jammu–180009, Jammu & Kashmir, India

Prem Lal Kashyap
ICAR-Indian Institute of Wheat and Barley, Karnal–132 001, Haryana, India

Bhupendra Singh Kharayat
Department of Plant Pathology, College of Agriculture, GBPUA&T, Pantnagar, Uttarakhand, India

Bijendra Kumar
College of Agriculture, Department of Plant Pathology, G.B. Pant University of Agriculture and Technology, Pantnagar–263145, Udham Singh Nagar, Uttarakhand, India

Deo Kumar
Department of Soil Science, Banda University of Agriculture, Science and Technology, Banda–210001, Uttar Pradesh, India

Sudheer Kumar
ICAR-Indian Institute of Wheat and Barley, Karnal–132 001, Haryana, India

Ranganathswamy Math
Department of Plant Pathology, College of Agriculture, Jabugam,
Anand Agricultural University–391155, Gujarat, India

Kailash Behari Mohapatra
Department of Plant Pathology, College of Agriculture, Odisha University of Agriculture and Technology, Bhubaneswar–751 003, Odisha, India

Sudha Nandni
Department of Plant Pathology, College of Agriculture, G.B. Pant University of Agriculture and Technology, Pantnagar, Uttarakhand–263145, India

J. Krishna Prasadji
Department of Plant Pathology, Andhra Pradesh Rice Research Station and Regional Agricultural Research Station, Maruteru–534 122, West Godavari District, Andhra Pradesh, India

S. Krishnam Raju
Department of Plant Pathology, Andhra Pradesh Rice Research Station and Regional Agricultural Research Station, Maruteru–534 122, West Godavari District, Andhra Pradesh, India

P. V. Satyanarayana
Director, Andhra Pradesh Rice Research Institute and Regional Agricultural Research Station, Maruteru–534 122, West Godavari District, Andhra Pradesh ANGRAU, Andhra Pradesh, India

Divya Sharma
Department of Plant Pathology, College of Agriculture, GBPUA&T, Pantnagar, Uttarakhand, India

A. K. Singh
Division of Plant Pathology, Sher-E-Kashmir University of Agricultural Sciences and Technology (SKUAST-J), Chatha, Jammu–180009, Jammu & Kashmir, India

D. P. Singh
ICAR-Indian Institute of Wheat and Barley, Karnal–132 001, Haryana, India

K. P. Singh
Department of Plant Pathology, College of Agriculture, G.B. Pant University of Agriculture and Technology, Pantnagar, Uttarakhand–263145, India

Kunal Pratap Singh
Department of Plant Pathology, Bihar Agricultural College, Sabour, Bihar Agricultural University, Sabour, Bhagalpur, Bihar, India

S. K. Singh
Division of Plant Pathology, Sher-E-Kashmir University of Agricultural Sciences and Technology (SKUAST-J), Chatha, Jammu–180009, Jammu & Kashmir, India

V. B. Singh
Rainfed Research Sub-Station for Sub-Tropical Fruits, Raya, Technology (SKUAST-J), Jammu–180009, Jammu & Kashmir, India

Virendra Kumar Singh
Department of Plant Pathology, Banda University of Agriculture, Science and Technology, Banda–210001, Uttar Pradesh, India

Vivek Singh
Department of Plant Pathology, Banda University of Agriculture, Science and Technology, Banda–210001, Uttar Pradesh, India

Yogendra Singh
Department of Plant Pathology, College of Agriculture, GBPUA&T, Pantnagar, Uttarakhand, India

J. N. Srivastava
Department of Plant Pathology, Bihar Agricultural University, Sabour–813210, Bhagalpur, Bihar, India

Abbreviations

BGRI	Borlaug Global Rust Initiative
BLB	bacterial leaf blight
BLSB	banded leaf and sheath blight
BSDM	brown stripe downy mildew
CoYVV	Corchorus yellow vein virus
CoYVYuV	Corchorus yellow vein Yucatan virus
CZ	Central Zone
DAE	days after emergence
dPCR	direct PCR
dsRNA	double-stranded RNA
EEI	expected economic impact
ESR	erwinia stalk rot
FAO	Food and Agriculture Organization
FAS	Foreign Agricultural Service
GLH	green leaf hoppers
IGP	Indo-Gangetic plains
LTR	long terminal repeats
MDMV	maize dwarf mosaic virus
MMV	maize mosaic virus
MStV	maize stripe virus
NEPZ	North Eastern Plains Zone
NEZ	Northern Hills Zone
NWPZ	North Western Plains Zone
PAGE	poly acrylamide gel electrophoresis
PAL	phenylalanine ammonia lyase
PAU	Punjab Agricultural University
PCR	polymerized chain reaction
PDI	percent disease index
PFSR	post flowering stalk rot
PGPR	plant growth promoting rhizobacteria
PPO	polyphenol oxidase
PZ	peninsular zone
QTL	quantitative trait loci
RAPD	random amplified polymorphic DNA

RFLYA	Ragi flour lactose yeast agar
RTBV	rice tungro bacilliform virus
RTSV	rice tungro spherical virus
SBP	stable bleaching powder
SCMV	sugarcane mosaic virus
SrMV	sorghum mosaic virus
TDP	thermal death point
VAM	vesicular arbuscular mycorrhyza

Preface

There has been a long-felt need by the students, researchers, and teachers for a comprehensive book on plant diseases. Plant diseases constitute important factors for causing yield loss in crop production. In order to sustain higher levels of agricultural productivity under the ever-changing agricultural scenario, the focus of attention has been the ways to manage plant diseases and reduce crop losses in order to avoid wide fluctuations in the production. The global change in climatic conditions has also contributed towards influencing infection, development, perpetuation, and severity of many plant diseases, directly affecting crop production. The multiplicity of crops, largely in a tropical and subtropical climate, overlapping growing seasons, a large number of varieties, production systems, and various cultural practices favor the occurrence of a number of plant diseases. A few of them are quite serious and cause yield reduction and poor quality of the produce, resulting in great economic losses. The subject of plant pathology has tremendously expanded pathogenesis, and the ways by which plants defend themselves are now relatively better understood. Advances in molecular plant pathology are providing appropriate diagnostic techniques for the identification of causal agents and diseases.

In view of the above facts and keeping in mind the implementation of recommendations of the Indian Council of Agricultural Research, New Delhi, the immediate need for crop-wise book pertaining to plant diseases is the need of the hour.

This book, *Diseases of Field Crops: Diagnosis and Management, Volume 1,* covers cereals, small millets, and fiber crops and is comprised of eighteen chapters. Each chapter has a detailed account that includes an introduction, symptoms, causal organisms, disease cycles, epidemiology, and management of economically important diseases. The book chapters have been contributed by authors who are engaged in teaching, research, and extension services and are well known national scientists in their respective fields. The authors, while writing the chapters, have incorporated their experience and knowledge with the recent development in the field of plant diseases.

It is hoped that the book will cater to the needs of students studying at undergraduate and postgraduate levels, researchers, teachers, planners, administrators, and growers, not only in the discipline of plant pathology but also in the other fields of agriculture.

We would highly appreciate receiving your comments, suggestions, and research contributions that relates to the different themes of the book. We will be happy to have your support in making valuable contributions to the scientific community.

We sincerely acknowledge our thanks and gratitude to the esteemed scientists who have spared their time and contributed valuable chapters for this book. We are also thankful to Apple Academic Press for publishing this book.

—J. N. Srivastava, Ph. D

A. K. Singh, Ph. D

Introduction

Among the biotic stress plant diseases are causing considerable yield loss in economically important crops. The disease may attack at juvenile stage to crop maturity or harvesting of the crops. They may affect different parts of the plants, such as foliage, stem, root, flowers and seed thatinduce various types of symptoms. However, vascular system infecting pathogens causing wilt and affect the entire plant. Many pathogens survive on the stored grains which may be onward transmitted or causes spoilage. Several pathogens causing complex symptom e.g. root rot, wilting leaf spot fruit rot which is difficult to distinguish each other.

The crop losses can be reduced by suitable control measuresbytargeting specific pathogens if theyaccurately diagnosed.Need-based application of fungicides will improve environment and economic gains. In plant disease, visual observation of infected plant continues to be the dominant methods. Several sophisticated tools are being used for disease diagnosis, which include microscopy, isolation, immunological, biochemical assay and genome analysis.

Adaptation of single practice like fungicide application will leads development of resistance in fungal pathogens. Integrated approach will helpful in order to sustainable management of crop disease. Conventional cultural practices are effective to control various soil borne plant pathogens e.g.*Pythium*spp, *Fusarium*spp, *Rhizoctonia*spp, *Sclerotium*sppetc. Soil solarization and mulching provides the effective efforts to manage the nematode and root infecting pathogens. Antagonistic microbial bioagents like *Trichoderma*, *Pseudomonas fluorescence* found effective to control the soilborne plant pathogen. Application of different group fungicides (systemic and non-systemic) provide effective control of diseases in crops.Integration of crop varieties with different genetic makeup is always pronouncing the safest way to manage the disease with high yield output.

The compiled book *"Diseases of field crops: diagnosis and management"* volume 1 deal on several aspects in different crops like cereals, small millets, and fiber crops.The chapters of the book focused on economic importance, symptom, causal organism, disease cycle, epidemiology and suitable disease management options.

The book containing different disease of rice including common and minor disease having greater significance.Wheat and barley are commonly grown in norther part of India which are affected by different disease e.g. rust, bunts, smut, powdery mildew spot blotch covered under the chapter. Maize is commonly grown throughout the year in Indian which affected by several diseases and causing considerable yield loss.The crop is majorly hampered by *Helminthosporium* leaf blight, banded leaf and sheath blight, downey mildew, common rust, stalk rot, charcoal rot, fusarium rot, late wilt, smut, maize chlorotic dwarf and maize dwarf disease.Sorghum is commonly grown for food as well as fodder crop. The crop is severely challenged by various pathogens inducing anthracnose, leaf blight, zonate leaf spot, downey mildew, Cercospora leaf spot, target leaf spot, shooty stripe, rough leaf spot, rust, ergot and smut.

Downey mildew, ergot, smut, rust, blast targeting the bajra crop in rainfed and drought ecology which discussed in the specified chapter.The emphasis also given on small millets, common millets diseasese.g. smut, leaf spot, sheath blight, rust, downy mildew, blast, udabatta and bacterial stripe. Additionally, association of nematode with millets also described in the text. Cotton and jute are commercially grown which provides the raw material to textile industries. In the book, several diseases of cotton and jute fiber are also described with various aspects of management tactics. Common fungal disease like root rot, seedling blight, vascular wilt, anthracnose, leaf spot, Myrothecium leaf blight, rust, grey mildew described in the chapter. Moreover, both crops are also affected by whitefly transmitted geminivirus like leaf curl and Jute mosaic virus, okra mosaic virusalso explained in respective chapter.

The compiled chaptersare substantiated with novelfigures, which are of excellent quality labelled in the book.Each chapter also provides suitable and recent references. Information provided on different aspect of disease will helpfulforacademician and researchers. Moreover, the book will also provide sufficient information to undergraduate and postgraduate studies in colleges and universities.

—Editors

CHAPTER 1

Present Scenario of Diseases in Rice (*Oryza sativa* L.) and Their Management

S. KRISHNAM RAJU,[1] V. BHUVANESWARI,[2] J. KRISHNA PRASADJI,[1] and P. V. SATYANARAYANA[3]

[1]*Department of Plant Pathology, Andhra Pradesh Rice Research Station and Regional Agricultural Research Station, Maruteru–534122, West Godavari District, Andhra Pradesh, India, E-mail: rajupathol@yahoo.co.in*

[2]*Department of Plant Pathology, Andhra Pradesh Rice Research Station and Regional Agricultural Research Station, Maruteru–534122, West Godavari District, Andhra Pradesh, India*

[3]*Andhra Pradesh Rice Research Institute and Regional Agricultural Research Station, Maruteru–534122, West Godavari District, Andhra Pradesh ANGRAU, Andhra Pradesh, India*

1.1 INTRODUCTION

Rice (*Oryza sativa* L.) is a consequential staple victuals crop for a more sizably voluminous part of the world's population and is engendered around the globe. More than 3 billion people, half of humanity, eat rice as their staple food. Globally rice is cultivated on approximately 158.8 M ha on 11% of the world's cultivated land and 685 million tonnes produced annually (FAO, 2009). More than 90% of the rice production of the world is grown in Asia. World production of rice has been steadily rose from 200 MT of raw rice in 1960 to over 472.09 MT in 2015. China was the leading rice producer followed by India, Indonesia, and Bangladesh in 2015–16. India is the 2nd largest producer of rice in the world with an area of 43.46 M ha. India's rice production was at around 103.5 MT during 2015–16 and it was projected to 104.82 MT during August 2016 (USDA, Foreign Agricultural Service (FAS) (m.world-grain.com)). It is grown in almost all the states of India contributing about 42% of the country's foodgrain production and provides the livelihood for about 70% of the population. India was the largest exporter of rice in 2015–16 followed by Thailand, Vietnam, and Pakistan. The world needs 8–10 MT more rice each year to meet people's needs and keep rice affordable. The population of rice eaters is increasing at an exceptionally fast rate and the number of rice consumers will probably double by 2020. The country needs to increase its rice production at a rate of 3.75 MT per year until 2050 to meet its food security. The requirement of rice by 2025 is estimated to be around 125–127 MT.

The demand for food and processed commodities is increasing due to growing population and rising per capita income. There are projections that demand for food grains would increase from 192 million tonnes in 2000 to 345 million tonnes in 2030. Hence, in the next 20 years production of food grains needs to be increased at the rate of 5.5 million tonnes annually.

Although, India is having the maximum area under rice cultivation but several biotic and abiotic factors are mainly responsible for low production and productivity of rice. Due to apparent changes in climatic conditions, change of genotypes and cultivation practices, the profile of diseases in rice has changed over a period of time. Several pathogenic diseases have been found to occur on the rice crop resulting in extensive damage to grain and straw yield. Among diseases blast, sheath blight, bacterial leaf blight (BLB), sheath rot, brown spot, stem rot, false smut, and rice tungro cause severe yield losses in rice.

1.2 BLAST DISEASE

1.2.1 ECONOMIC IMPORTANCE

It is also known as rice blast disease, rice rotten neck, rice seedling blight, blast of rice, oval leaf spot of graminea, pitting disease, ryegrass blast, Johnson spot. Blast is considered a major disease of rice because of its wide distribution and destructiveness under favorable conditions. Blast occurs over 85 rice growing countries of the world causing considerable reduction in yield. The disease was first recorded as 'rice fever disease' in China by Soong Ying-shin in 1637. In Japan, it is believed to have occurred in *Imochi-byo* as early as in 1704. In Italy, the disease called 'brusone' was reported in 1828 and in U.S.A in 1876. The causal organism was first detected by Cavara in 1891 from Italy.

In temperate and subtropical Asia, blast is highly destructive in lowland rice and in tropical Asia, Latin America, and Africa, it affects upland rice.

In India, the occurrence of this disease was first reported from the Tanjore district of Tamil Nadu in 1918, Maharashtra in 1923 and subsequently in several other parts. Since then heavy destruction of rice crop due to this disease over large areas have been reported from Andhra Pradesh, Kerala, Maharashtra, Gujarat, Orissa, West Bengal and other states of the country. Now it occurs in almost all principal rice-growing states of the country. In severe cases, losses amounting to 70–80% of grain yield are reported (Ou and Nuque, 1985).

In India blast epidemics were reported from the Sub-Himalayan regions of Jammu and Kashmir, Andhra Pradesh, Tamil Nadu and Coorg regions of Karnataka and North Eastern region comprising the states of Arunachal Pradesh, Manipur, Mizoram, Meghalaya, Assam, and uplands of Bihar and Orissa.

Although blast is capable of causing very severe losses of up to 100%, little information exists on the extent and intensity of actual losses in farmers' fields. Losses of 5 to 10%, 8%, and 14% were recorded in India (1960–1961), Korea (mid-1970s), and in China (1980–1981), respectively. In the Philippines, yield losses ranging from 50 to 85% were reported.

1.2.2 SYMPTOMS

The fungus attacks all aerial parts of plants at all stages of growth and they can symptoms produce on plant parts such as leaf (leaf blast), leaf collar (collar blast), culm, culm nodes, panicle neck node (neck rot), and panicle (panicle blast).

1.2.2.1 LEAF BLAST

Symptoms on the leaves originate as small specks which subsequently enlarge into spindle-shaped spots varying in length from 0.5 cm to several cms. The center of a well-developed spot is whitish-grey with a brown margin. Under favorable conditions, lesions on the leaves expand rapidly and tend to coalesce, leading to complete necrosis of infected leaves giving a burnt appearance from a distance. So the name rice blast is given to this disease. Similar spots are also formed on the sheath.

1.2.2.2 COLLAR BLAST

Collar blast occurs when the pathogen infects the collar that can ultimately kill the entire leaf blade.

1.2.2.3   NODE BLAST

The pathogen infects the culm node that turns grayish brown to blackish and breaks easily at the point of infection. This condition is referred as node blast.

1.2.2.4   NECK BLAST

The neck of the panicle is infected when the earhead emerges. This phase of infection commonly called phase is called black neck or neck blast or rotten neck or neck rot or panicle blast. The neck region turns grayish brown to blackish and shriveled. Grain set in earheads is completely or partially inhibited. When the grains are set, the panicle breaks at the neck due to weakening of the neck tissues. Such panicle hangs down and can be distinguished from a distance. Out of the symptoms, neck blast is more destructive and directly reduces the economic value of the produce. If neck blast occurs before the milky stage, the entire panicle may die prematurely; leaving it white and completely unfilled grains. The pathogen also causes brown lesions on the branches of the panicles and on the spikelet pedicels, resulting in panicle blast. Infection of the neck, panicle branches, and spikelet pedicels may occur together or may occur separately. Sometimes glumes also become infected and develop brown to black spots (Figures 1.1–1.6).

FIGURE 1.1   Field symptoms of leaf blast.

FIGURE 1.2   Symptoms of leaf blast in nursery.

FIGURE 1.3   Nursery giving burnt appearance due to severe blast incidence.

FIGURE 1.4   Symptoms of node blast.

FIGURE 1.5   Field symptoms of neck blast.

FIGURE 1.6   Rotting of neck tissue due to blast incidence.

1.2.3   CAUSAL ORGANISM

Magnaporthe oryzae (Herbert) Barr
Anamorph: *Pyricularia oryzae* (Cooke) Sacc.
Kingdom: Fungi
Phylum: Ascomycota
Order: Magnaporthales
Family: Magnaporthaceae

The fungus that causes rice blast is called *Magnaporthe oryzae* (formerly *Magnaporthe grisea*). It is an ascomycete because it produces sexual spores (ascospores) in structures called asci, and is classified in the newly constructed family Magnaporthaceae. The asexual stage of *Magnaporthe oryzae* is described by the name *Pyricularia oryzae* (formerly called *P. grisea*). The fungus produces septate, branched hyaline to slightly colored mycelia. The latter give rise to conidophores emerging either through stomata or by rupturing the epidermis and the cuticle. Conidiophores are produced singly or in groups, are simple, rarely branched, septate, slender, and denticulate, grayish color and show sympodial growth. Conidia are formed singly at the tip of the conidiophores in succession. They are ovate to obclavate, usually 3-celled (2-septate) with a small basal appendage, hyaline to pale olive in color. Conidia often form appressoria at the tip of germ tube. Mature conidia are usually three-celled or 2-septate, pyriform (pear-shaped), hyaline or colorless to pale olive, exhibit a basal appendage at the point of attachment to the conidiophore. Conidiophore usually germinates from the apical or basal cells.

1.2.4   DISEASE CYCLE

Mycelium and conidia in the infected straw and seeds are important sources of primary inoculums in tropical and temperate regions both. The mycelium and conidia of the fungus survives in the infected straw for one or two years under dry condition, but it is easily destroyed by moisture and microbial activities when buried in soil. The fungus hibernates in the seeds during storage and may help to some extent in the perpetuation of the fungus during off-season. Seed borne inoculum fails to initiate the disease in the plains due to high soil temperature in June. In the tropics, air-borne conidia are present all round the year because susceptible hosts are always present.

Magnaporthe oryzae also infects several grasses, including *Echinocloa colonum, Digitaria marginata, Brachiaria mutica, Echinochloa crus-galli, Setaria intermedia, Digitaria sanguinalis, Leersia hexandra, Panicum repens,* and *Dinebra retroflexa* which occur on field bunds in many parts of India; some of them likely to play a part in the epidemiology of this disease by acting as primary sources of inoculum. The conidia produced on the primary lesions become wind-borne and cause secondary spread of the disease.

Conidia lodged on leaves or neck of the panicle requires free water (dew, fog) for germination before penetrating into the host tissue. If such conditions are prevailing, it takes only 4–5 days for establishment of a visible symptom after infection and the secondary cycle of the disease is completed very quickly, such as several secondary infection cycles are completed by the pathogen in the same season of disease resulting in severe damage to the crop.

1.2.5   PHYSIOLOGICAL RACES

M. oryzae consists of many physiological races, which differ in their ability to infect paddy varieties. Thirteen races have been reported from Japan (Goto, 1963), Thirty from India (Chakrabarthi et al., 1966) ten from Korea (Lee and Matsumoto, 1966), Nineteen from Taiwan (Chiu et al., 1963).

Recently, 81 physiological races have been identified in the Philippines by the pathologists at IRRI, Philippines (Singh, 2013).

Magnaporthe oryzae exhibits an extreme degree of variability in pathogenicity and several physiologic races are recognized, based on the infection types developing on different hosts from artificial inoculation. The international differentials for distinguishing physiologic races of *P. oryzae* are rice varieties, Raminad Str-3, Zenith, NP-125, Usen, Dular, Kanto 51, Sha-tiao-tsao (S) and Caloro. At least 54 races of *P. oryzae* have been so far identified in India.

1.2.6 EPIDEMIOLOGY

- The environmental factors such as temperature, humidity, and moisture profoundly influence the initiation and development of the disease. A temperature of around 20°C particularly during the nights, alternate with a day temperature around 30°C with daylight for about 14 hours and darkness for about 10 hours predispose the plants to infection. The fungus invades the rice leaves most rapidly at 24–28°C.
- A high relative humidity (92%) and free water are required for conidial germination and infection.
- Sunshine inhibits the development of blast fungus and spores do not germinate in direct sunlight. Low solar radiation and overcast sky also favors spread of disease.
- Cloudy overcast weather is reported to favor spread of the fungus. Dewdrops on the leaves stimulate spore germination and infection. The fungal spores are deposited on the leaves more during night than during day time and the amount deposited depends on the angle of the leaf with the stem of the plant. Plants with leaves in vertical position trap fewer spores than those with slanting and horizontal leaves. Severe blast epidemics are usually associated with moist weather. Frequent periods of rain showers or drizzle are more favorable to infection than heavy rainfall of short duration.
- Susceptibility to blast is inversely related to soil moisture. Plants grown under lowland condition (i.e., flooded soil or with high soil moisture) become more resistant, while plants grown under upland conditions become more susceptible. The pathogen requires free moisture for spore penetration.
- High nitrogen supply favors heavy blast infection regardless of phosphorus or potassium supply. Rice plants that receive high levels of nitrogen are found to have fewer numbers of silicated epidermal cells.

1.2.7 MANAGEMENT

- Using seed obtained from disease-free crop.
- Using resistant or tolerant varieties in blast epidemic areas.
- Varieties possessing resistance or tolerance to blast: Ajaya, IR 36, IR 64, Jagannath, Jaya, Vijetha, Cottondora Sannalu, Prabhat, Nellore Mahsuri, Krishna Hamsa, Pusa 205, Pusa Basmati 1, Rasi, Ratna, Suraksha, Apurva, Chaitanya, Gautami, MTU 9993, Phalguna, Penna, Pinakini, Raja Vadlu, Sagar Samba, Simhapuri, Sriranga, Swarnamukhi, Tikkana, Vajram, Vasistha, Vikramarya, Vandana, Hemavathi, Madhuri, Kanchana, Swetha, Karjat 2, Ratnagiri-1, Sugandha, Indira, Indravati, Moti, Pratap, Rudra, ADT 36, ADT 39, CO 45, TKM 10, IR 24, and Pant Dhan 4 varieties were found to have resistance / tolerance genes to blast pathogen.
- Destruction of stubbles, weeds, and collateral hosts in and around rice fields and bunds.
- Applying sufficient quantity of farmyard manure and green manuring during land preparation.
- Adopting early sowing of *kharif* rice in blast endemic areas.
- Treating seeds with carbendazim @ 1.0 g or mixed fungicide of carbendazim 25% + mancozeb 50% – 75 WP @ 2.0 g/kg seed per liter of water as wet seed treatment.
- Balanced application of fertilizers.
- Need-based application of fungicides like tricyclazole 75 WP @ 0.6 g or isoprothiolane 40 EC @ 1.5 ml or kasugamycin 3 L @ 2.5 ml/l.

Recent work on fungicidal evaluation revealed that propiconazole + tricyclazole 52.5 SE (Filia) @ 2.5 ml/l, metiram 70 WG (Sanit) @ 1.0 g/l, propineb 70 WP (Antracol) @ 3.0 g/l, metaminostrobin 20 SC @ 0.5 ml/l were found superior in reducing the disease and recorded higher grain yield as that of tricyclazole @ 0.6 g/l.

1.2.8 HOST PLANT RESISTANCE

In 1924, Co-4 (Anaikomban), a local race of rice was selected as resistant to blast. Later TKM 1 was recognized as resistant. Further S67 and BJ1 were identified as resistant to blast disease. Tadukan, Rasi, Tetep, and IR-64 were also potential sources for blast resistance breeding. Rice has the genome size of 390 Mb and the blast fungus has the genome size of 40 Mb and both genomes were sequenced and publicly available. More than 90 genes and 347 quantitative trait loci (QTLs) have been detected (Yohei et al., 2009). Many reports mentioned that genes affecting blast resistance are co-localized on chromosomes 6, 11 and 12. On chromosome 6, at least 14 genes and / or alleles (*Pi2, Piz, Piz-t, Piz-5, Pi 8(t), Pi9, Pi13, Pi 13(t), Pi25(t), Pi26(t), Pi27(t), Pid2, Pigm(t)*, and *Pi40(t)* have been mapped in the region near the centromere. On the long arm of chromosome 11, at least nine genes (*Pi1, Pi7, Pi18, Pif, Pi34, Pi38, Pi44 (t), PBR*, and *Pilm2*) and six alleles at the Pik locus (*Pik, Pik-s, Pik-p, pik-m, pik-h,* and *pik-g*) have been mapped. On chromosome 12, at least 17 resistance genes and / or alleles *Pita, Pita-2, Pitq6, Pi6(t), Pil2(t), Pi19(t), Pi20(t), Pi21(t), Pi24(t), Pi31(t), Pi32(t), Pi39(t), Pi62(t), Pi157(t), IPi*, and *IPi3* have been mapped in the region near the centromere.

Koide et al. (2010) developed pyramided lines with two major blast resistance genes *Pish* and *Pib* in the genetic background of blast susceptible variety, CO-39. Madhavi et al. (2012) stacked the blast resistance gene, *Pi54* into the genetic background of the elite fine-grain type, high yielding, bacterial blight resistant rice variety, improved Samba Mahsuri through MAS. Hari et al. (2013) introgressed the blast-resistant gene *Pi54* and a bacterial blight resistance gene *Xa21* into the genetic background of the elite maintainer line IR58025B with the help of markers. Attempts are in progress to introgress *Pi-1, Pi-2* and *Pi-54* which are the promising R genes in Indian agro-climatic conditions from the donor source C101LAC, C101A51 and Tetep which are highly resistant to the rice blast fungus *Magnaporthe oryzae* with bacterial blight resistance in the rice cultivar Improved Samba Mahsuri (Ratna Madhavi et al., 2013). Akanksha Srivastava et al. (2013) introgressed multiple genes for BB (*Xa21, Xa33*) and blast (*Pi54, Pi2*) resistance into the genetic background of Sampada. Aruna Kumari et al. (2013) reported marker-assisted gene pyramiding of bacterial blight and blast resistance genes *xa13, Xa21* and *Pikh, Pi-1* in the elite Indica rice variety MTU-1010 (Cottondora Sannalu).

1.3 BROWN SPOT OR SESAME LEAF SPOT OR HELMINTHOSPORIOSE

1.3.1 ECONOMIC IMPORTANCE

It is also called 'nai-yake' i.e., seedling blight, sesame leaf spot and Helminthosporiosis. The disease known in Japan since 1900 and it has been to occur in all the rice-growing countries including Japan, China, Burma, Sri Lanka, Bangladesh, Iran, Africa, South America, Russia, North America, Philippines, Saudi Arabia, Australia, Malaya, and Thailand (Ou, 1985; Khalili et al., 2012).

In India, it is known to occur in all the rice-growing states (Gangopadhyay, 1983; Ou, 1985) since its first report from Madras in 1919 by Sundraraman. The disease is more severe in dry / direct-seeded rice in the states of Bihar, Chhattisgarh, Madhya Pradesh, Orissa, Assam, Jharkhand, and West Bengal. In India, this disease is the principal cause of Bengal famine of 1942–43. Under highly favorable conditions, the disease causes a reduction in yield ranging up to 90%. The disease is of great importance in several countries and has been reported to cause enormous losses in grain yield (up to 90%) particularly when leaf spotting phase assumes epiphytotic proportions as observed in Great Bengal Famine during 1942 (Ghose et al., 1960).

1.3.2 SYMPTOMS

The pathogen attacks the crop from seedling to milk stage. The symptoms appear as minute spots on the coleoptile, leaf blade, leaf sheath, and glume, being most prominent on leaf blades and glumes.

On leaves, typical spots are brown in color with grey or whitish center, cylindrical or oval in shape resembling sesame seeds usually with yellow halo while young spots are small, circular, and may appear as dark brown or purplish brown dots. Several spots coalesce and the leaf dries up. The affected nursery can often be recognized from a distance by scorched appearance due to death of the seedlings.

The pathogen has also been reported to cause brown to dark brown lesions on panicle stalk at the joint of flag leaf to stalk. These lesions usually extend downward beneath the sheath resulting in severe wet rotting, partially filled to chaffy dull grains and occasional hanging down of panicles. In severe cases, grayish mycelial growth is seen between sheath and stalk (Sunder et al., 2005).

On glumes, black or dark brown spots are produced resulting in discolored and shriveled grains. Under favorable conditions, dark brown conidiophores and conidia develop on the spots giving a velvety appearance. The fungus may penetrate the glumes and leave blackish spots on the endosperm. Severe infection of grains has been reported to prevent germination (Ranganathaiah, 1985) and to cause seed rotting and pre-emergence damping off (Kulkarni et al., 1980b; Hiremath and Hegde, 1981). Young roots may also show blackish lesions. Nodes and internodes are rarely infected.

1.3.3 CASUAL ORGANISM

Helminthosporium oryzae
Syn: *Drechslera oryzae and bipolaris oryzae*
Sexual stage: *Cochliobolus miyabeanus*

1.3.4 DISEASE CYCLE

The fungus is reported to survive in soil and infected plant parts including stubbles, straw, and grains for 2–3 years, which act as the primary source of inoculum (Ou, 1985). The pathogen has been reported to survive in seed also (Wesely et al., 1996). The primary infection is usually initiated by the infected seed (Bernaux, 1981). Diseased seeds (externally seed borne) may give rise to the seedling blight, the first phase of the disease. The environmental factors influence survival of fungus in seed and soil. Some weed hosts have also been reported as inoculum reservoirs (Biswas et al., 2008) such as collateral hosts like *Digitaria sanguinalis, Leersia hexandra, Echinochloa colonum, Pennisetum typhoides, Setaria italica,* and *Cynodon dactylon.* Brown spot is known to occur in nutrient deficient soils (Ou, 1985).

1.3.5 EPIDEMIOLOGY

Temperature and relative humidity significantly affect the disease progress. The temperature of 25–300°C with relative humidity above 80% are highly favorable. Excess of nitrogen aggravates the disease incidence.

1.3.6 MANAGEMENT

- Use of seeds from a disease free crop.
- Grow resistant varieties like Bala, BAM 10, IR-20, Jaya, Ratna, Tellahamsa, and Kakatiya.
- Give three to four applications of fungicides like Dihane Z-78 or Dihane M-45 @ 2.5 g/L water at 10 days interval just before the appearance of initial symptoms of disease.

1.4 SHEATH BLIGHT

1.4.1 INTRODUCTION

Sheath blight is an economically significant disease of rice in all growing areas of the world. Sheath blight was first reported from Japan in 1910 by Miyake, but was first noted by Yano in 1901. Reinking (1918) and Paolo (1926) found a very similar disease in the Philippines. Park and Bertus (1932) reported the disease in Sri Lanka and Wei (1934) in China and are now known to occur in most rice growing countries. In the Philippines, it has been estimated that the damage caused by this disease may affect 25–50% of rice production. Yield losses of up to 50% are reported when susceptible varieties are grown. A modest estimation of losses due to sheath blight disease alone in India has been up to 54.3% (Rajan, 1987; Roy, 1993). Sheath blight disease of rice occurs in all rice production areas world-wide (Ou, 1985; Teng et al., 1990; Savary et al., 2000, 2006). The disease is particularly important in intensive rice production systems (Savary and Mew, 1996). Yield losses of 5–10% have been estimated for tropical lowland rice in Asia (Savary et al., 2000). The disease also occurs in Kerala and parts of Tamil Nadu and Andhra Pradesh. As most of the high yielding semi-dwarf varieties are susceptible to this disease, it has become one of the important diseases of rice during *kharif* season. Rice sheath blight is an increasing concern for rice production especially in intensified production systems. There is a 20–25% yield reduction if the disease develops up to flag leaf. Sheath blight has assumed economic importance in the last two decades when modern, semi-dwarf nitrogen responsive cultivars have been commercially grown.

1.4.2 SYMPTOMS

The pathogen causes spots or lesions mostly on the leaf sheath, extending to the leaf blades under favorable conditions. The initial lesions are small, ellipsoid or ovoid, and greenish-gray and usually develop near the water line in lowland fields. Under favorable conditions, they enlarge and may coalesce forming bigger lesions with irregular outline and grayish-white center with dark brown borders. The presence of several large spots on a leaf sheath usually causes the death of the whole leaf. In the advanced stages, brown sclerotia are formed, which are easily detached from these spots. Under humid conditions, the fungal mycelium spreads to other leaf sheaths and blades. Eventually, the whole sheath rots and the affected leaf can easily be pulled off from the plant. In severe cases, all the leaves of a plant are blighted, resulting in death of the plant. Plants are usually attacked at the tillering stage, when the leaf sheaths become discolored at or above water level (Figures 1.7–1.12).

FIGURE 1.7   Symptoms of sheath blight disease at early stages of crop growth.

FIGURE 1.8   Greyish brown lesions developed at the base of the tillers.

FIGURE 1.9   Greyish brown lesions developed on leaf blade.

FIGURE 1.10   Symptoms of sheath blight extending up to panicle.

FIGURE 1.11   Sclerotial bodies on the infected dead tissue of rice plant.

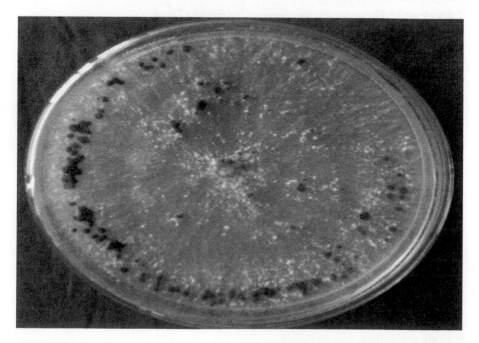

FIGURE 1.12   *Rhizoctonia solani* fungus grown on PDA medium.

Instead of spores, the rice sheath blight fungus produces sclerotia, usually measuring 1 to 3 mm in diameter which is relatively spherical. Sclerotia are formed on or near the spots and can be easily detached from the plant. Under natural conditions, sclerotia usually occur singly but may sometimes coalesce to form larger masses. They are whitish when young and turn brown or dark brown when older. Sclerotia on the soil or floating on the water are assumed to be the sole source of inoculum in most temperate countries. In a subsequent cropping season, they float to the surface of the paddy water during soil puddling, leveling, and other farm operations. Sclerotial densities are much lower in the tropics than in temperate and subtropical areas. Mycelium in colonized crop residues is relatively persistent and acts as a source of primary inoculum. It may be a more important source of primary inoculum than sclerotia in the tropics because climatic conditions are more favorable for its survival. The pathogen is soil-borne but it can also infect seeds.

1.4.3   CAUSAL ORGANISM

Rhizoctonia solani Kuhn.
Teleomorph: *Thanatephorus cucumeris* (Frank) Donk.

Thanatephorus cucumeris (Frank) Donk is the perfect stage and AG-1 as anastomosis group. The fungus grows over a wide range of temperature, the optimum being 28–30°C. The size and number of sclerotia on culture media depend on the nitrogen source and the concentration. Isolates differ in their virulence and produce a toxin called phenylacetic acid.

1.4.4   DISEASE CYCLE

The sclerotia are the main source of infection and they survive in the soil for several months, depending on temperature and moisture conditions. The sclerotia float to the surface of the water during soil puddling, leveling, weeding, and other operations and infect the plants with which they come into contact. The mycelium grows inside the tissues in all directions, initiating secondary spots, in turn producing sclerotia on the spots. The mycelium enters into the host plant through the stomata or it penetrates directly through the cuticle. The mycelium is most active and infectious when the lesions are young.

1.4.5 EPIDEMIOLOGY

- The disease is highly destructive in highly humid and warm temperatures. Critical factors for rice sheath blight infection are relative humidity and temperature. The pathogen thrives when the canopy humidity is 96-97%. High temperature (28–32°C) was reported to favor infection. Frequent rainfall favors disease development. Therefore, the disease is more common during the rainy than in the dry season in the tropics.
- High rates of nitrogenous fertilizer also make the tissue more susceptible to the disease while high rates of potassium induce resistance to the disease. High rate of nitrogen fertilizer promote luxuriant crop growth with dense canopies. This type of canopy structure has high relative humidity which provides a favorable microclimate for rice sheath blight. Nitrogen supply to the crop indirectly affects disease spread by increasing tissue contacts and leaf wetness in the canopy.
- High seeding rates and close plant spacing favor the spread of disease because, aside from creating a favorable microclimate, it allows more tissue contacts and longer durations of leaf wetness.
- Leaf wetness plays a major role in the development of rice sheath blight and is a critical factor affecting infection, i.e., infection efficiency was higher in interrupted leaf wetness regimes than in permanently dry or wet regimes.
- Primary infection is closely related to the number of sclerotia that come into contact with the plant and subsequent disease development is greatly influenced by environmental conditions and susceptibility of plants.
- Increased cropping frequency is assumed to provide a continuous availability of host tissues for the pathogen and may favor inoculum survival across cropping seasons.

1.4.6 MANAGEMENT

- Selection of healthy seeds.
- Treatment seed with carbendazim @ 1.0 g/kg seed per liter water as wet seed treatment or carbendazim 3 grams per 1 kg of seed for dry seed treatment.

- Adopting optimum spacing as per the season and as per the method of planting.
- Timely weed control in the field and on bunds.
- Avoiding excess nitrogen application.
- Flooding and draining the field once at 10 days before puddling is useful to let out the sclerotia overwintering on crop residue as well as in the soil.
- In heavily infected soils spraying with hexaconazole 5 EC @ 2.0 ml or validamycin 3 L @ 2.0 ml or propiconazole 25 EC @ 1.0 ml or trifloxystrobin 25%+ tebuconazole 50% WG @ 0.4 g or hexaconazole 75 WG @ 0.13 g/l or azoxysytrobin 25 SC @ 1.0 ml/liter of water helps to control the disease.

Studies on evaluation of fungicides against sheath blight disease indicated that thifluzamide 24 SC (Spencer) @ 0.75 ml/l, captan + hexaconazole 75 WP (Taqat) @ 1.5 g/l, hexaconazole 75 WG (Epic) @ 0.13 g/l, azoxystrobin 18.2% + difenoconazole 11.4% SC (Amistar top) @ 1.0 ml/l, Picoxystrobin 25% SC @ 1.2 ml/l, picoxystrobin + propiconazole 20% SC @ 2.0 ml/l and Azoxystrobin 11% + tebuconazole 18.3% SC @ 1.0 ml/l were found most effective in reducing disease incidence and severity.

1.4.7 HOST PLANT RESISTANCE

Almost all the popular rice cultivars which are under cultivation are susceptible to sheath blight disease and it is very difficult to find genotypes which are completely resistant to *R. solani* which is more of a saprophytic in nature. However, a few lines *viz.*, CR 3608-11-1-1-1 (IET No 23145), CR 3598-1-4-2-1-1 (IET No 23161) and CR 2683-45-1-2-1-1 (IET No. 23175) were found moderately resistant.

1.5 SHEATH ROT

1.5.1 INTRODUCTION

Sheath rot was first described in Taiwan in 1922. It is reported in all countries in South Asia. Earlier this disease was considered as minor. But now it has assumed importance with its sporadic occurrence in some places of Andhra Pradesh. In India, a yield reduction of 9.6% to 26% with an average of 14.5%

was reported in 1978. Crop losses ranging from 3 to 20–60% were reported in Taiwan in 1980. It has become prominent in Andhra Pradesh, Kerala, Orissa, Tamil Nadu, Bihar, and North Eastern states causing losses to yield.

1.5.2 SYMPTOMS

The disease appears during heading to maturity stages. Small water soaked lesions occur on the upper most leaf sheaths (boot leaf heath). Lesions gradually develop into grey color irregular shaped spots which are surrounded by brown colored margins. As the disease progresses, the lesions enlarge and coalesce and may cover entire leaf sheath. Lesions may also consist of diffuse reddish brown discolorations in the sheath. In severe cases, the infection reaches the panicle and glumes. An abundant whitish powdery growth may be found inside the affected sheaths, although the leaf sheath may look normal from the outside. With early or severe infection, the panicle may fail to emerge completely or not at all; the young panicles remain within the sheath or only partially emerge. Panicles that have not emerged tend to rot and florets turn red-brown to dark brown. Most grains are sterile, partially or unfilled and discolored (Figures 1.13–1.16).

FIGURE 1.13   Initial symptoms of sheath rot disease.

FIGURE 1.14   Lesions spreading entire leaf sheath.

FIGURE 1.15   Panicle rotting due to sheath rot infection.

FIGURE 1.16   Discolored panicle with chaffy grains due to Sheath rot incidence.

1.5.3   CAUSAL ORGANISM

Sarocladium oryzae
Division: Deuteromycetes
Order: Moniliales
Family: Moniliaceae

1.5.4   DISEASE CYCLE

The primary source of inoculum is by means of infected plant debris. The secondary spread is by means of airborne conidia produced on the leaf sheath. Conidia are colorless, single-celled and cylindrical in shape. The fungus produces sparsely branched, septate white mycelium. Conidia are cylindrical to slightly fusiform, often somewhat curved, hyaline, smooth, and single-celled. The fungus is a weak pathogen and most often associated with the presence of panicle mite, stem borers, and other forms of injury such as insect damage of the flag leaf sheath. This suggests that the fungus attacks the leaf sheaths enclosing the young panicles more easily when there is injury. The fungus invades rice through the plants, stomata, and wounds and grows intercellularly in the vascular bundles and mesophyll tissue. The sheath rot fungus survives as a mycelium in infected residue and on seeds.

1.5.5   EPIDEMIOLOGY

- Night temperatures of 20°C, dew (mist), cold weather and high humidity.
- More use of Nitrogen fertilizers.
- Incidence of panicle mite, stem borer favors the disease development.

1.5.6   MANAGEMENT

- Destruction of the infected plant debris by burning.
- Seed treatment with carbendazim at 1.0 g/kg seed.
- Use of balanced fertilizers and application of nitrogen in three split doses.
- Avoiding excess seed rate as close spacing favors the disease spread.
- Adopting spacing of 20 × 20 cm in endemic areas.
- Spraying with carbendazim 50 WP @ 1.0 g or propiconazole 25 EC @ 1.0 ml/l once at the appearance of disease and second spray at 10 days after first spray controls the disease.
- The new fungicide, trifloxystrobin + tebuconazole 75 WG @ 0.4 g/l was found effective against sheath rot of rice.

1.6   FALSE SMUT

1.6.1   INTRODUCTION

False smut disease, also known as orange or green smut, has been in the United States for many years but rice false smut was first reported in the United States in 1997 in Arkansas. False smut caused widespread concern in 1998, especially in northeast Arkansas, where environmental conditions favored disease development. False smut disease was reported for the first time from Tamil Nadu, India False smut is caused by the fungus *Ustilaginoidea virens* and survives in the soil or contaminated rice grain as spore balls. Similar to kernel smut, false smut spores replace the developing rice kernel. Under highly favorable conditions, the disease causes a reduction in yield ranging up to 50%.

1.6.2 SYMPTOMS

Spore balls in the soil are believed to germinate late in the growing season and release spores into the air. When the spores land on and infect rice flowers, primary infection occurs. The fungus transforms infected individual grains into greenish spore balls or yellow spore balls of velvety appearance which are small at first and 1 cm or longer at later stages. When mature, the spore balls are 1/4 to 1/2 inch in diameter and are orange. At early stages, the spore balls are covered by a membrane which bursts with further growth. Due to the development of the fructification of the pathogen, the ovaries are transformed into large velvety green masses. Infection usually occurs during the reproductive and ripening stages, infecting a few grains in the panicle and leaving the rest healthy.

Yellow or greenish spore balls

1.6.3 CAUSAL ORGANISM

Ustilaginoidea virens
(P.S: *Claviceps oryzae – sativa*)
Kingdom: Fungi
Division: Ascomycota
Class: Sordariomyycetes
Order: Hypocreales
Family: Incertaesedis

Genus: *Ustilaginoidea*
Species: *virens*

Chlamydospores are formed on the spore balls; they are spherical to elliptical, waxy, and olivaceous.

1.6.4 EPIDEMIOLOGY

- The disease can occur in areas with high relative humidity (>90%) and temperature ranging from 25–35°C.
- Rain, high humidity, and soils with high nitrogen content also favors disease development. Wind can spread the fungal spores from plant to plant.
- False smut is visible only after panicle exsertion. It can infect the plant during flowering stage.

1.6.5 DISEASE CYCLE

In temperate regions, the fungus survives the winter through sclerotia as well as through chlamydospores. Ascospores produced on the over wintered *sclerotia* apparently start primary infection. Chlamydospores are important in secondary infection which is a major part of the disease cycle. Infection usually occurs at the booting stage of rice plants. Chlamydospores are borne, but do not free them from spore ball easily because of the presence of sticky material.

1.6.6 MANAGEMENT

1.6.6.1 CULTURAL METHODS

- Use cleaned, fungicide-treated seed free of false smut spore balls to minimize the introduction of false smut to previously unaffected fields.
- Among the cultural control, destruction of straw and stubble from infected plants is recommended to reduce the disease.
- Use varieties that are found to be resistant or tolerant against the disease in India.
- Avoid field activities when the plants are wet.
- Early planted crop has less smut balls than the late planted crop.

- At the time of harvesting, diseased plants should be removed and destroyed so that sclerotia do not fall in the field. This will reduce primary inoculum for the next crop.
- Field bunds and irrigation channels should be kept clean to eliminate alternate hosts.
- Excess application of nitrogenous fertilizer should be avoided.
- Regular monitoring of disease incidence during Rabi season is very essential.
- Proper Destruction of straw and stubble.

1.6.6.2 CHEMICAL METHODS

- Spray copper oxychloride@0.3% or carbendazim@0.1% at panicle emergence stage.
- Spray Propiconazole at 1.0 ml/liter at boot leaf and milky stages will be more useful to prevent the fungal infection.
- Seed treatment with carbendazim 2.0 g/kg of seeds.
- Treat seeds at 52°C for 10 min.
- At tillering and preflowering stages, spray Hexaconazole @ 1 ml/lit or Chlorothalonil 2 g/lit.
- In areas where the disease may cause yield loss, applying captan, captafol, fentin hydroxide, and mancozeb can be inhibited conidial germination.
- At tillering and preflowering stages, spraying of carbendazim fungicide and copper base fungicide can effectively control the disease.

1.7 FOOT ROT OR BAKANAE DISEASE

1.7.1 INTRODUCTION

The disease is also known as Foolish-seedling disease. It was first described from Japan by Hori in 1898 and later in detail by Kurosawa in 1926. The disease was reported for the first time in India by Thomas in 1931 from the Godavari delta of old Madras State, now in Andhra Pradesh. It is also found in the Philippines, China, British Guiana, Uganda, Italy, and other countries. The disease is a minor one in India. It occurs in both in upland and low land cultivated rice. It is more predominant in Haryana, Punjab, and Uttar Pradesh. It causes up to 20% yield loss.

1.7.2   SYMPTOMS

The disease occurs in the nursery and in the main field. The infected seedlings are lean and lanky, much taller than healthy seedlings, abnormally elongated and die after some time. In the main field, the peculiar symptom is the appearance of tall, lanky tillers which are abnormally elongated and develop into flower earlier than the healthy plants, show the symptoms of fungus infection at the collar region and die within 2–6 weeks. There is profuse branching of the roots. Sometimes the fungus causes stunting of plants. Adventitious roots are also produced from the first two or three nodes above the ground level. A pinkish bloom / white powdery growth may be present on the base of the plant. The plants are killed before earhead formation or they produce only sterile spikelets. When the culm is, split open white mycelial growth can be seen.

Early infection can cause seedlings to die at early tillering stage. Later infection results in plants that develop few tillers and have dry leaves. If the plants survive to maturity stage, they develop partially filled grain, sterile or empty grains.

1.7.3   CAUSAL ORGANISM

Fusarium fujikuroi
Teleomorph: *Gibberella fujikuroi* (Saw.) Wr.

The perfect stage is an ascomycete, *Gibberella fujikuroi*. The mycelium is yellow to rosy white and present inter and intra-cellularly in the host tissue, but concentrated in the xylem vessels. Fungus produces both macro and microconidia. Micro conidia are hyaline, single-celled and oval in shape. Macro conidia are slightly sickle-shaped narrow at both ends and two to five celled. The fungus is systemic and produces growth stimulating substances *viz.*, gibberellic acid in the plant. Excessive elongation of the diseased plants is due to gibberellic acid. The fungus is externally seed-borne (Figures 1.17 and 1.18).

1.7.4   MANAGEMENT

- Use certified seeds free from disease. The varieties, Co.18, C0.22, ADT 8 are found resistant to this disease.
- Treat the seeds with carbendazim @ 2 g/kg of the seeds.
- Steeping the seeds in 1% copper sulfate solution.

- Avoid draining fields early, as the aerobic fungus may reproduce rapidly in the presence of oxygen.
- Avoid application of nitrogen as it favors the development of the disease.

FIGURE 1.17   Field symptoms of Bakanae disease.

FIGURE 1.18   Adventitious roots produced from nodes above the ground level.

1.8 RED STRIPE

1.8.1 INTRODUCTION

Red stripe is an emerging disease that has been observed in the recent years in intensive rice production areas of the world especially in tropical countries such as Vietnam, Thailand, and the Philippines. The disease was first reported in Indonesia in 1988 (Mogi et al., 1988). Since then it has been reported in Vietnam (Du et al., 1991), the Philippines (Barroga and Mew, 1994), Malaysia (Yazid et al., 1996), Thailand (Dhitikiattipong et al., 1999) and in Cambodia (Du et al., 2001). The disease is also described as bacterial red stripe, bacterial leaf stripe, and yellow leaf syndrome and rice leaf yellowing (Mogi et al., 1988).

In India, the disease was first observed during the year 2000 in a sporadic manner on the rice varieties viz., PLA-1100 and MTU-1001 in East and West Godavari districts of Andhra Pradesh (Rajamannar et al., 2007). Subsequently, it appeared during the years 2001 and 2002 at moderate levels in these districts. The symptoms of the disease were observed throughout the infected field. The infected panicles were abnormal and had more unfilled grains at the time of harvest. Number of grains per panicle decreased from 159.69 to 153.99, grain weight per 30 panicles from 109.27 to 101.88 g, healthy grains from 67.5 to 61.41%. Chaffy grains increased from 28.91 to 33.48% and discolored grains from 3.59 to 5.12% due to red stripe disease of four rice varieties (Krishnam et al., 2012).

1.8.2 SYMPTOMS

Red stripe disease occurs as primary lesions on any leaf at different positions of the plant under high humidity and that it can easily infect rice seedlings (Bien et al., 1992). The symptoms of the disease were more severe on seedlings and immature plants and sometimes they lead to seed discoloration. The leaf streaks and stripes extend into the sheaths and occasionally a red stalk rot develop (Saddler, 1998). The red stripe lesions are peculiar sometimes they resemble foliar diseases caused by fungal pathogens. Initial lesions are pin point sized spots often light yellowish green to light orange at the base of the leaf blades, ultimately blighting is common on the leaves. The lesions also extend on the sheath. The symptoms occur from flowering to ripening stage of the crop (Mew et al., 2001). Typical symptoms of this disease are usually manifested from flowering to ripening stage of the rice. The characteristic symptoms are yellow orange spots which gradually turn to streaks advancing towards the leaf tips. The symptoms are more common on the leaves while, the lesions also extend on to leaf sheaths and culms (Elazegui et al., 2004) (Figures 1.19–1.22).

FIGURE 1.19   Initial symptoms of red stripe disease.

FIGURE 1.20   Orange colored leaves due to red stripe infection.

FIGURE 1.21   Field symptoms of red stripe disease.

FIGURE 1.22   Lesion extending upwards from the point of infection.

1.8.3   ETIOLOGY

Red stripe etiology is not yet confirmed and still control measures are to be developed to manage the disease. However, some reports are available. Du et al. (1991) isolated the causal organism and later identified it as *Curvularia lunata* (*Cochliobolus lunatus*) and confirmed its pathogenicity. Several fungal species isolated from leaf tissues with stripe lesions on nutrient media included *Curvularia lunata, Nigrospora oryzae, Cercospora* spp., *Alternaria* spp., *Helminthosporium* spp., *Colletotrichum* spp., and *Fusarium* sp. However, pathogenicity could not be proved with any of these fungal isolates (Du et al., 1991; Vinh, 1997; Wakimoto et al., 1998; Vinh et al., 1999; Saad, 2001). Elazequi and Castilla (2004) included several natural and synthetic media to isolate the causal agent. They isolated a fungus, *Gonatophragmium* consistently from the leaves with red stripe symptoms using a blotter method and subsequently Koch's postulates were proved.

1.8.4 MANAGEMENT

- Spraying with carbendazim @ 1.0 g/l and benomyl @ 1.0 g/l immediately after disease appearance in the field is useful to prevent spread of disease.

Among the different, new fungicides tested for their efficacy against red stripe disease, trifloxystrobin, + tebuconazole 75 WG @ 0.4 g/l, Propiconazole 25 EC @ 1.0 ml/l, and azoxystrobin 18.2% + difenoconazole 11.4% SC @ 1.0 ml/l could effectively control the disease.

1.9 BUNT DISEASE

1.9.1 INTRODUCTION

Bunt disease of rice is also known as 'Kernel smut' or 'black smut' of rice and is worldwide in distribution. It has been first reported and described from Japan. In India, the disease occurs in Bengal, Assam, Uttar Pradesh and parts of Kerala.

1.9.2 SYMPTOMS

Only a few grains in the panicle are infected either partly or wholly, because of localized infection, the disease not of systemic in nature. The symptoms appear first as minute black streaks bursting through the glumes at the time of ripening. If the infected grain is crushed between the fingers, a black powdery mass of spores comes out. The spores, however, shed from the infected grains in the field and settle on the leaves forming a characteristic black covering.

1.9.3   CAUSAL ORGANISM

Neovossia horrida
Kingdom: Fungi
Phylum: Basidiomycota
Class: Ustomycetes
Order: Ustilaginales
Family: Tilletiaceae

The teleutospores are spherical, black with a spiny epispore (Exospore).

1.9.4   EPIDEMIOLOGY

- Temperature of 25–30°C and relative humidity of 85% or intermittent light showers at the time of panicle-emergence are favorable for infection.
- High doses of nitrogen fertilizers favor disease.

1.9.5   MANAGEMENT

- Field sanitation, crop rotation, and use of resistant varieties.

1.10   LEAF SCALD

1.10.1   INTRODUCTION

Leaf scald commonly occurs in Central and South America, resulting in significant yield losses. It also occurs in Asia, Africa, and the USA. Instances are there in Latin America and West Africa, where caused considerable losses. The disease is found in upland, rainfed, and irrigated conditions. Disease development usually occurs late in the season on mature leaves and is favored by wet weather, high nitrogen fertilizers and close spacing. It develops faster in wounded leaves than in unwounded leaves.

1.10.2 SYMPTOMS

Typical symptoms appear on the upper part of the leaves. Lesions either start from the tip or along the margin and are oblong with characteristic alternate zonations of the dark-brown and light yellow areas, leaf tips, and margins are translucent. Individual lesions are 1–5 cm long and 0.5–1 cm wide or may cover the entire leaf. Enlargement and coalescing of lesions result in blighting of a large part of the leaf blade. The affected areas dry out giving the leaf scalded appearance. In some countries, lesions rarely develop the zonate pattern and only the scalding symptom is prominent.

1.10.3 CAUSAL ORGANISM

Microdochium oryzae

The pathogen produces conidia which are hyaline, two-celled, and unequal with a short lateral beak on the apical cell.

1.10.4 DISEASE CYCLE

The sources of infection are seeds and crop stubbles. Wet weather and high doses of nitrogenous fertilizer favor the disease.

1.10.5 MANAGEMENT

- Use of resistant varieties.
- Avoid high doses of fertilizers. Application of Nitrogen in split doses is useful.
- Remove weeds and infected rice ratoons.
- Spray 0.1% carbendazim or benomyl or 0.1% or thiophanate-methyl after the appearance of disease in the field.

1.11 UDBATTA DISEASE

1.11.1 INTRODUCTION

Udbatta disease is a minor disease of rice crop. The other vernacular names are Agar batti, Mathapukaddi roga, and Kari kaddi roga. Udbatta disease caused by *Blansia oryzae sativa* was first described by Sydow (1914) from India.

1.11.2 SYMPTOMS

The panicle emerges from the leaf sheath as a straight, dirty colored, hard, cylindrical spike, reduced in size much resembling an 'agarbatti' or Udbatta hence the name comes for this disease. No grain is formed on the affected ear and causes 100% sterility of the panicle. The fungus perennates through the sclerotium in the soil.

1.11.3 CAUSAL ORGANISM

Ephelis oryzae Syd *(Balansia oryzae)*.

The fungus forms a stroma over the entire length and girth of the inflorescence and black, convex pycnidia are formed in the stroma. Pycnidiospores are hyaline, 4–5 celled, and needle shaped.

1.11.4 MANAGEMENT

- Hot water seed treatment at 45°C for 10 min or seed treatment with 0.1% Carbendazim.
- Removal of collateral hosts like *Cynadon dactylon.*
- Spraying of carbendazim @ 1.0 g/l at the panicle initiation stage.

1.12 GRAIN DISCOLORATION

1.12.1 INTRODUCTION

Grain discoloration occurs at the time of panicle initiation stage. At this time normal grain color changes to brownish white color.

1.12.2 SYMPTOMS

The disease appears on the grains during maturity stage when there is a continuous rain or dew for more number of days. The spike let's turn to brown to dark brown color and later seed setting will not be there.

1.12.3 CAUSAL ORGANISM

Drechslera oryzae, Fusarium sp., Nigrospora oryzae, Penicillium sp. Curvularia sp., Sarocladium oryzae and *Rhizophus sp.*

1.12.4 EPIDEMIOLOGY

- Grain discoloration disease break out by high temperature at night and high rain condition.
- The high humidity coincides with panicle emergence stage of rice crop leads to heavy yield loss.
- Heavy nitrogenous fertilizers increase the discoloration of grains.

1.12.5 MANAGEMENT

- Dry seed treatment with carbendazim 3 g/l or Sprint 75 WP (carbendazin 25% + mancozeb 50%) @ 2.0 g/L water or vitavax 2.0 g/L water for 6 hours.
- Spraying of 0.1% carbendazim at the time of panicle emergence.

1.13 BACTERIAL LEAF BLIGHT (BLB)

1.13.1 INTRODUCTION

Bacterial leaf blight (BLB) of rice caused by *Xanthomonas oryzae* pv. *oryzae* a major production constraint in rice cultivation in many countries. In recent years, this disease has been reported to appear regularly at an alarming intensity in many areas, which were considered earlier as non-endemic to this disease.

It is one of the oldest known and most destructive diseases of rice in majority of the rice growing countries especially in Asia *viz.*, Japan, China,

Philippines, Korea, Mexico, Malaysia, and Indonesia. The disease accounts for annual yield loss of 20–30% in Asia and Africa. The disease and causal organism were described in detail from Japan by Udeya and Ishiyama in 1922, who named the organism *Pseudomonas oryzae*. Subsequently, the disease was reported from most of the rice growing countries. Dowson (1949) named it *Xanthomonas oryzae*. Dye (1982) renamed it as *Xanthomonas campestris* pv. *oryzae*. Now it is called *X. oryzae* pv. *oryzae*. In India, the occurrence was first reported by Srinivasan et al. in 1959 from Maharashtra. First authentic report of typical bacterial blight symptoms was made by Bhapkar and co-workers in 1960. The disease broke out in epidemic form in Shahabad district of Bihar during 1963 and since then it has spread fast to other rice growing regions of the country often causing considerable yield loss especially in high yielding varieties during rainy season.

In India, bacterial blight is considered as a serious production constraint especially in irrigated and rainfed lowland ecosystem. In Punjab and Haryana states of India, major epidemics occurred in 1979, 1980; severe kresek was observed and total crop failure was reported (Production Oriented Survey, 1979, 1980; Mew, 1987). The disease was again reported in epidemic form during 1998 in Pallakad district of Kerala and since then it has become endemic in that region (Priyadarishini and Gnanamanickam, 1999). This is a serious problem in the terai belt of Uttar Pradesh and delta regions of Andhra Pradesh wherever highly susceptible varieties are grown.

The disease is a major problem in *kharif* season (wet season) crop in rice growing regions of North Western Parts of India, Eastern India, Parts of North Eastern India, entire eastern coast, parts of Kerala and Karnataka and Konkan region of Maharashtra. During *kharif* 2010 and 2016, due to heavy rains over a period of 4 months from June to September the bacterial blight was appeared in moderate to severe form in rice growing areas of East and West Godavari districts. Severe incidence of disease was observed on Swarna variety in certain pockets of East Godavari district where the crop was subjected to inundation for five days.

1.13.2   ECONOMIC IMPORTANCE

The disease is known to occur in epidemic proportions in many parts of the world, incurring severe crop loss of up to 50%. Crop loss assessment studies have revealed that this disease reduces grain yield to varying levels, depending upon the stage of the crop, degree of cultivar susceptibility and to a great extent, the conduciveness of the environment. Damage may be due to partial or total blighting of the leaves (leaf blight phase) or due to complete wilting of the affected tillers (Kresek phase) leading to unfilled grains. This disease is a problem of rainy season (wet season). Generally, the stage between maximum tillering and booting is highly sensitive to disease infection, as it affects the yield significantly in terms of filled grain weight per hill and total yield. Depending on the stage of infection and severity of the disease under natural condition, the extent of loss has been reported to vary from 6–60% (Srivastava et al., 1966). Rao and Kauffman (1971) have reported yield loss up to 50% depending on the variety, severity, and stage of infection.

1.13.3 SYMPTOMS

The disease appears in two phases viz., wilt or 'Kresek' phase and leaf blight phase.

1.13.3.1 KRESEK PHASE

Wilting syndrome known as 'Kresek' occurs sporadically in the fields causing serious damage. It commonly occurs within 3–4 weeks after transplantation of the crop. The leaves roll completely, turn yellow or grey and finally the tillers wither away. Kresek results either in the death of whole plant or wilting of only a few leaves.

1.13.3.2 LEAF BLIGHT PHASE

Leaf blight phase is the most predominant form of the disease occurring between tillering and heading stages of the crop. The symptom starts as dull greenish water-soaked or yellowish lesions on the tip of the leaves and increases in length downwards. Initially, the lesions are pale green in color and later turn into yellow to straw colored stripes with wavy margins. The lesions adjoining the healthy part show water soaking. Lesions may start at one or both edges of the leaves. Occasionally, the linear stripes may develop anywhere on the leaf lamina or along the midrib with or without marginal stripes. As the disease advances, the lesion covers the entire leaf blade, turns white, and later becomes grayish (Figures 1.23–1.26).

FIGURE 1.23   Field symptoms of bacterial leaf blight of rice caused by *X. oryzae* pv. *oryzae.*

FIGURE 1.24   Yellowish, opaque, and turbid drops of bacterial ooze on infected leaves.

FIGURE 1.25   Water becoming yellowish and turbid due to release of bacteria.

FIGURE 1.26 Microscopic view of cloudy mass of bacteria oozing out from cut ends.

In humid areas, on the surface of the young lesions, pale amber in color or yellowish, opaque, and turbid drops of bacterial ooze may be observed during early morning hours. They dry up to form small, yellowish, spherical beads on the lesions and these will fall in water streams and spread to other fields. In severe infections, all the leaves are attacked and premature drying results. When the affected leaves are cut and immersed in clear water in a test tube, a turbid ooze of the bacterium streaming from the vascular bundles can be observed.

1.13.3.3 YELLOWING OF LEAVES

A third type of less conspicuous symptom caused by the bacterium is yellowing of leaves. Such leaves show blighted appearance.

1.13.4 TESTS FOR RAPID DIAGNOSIS OF THE DISEASE

These tests are based on bacterial exudation from the cut ends of vascular system of diseased leaves.

1. **Dipping Method:** When fresh bacterial blight infected leaves are cut into small pieces with a scissor across the yellow lesion and placed in a

test tube filled with water, whitish turbid substance can be seen coming down from the cut ends of the leaf bits into water within few minutes. This substance is the causal bacterium of the disease. After 30–40 minutes, the entire water in the test tube becomes yellowish turbid.

2. **Microscopic Observation:** A small section of fresh lesion is placed in between glass slide and the cover slip with a few drops of water. The slide is then observed under microscope in low power. A large amount of cloudy mass of bacteria (ooze) streaming out from the cut ends of the leaves is observed.

1.13.5   ISOLATION OF THE PATHOGEN

The bacterium can be readily isolated from infected leaf samples on culture medium. Generally, the infected leaf samples are checked for bacterial ooze under microscope. The positive samples are then surface sterilized with 0.1% mercuric chloride or 95% ethanol for 30 seconds followed by 2–3 times rinsing with sterile distilled water. The infected leaf is then cut into small sections and put in a vial containing 2 ml sterile distilled water. After 5–10 minutes, when the bacterial ooze comes out from the cut ends of infected leaf bits into water, a loopful of water containing the bacteria can be streaked on to a suitable medium. The bacterium can be isolated on a number of culture media *viz.*, Nutrient Agar, Wakimoto's medium, Peptone sucrose agar. After 4–5 days of incubation at 28+/–2°C, pin head sized *Xoo* colonies can be observed in culture plates.

The presence of bacteria in the plants can be confirmed using modern molecular tools like polymerized chain reaction (PCR). Total genomic DNA is isolated from the suspected BB infected leaves and used for PCR assay using primers designated as TXT (5-GTCAAGCCAACTGTGTA-3) and TXT4R (5-CGTTCGGCACAGTTG-3). These primers amplify a 964-bp fragment of an insertion sequence (IS1113). A pure culture of *Xoo* should be used as positive control and DNA sample from a healthy plant should be used as a negative control (Sakthivel et al., 2001).

1.13.6   CAUSAL ORGANISM

Bacterial blight of rice is caused by *Xanthomonas oryzae* pv. *oryzae* (Ishiyama). According to the new classification system, the bacterium has been placed in:

1. **Family:** Xanthomonadaceae;
2. **Order:** Xanthomonadales;
3. **Class:** Gammaproteobacteria;
4. **Phylum:** Proteobacteria;
5. **Domain:** Bacteria.

X. oryzae pv. *oryzae* is a rod-shaped gram-negative and non-spore forming bacterium with a single polar flagellum. The bacteria do not produce spores or chains. *X. oryzae* colonies developed on nutrient agar medium are waxy yellow, round, smooth, and glistening. The mucous capsule is soluble in water, precipitated by acetone and is a polysaccharide, called Xanthomonadin.

1.13.7 DISEASE CYCLE

Primary infection may result from the inoculum overwintering in the seed, being present in the husk as well as in the endosperm. It also survives in soil or plant stubbles and debris, and the initial inoculum may be built upon the nursery seedlings. The bacterium is reported to infect some grasses like *Leersia* spp., *Cyperus rotundus*, which might play a role in the spread of the disease. It is disseminated through irrigation water and wind-borne rain, while hydathodes and wounds are its portals of entry. Once inside the host, the bacterium becomes systemic in the vascular bundles. After seedlings are transplanted, the disease symptoms may not show up until a few weeks later and become more severe at the time of flowering.

1.13.8 EPIDEMIOLOGY

- A combination of rainy weather, dull windy days and atmospheric temperature of 20–26°C are conducive for the development of the disease.
- Plants in the shade and close planted crops supplied with high doses of nitrogen show more disease incidence.

1.13.9 VARIETIES POSSESSING RESISTANCE OR TOLERANCE TO BACTERIAL BLIGHT

Ajaya, IR 20, IR 64, Pant Dhan 19, Swarnadhan, Improved Pusa Basmati-1, Improved Samba Mahsuri, Swarna, Indra, Tholakari, Mahsuri, Tikkana,

Pinakini, Deepti, Badava Mahsuri, MTU-9992, Godavari Ranjeet, Jayashri, Rajendran-201, Madhuri, Karjat-1, Kanchan, PR4141, BK79, ADT36, CO45, Sarjoo 52 and Pant Dhan 4 etc.

1.13.10 MANAGEMENT

- Host plant resistance offers the best solution for management of the disease. Cultural practices, host nutrition, and limited chemical control measures help in the reduction of the initial inocula and the secondary spread of the disease.
- Growing resistant / tolerant varieties.
- Destruction of wild collateral hosts. Species of *Cyperus* and *Leersia* must be removed from the field.
- Infected straw or chaff should not be let *in situ* or applied in the field. Rather, it should be burnt.
- Infected plant debris, self-sown rice plants have to be plowed down and the field is to be irrigated a month before sowing and transplanting to bring down the inoculum potential.
- Secure disease free seed. Seed treatment with plantomycin @ 1.0 g/liter/kg seed is useful.
- Avoiding flow of irrigation water from affected field to healthy field.
- Growing of rice crop under the shade should be avoided.
- Nitrogenous fertilizer should be applied in 3–4 split doses.
- Application of MOP @ 15 kg/acre at panicle initiation is useful.
- Streptomycin sulfate @ 1000 ppm in combination with copper hydroxide (Kocide) @ 3.0 g/l was found moderately effective against this disease.

1.13.10.1 HOST PLANT RESISTANCE

Deployment of varieties carrying one or more major resistant genes is the most effective approach for managing the disease. Resistance to bacterial blight is mostly qualitative in nature and there are only few reports about quantitative resistance to the disease (Nino-Liu et al., 2006). Till date, more than 30 BB resistance genes have been identified from diverse sources (Nino-Liu et al., 2006; Sundaram et al., 2011; Bhasin et al., 2012; Natarajkumar et al., 2012). The resistance genes have been designated as *Xa1* to *Xa38* with eight of them being recessive (*Xa5, Xa8, Xa13, Xa19, Xa20, Xa24,* and *Xa2*); six have been cloned (*Xa*1, *Xa*5, *Xa*13, *Xa*21, *Xa*3/*Xa*26, and *Xa*27) six have been physically mapped (*Xa*2, *Xa*4, *Xa*7, *Xa*30, *Xa*33, and *Xa*38).

The first successful report of marker-assisted gene pyramiding for BB resistance in India was made by Singh et al. (2001) from Punjab Agricultural University (PAU), Ludhiana, where three genes viz., *Xa*21, *Xa*13, and *xa*5 were successfully pyramided into the genetic background of an elite high yielding rice cultivar, PR 106 through marker-assisted backcross breeding. Davierwala et al. (2001) introgressed three BB resistance genes viz., *Xa*21, *Xa*13, and *Xa*5 into the genetic background of IR 64, singly or in combination through marker-assisted breeding. Leung et al. (2003) introgressed the gene combinations *Xa*4+ *Xa*5 and *Xa*4+ *Xa*7 into the genetic background of IR 64 and developed two new varieties 'Angke' and 'Conde' which were released by the Indonesian Government for commercial cultivation in 2002.

Joseph et al. (2004) introgressed *Xa*21 and *Xa*13 into the genetic background of the elite Basmati cultivar, Pusa Basmati-1 through restricted backcrossing involving one backcross cycle followed by five cycles of selfing. Pusa 1460-01-32-6-7-67, possessing maximum genomic background and quality characteristics of Pusa Basmati-1 gave resistance reaction against BB, similar to that of non-basmati resistant check variety (Ajaya) was released as a new variety in the name of 'Improved Pusa Basmati-1 for commercial cultivation in India in the year 2007 (Gopalakrishnan et al., 2008).

The research team at the Indian Institute of Rice Research-Hyderabad and CCMB-Hyderabad ventured to introgress three BB resistance genes *Xa*21, *Xa*13, and *Xa*5 into the genetic background of an elite, fine grain quality rice cultivar, Samba Mahsuri through marker-assisted backcross breeding strategy. They used the PCR-based linked markers pTA248, RG136 and RG556 in a backcross-breeding program to introgress three major BB resistance genes, viz., *Xa*21, *Xa*13, and *Xa*5, respectively into Samba Mahsuri from a donor line, named SS1113, which was earlier developed by Singh et al. (2001) and in which all the three resistance genes are present in a homozygous condition. Finally, the promising entry nominated for multi-location trials, RPBio-226 (IET 19046) was identified by the Varietal Identification Committee of ICAR as a new variety in the year 2007 and was subsequently notified and released under the name 'Improved Samba Mahsuri' for commercial cultivation in the year 2008 by CSCCSNRV constituted by Ministry of Agriculture, Govt. of India (Sundaram et al., 2008). The BB resistant genes *Xa*13 and *Xa*21 from IPB1; *Xa*33 from FBR15-1 and *Xa38* from PR114-*Xa38* are being incorporated into the genetic background of excellent quality rice varieties Pusa Basmati 1121 and Pusa Basmati 6 through marker-assisted simultaneous but step wise-backcross breeding program (Gopalakrishnan et al., 2013).

Swathi et al. (2013) introgressed two BLB resistance genes (*Xa13* and *Xa21* through marker-assisted backcross breeding in Jagtial Sannalu (JGL 1798). Rekha Malik et al. (2013) attempted to introgress major BB resistance genes (*Xa21, Xa13,* and *Xa5*) into susceptible basmati rice variety CSR-30 from BB resistant donor variety IRBB-60 through marker-assisted selection. Rao et al. (2013) reported the successful pyramiding of four bacterial blight resistant genes (*Xa4, Xa5, Xa13,* and *Xa21*) into Lalat and Tapaswini, two highly popular cultivars of Eastern India. Two gene pyramids of Lalat and Tapaswini were already released for cultivation while the gene pyramids of the mega varieties Swarna and IR 64 were identified for release (Mohapatra et al., 2013). Sujein Chang et al. (2013) reported that *xa5, Xa7,* and *Xa21* was effective genes for the dynamics of *Xanthomonas oryzae* pv. *oryzae* of Taiwan. Three R genes, *Xa4, Xa5,* and *Xa21* have been pyramided into an elite high-quality japonica cultivar by marker-assisted selection and the advanced progenies were challenged with the new virulent race K3a of *Xoo* and promising breeding lines produced by three R-gene pyramids were free from linkage drag and devoid of penalty on yield (Jena et al., 2013). About 1,442 introgression lines (ILs) (genetic background of Pusa 44 and PR 114) derived from the cross of various accessions of six different wild rice species viz., *O. rufipogon, O. nivara, O. glaberrima, O. barthii, O. glum-aepatula* and *O. longistaminata* were evaluated for BB resistance. Many cultures were found resistant to three or more strains. Eight cultures viz., PAU-547, PAU-549, PAU-550, PAU-695, PAU-747, PAU-1061, PAU-1077, and PAU-1195 were found to have broad spectrum resistance against all the BB strains used. These lines may provide novel sources of resistance (Laha et al., 2013).

Based on the virulence pattern, the *Xoo* isolates were categorized into 22 pathotypes. Among these, pathotypes 18, 20, 21, and 22 were highly virulent and were also individually virulent on *Xa*13 and *Xa*21 (Yugander et al., 2013).

1.14 BACTERIAL LEAF STREAK

1.14.1 INTRODUCTION

Bacterial leaf streak is common in tropical and subtropical regions of Asia, Africa, South-America, and Australia. It can affect the plant during early stages from maximum tillering to panicle initiation. Generally, it occurs in areas with high temperature and high humidity. Yield loss caused by bacterial leaf streak can range from 8–17% in the wet season and 1–3% in dry season. Mature rice plants can easily recover from leaf streak and have minimal grain yield losses.

1.14.2 SYMPTOMS

Initially symptoms appear as small, water-soaked, linear lesions between leaf veins. These streaks are initially dark green and later become light brown to yellowish gray. The lesions are translucent when held against the light. Entire leaves may become brown and die when the disease is very severe. Infected plants show browning and drying of leaves. Under humid conditions, yellow droplets of bacterial ooze, which contain masses of bacterial cells, may be observed on the surface of leaves. Disease incidence from booting to panicle emergence stage and further aggravation by favorable weather conditions lead to lead to reduced grain weight due to loss of photosynthetic area.

When the affected leaf streaks are cut and placed in a glass with water, a mass of bacterial cells can usually be seen oozing out of the leaf, which makes the water turbid after 5 minutes.

1.14.3 CAUSAL ORGANISM

Xanthomonas oryzae pv. *oryzicola.*

1.14.4   DISEASE CYCLE

It is transmitted through seeds and infected stubbles to the next season. It can occur in fields where *X. oryzae* pv. *oryzicola* bacteria is present on leaves, in the water, or in the debris left after harvest.

1.14.5   MANAGEMENT

- Use of resistant varieties. Treat seeds with hot water.
- Dry the field during the fallow period to kill the bacteria in the soil and in plant debris. Field sanitation, removal of weed hosts and destruction of rice stubble and straw.
- Use balanced fertilization especially nitrogen.
- Ensure good drainage of fields and nurseries. Drain the field during severe flood. Avoid flow of irrigation water from affected field to healthy field.
- Application of potash @ 15 kg/acre at panicle initiation is useful.
- Streptomycin sulphate @ 1000 ppm in combination with copper hydroxide (Kocide) @ 3.0 g/l is useful.

1.15   RICE TUNGRO VIRUS

1.15.1   INTRODUCTION

Rice is affected by a number of viral diseases of which rice tungro disease is considered to be the most devastating, economically important and wide spread viral disease in South and South East Asia from Pakistan to the Philippines. Outbreaks of the tungro disease occur in Andhra Pradesh, Tamil Nadu, Uttar Pradesh, Bihar, Orissa, and West Bengal thus, climatologically tropical sub-humid rice fields are more vulnerable to this disease. Field surveys in the years 2006–2008 indicated that Rice Tungro Disease has become prevalent in certain districts of Andhra Pradesh, Tamil Nadu, and West Bengal.

Tungro causes an annual yield loss worth of about US $ 1.5 billion worldwide and substantial yield losses in India as well. Yield losses were 38–71% (less susceptible varieties) and 84–100% (highly susceptible varieties).

1.15.2 SYMPTOMS

Tungro is characterized by stunting of the plant and discoloration of leaves, ranging from various shades of yellow to orange and rusty blotches spreading downwards from the leaf tip and twisting of leaf tips. The young leaves show a mottled appearance and slightly twisted, whereas the older leaves appear rusty colored. In less susceptible varieties tungro virus infection delays flowering. If infection of highly susceptible varieties takes place at very early stages, the plants may die before flowering. The tungro virus is transmitted by green leaf-hopper *Nephotettix virescens.*

1.15.3 CAUSAL ORGANISM

Rice tungro disease is caused by two morphologically and genomically as dissimilar viruses viz., Rice tungro spherical virus (RTSV) is a plant picorna virus (Family: Sequiviridae, Genus: Waikavirus, Type species: RTSV) which is a polyadenylated single stranded RNA virus, measures 30 nm in diameter and Rice tungro bacilliform virus (RTBV) is pararetrovirus (Family: Caulimoviridae, Genus: Tungro virus, Type species: Rice tungro bacilliform virus) a double stranded DNA virus, replicating via RNA intermediate. RTBV measures about 103–224 nm in length and 18–23 nm in diameter (Rice Tungro viruses are possibly the member of machloviruses group). Plants infected with both the tungro viruses show severe symptoms of tungro disease, while RTBV-infected plants show mild stunting and yellowing.

Tungro viruses are transmitted in a semi-persistent manner by green leaf hoppers (GLH) *Nephotettix virescens, N. nigropictus, N. cincticeps, N. malayanus, N. parvus* and *Recilia dorsalis* of which *Nephotettix virescens* (Distant) is the principal vector. Once virulified, the vector becomes infective immediately i.e., no incubation period is required. The infectivity of the vector decreases gradually and eventually is lost within 2–6 days. This is referred as 'transitory." The activity of the vector is related to weather conditions mainly temperature, rainfall, and duration.

1.15.4 DISEASE CYCLE

- Wild collateral grasses – *Eleusine indica, Echinochloa colonum* are the primary sources of inoculum.

- GLH are the secondary source of infection (female hopper is more efficient over male hopper).

1.15.5   EPIDEMIOLOGY

- Mineral nutrition and N-fertilization had marked influence on development of disease.
- September to November and March to April the insect vector is more active and thus the disease is more prevalent.

1.15.6   MANAGEMENT

- The isolated plants having virus infection symptoms in the beginning and destroy them by burning so that the insect does not get inoculum to spread the disease.
- The green stubbles, voluntary plants should be uprooted and burnt after harvest.
- Adopt balance fertilizer application.
- Destroy weeds both in field and on bunds.
- Grow tolerant varieties like Vikramarya (IET 7302), Bharani (NLR 30491), Deepti (MTU 4870), NLR 34242, Vasundhara (RGL 2538) and Srinivas (IET 2508).
- In endemic areas follow rotation with pulses or oil seeds.
- Green jassids acting as vectors are to be controlled effectively in time by spraying monocrotophos @ 2.2 ml or ethofenprox @ 2.0 ml or acephate @ 1.5 g/l or by applying carbofuran 3G @ 10 kg/acre.

Recently, Utri Merah, Balimau Putih, Habiganj DW8, and *O. rufipogon* were reported as donor for resistance. At IRRI, a tropical Japonica rice cultivar, Japonica 1 is being introgressed with RTSV resistance gene tsv1 using RM336 marker (Shim et al., 2013).

1.16   RICE STRIPE VIRUS

1.16.1   INTRODUCTION

Red stripe virus disease occurs in the temperate regions of East Asia i.e., China, Japan, Korea, and Taiwan. It has also been reported in far-eastern

Russia. It can cause high yield losses when severe epidemics occur. Severe infection at the seedling to early tillering stage was reported to cause yield losses of 50–100%.

1.16.2 SYMPTOMS

- Chlorotic to yellowish white stripes, mottling, and necrotic streaks on the leaves.
- Infected plants at the seedling stage are having folded, twisted, wilted, and droopy leaves. Plants are stunted, have few tillers, may produce few panicles, and may die prematurely.
- Panicles produced by infected plants have whitish to brown and deformed and unfilled spikelets, not fully exerted. Leaves of infected plants have less severe chlorosis or mottling. Panicle exertion and ripening of plants may be delayed.

1.16.3 CAUSAL ORGANISM

Rice stripe virus

The virus is transmitted in a persistent, circulative-propagative manner mainly by the small brown plant hopper. *Laodelphax striatellus* Fallen. It is also transmitted by three other plant hopper species, *Unkanodes sapporona* (Matsumura), *U. albifascia* (Matsumura), and *Terthron albovittatum* (Matsumura).

1.16.4 MANAGEMENT

- Grow resistant varieties. Resistance to the virus is more effective than resistance to the vector.
- Adjust planting time so that the crop will be at stem elongation stage or older during the peak of immigration of viruliferous insects from winter crops.
- Spraying of insecticides judiciously to reduce the population of viruliferous vectors.
- Remove ratoon or stubbles of the previous crop and weeds to reduce the virus and the population of the vector.

There are also a few non-parasitic diseases of rice in India. The important ones are Khaira disease due to zinc deficiency found in parts of Uttar Pradesh and Karnataka, sulfide injury due to excess production of hydrogen sulfide in some soils of Kerala, pan-sukh or dry leaf disease due to some unknown physiological disorder found in some parts of North India.

KEYWORDS

- **bacterial leaf blight**
- **green leafhoppers**
- **polymerized chain reaction**
- **quantitative trait loci**
- **rice tungro spherical virus**

REFERENCES

Al-Heeti, M. B., & El-Bahadli, A. H., (1982). *Estimation of Yield Losses Caused by Sclerotium Oryzae* (pp. 113, 114). Catt. on rice in Iraq. College Agric. Univ. Baghdad, Abu Gharid, Iraq.

Aruna, K. K., Durga, R. C. V., Sundaram, R. M., Vanisree, S., Seshumadhav, M., Arun, P. K. N., Swathi, G., & Jamaloddin, M. D., (2013). Marker assisted pyramiding of BB and blast resistance genes xa13, Xa21 and Pikh, Pi-1 in the elite indica rice variety MTU 1010. *Proceedings of 4th International Conference on Bacterial Blight of Rice* (p. 36). Hyderabad, India from 2nd–4th December, 2013.

Barroga, J. F., & Mew, T. W., (1994). "Red stripe" a new disease of rice in the Philippines. *Paper Presented at the 25th Anniversary and Annual Scientific Convention of the Pest Management Council of the Philippines.* Cagayan de Oro City, Philippines.

Bhasin, H., Bhatia, D., Raghuvanshi, S., Lore, J. S., Sahi, G. K., Kaur, B., Vikal, Y., & Singh, K., (2012). New PCR-based sequence-tagged site marker for bacterial blight resistance gene Xa38 of rice. *Molecular Breeding, 30,* 607–611.

Bhuvaneswari, V., Krishnam, R. S., Reddy, A. V., & Satyanarayana, P. V., (2014). *New Fungicides for the Management of False Smut Disease in Rice.* Abstract accepted for poster presentation at 4th International Rice Congress held at Thailand, Bangkok.

Bien, P. V., Sang, P. M., Minh, P. N., Chen, H. Q., & Vinh, M. T., (1992). *Yellow Leaf Syndrome in South Vietnam and Some Methods of Controlling It.* Annual Scientific Report, Vietnam: Institute of Agricultural Sciences (Unpub).

Clother, E., & Nicol, H., (1999). Susceptibility of Australian rice cultivars to stem rot fungus *Sclerotium oryzae. Australasian Pl. Pathol., 28,* 85–91.

Davierwala, A. P., Reddy, A. P. K., Lagu, M. D., Ranjekar, P. K., & Gupta, V. S., (2001). Marker assisted selection of bacterial blight resistance genes in rice. *Biochemical Genetics, 39,* 261–278.

Dhitikiattipong, R., Nilpanit, N., Surin, A., Arunyanart, P., & Chettanachit, D., (1999). Study on rice red stripe in Thailand. *Paper Presented at the Planning Workshop on Red Stripe* (pp. 15–18). Ho Chi Minh City, Vietnam.

Du, P. V., Noda, T., & Lai, V. E., (2001). Studies on some aspects of red stripe disease of rice in the Mekong Delta. In: Mew, T. W., (ed.), *Proc. Planning Workshop Red Stripe* (pp. 15–20). International Rice Research Institute. Los Banos, Philippines.

Du, P. V., Lan, N. T. P., Dinh, H. D., & Van, D. P., (1991). Red Stripe, a newly reported disease of rice in Vietnam. *International Rice Research Newsletter, 16*(3), 25.

Elazegui, F. A., & Castilla, N. P., (2004). Causal agent of red stripe disease of rice. *Plant Disease, 88*(12), 1310–1317.

Ghose, R. L. M., Ghatge, M. B., & Subramanian, V., (1960). *Rice in India* (p. 474). New Delhi, ICAR.

Gopalakrishnan, S., Ranjith, K. E., & Singh, A. K., (2013). Marker assisted selection for development of bacterial blight resistance in basmati rice. *Proceedings of 4th International Conference on Bacterial Blight of Rice* (pp. S–22). Hyderabad, India.

Gopalakrishnan, S., Sharma, R. K., Anand, R. K., Joseph, M., Singh, V. P., Singh, A. K., Bhat, K. V., Singh, N. K., & Mohapatra, T., (2008). Integrating marker assisted backcross analysis with foreground selection for identification of superior bacterial blight resistant recombinants in Basmati rice. *Plant Breeding, 127,* 131–139.

Hari, Y., Srinivasarao, K., Viraktamath, B. C., Hari, P. A. S., Laha, G. S., Ahmed, M. I., et al., (2013). Marker-assisted introgression of bacterial blight and blast resistance into IR 58025B, an elite maintainer line of rice. *Plant Breeding, 132,* 586–594.

Hiremath, P. C., & Hegde, R. K., (1981). Role of seed-borne infection of *Drechslera oryzae* on the seedling vigor of rice. *Seed Res., 9,* 45–48.

Jena, K. K., (2013). Stacking and molecular characterization of major genes toward broad-spectrum resistance to virulent bacterial blight pathogen in rice. *Proceedings of 4th International Conference on Bacterial Blight of Rice* (p. S5). Hyderabad, India.

Joseph, M. S., Gopalakrishnan, R. K., Sharma, V. P., Singh, A. K., Singh, N. K., & Mohapatra, T., (2004). Combining bacterial blight resistance and Basmati quality characteristics by phenotypic and molecular marker-assisted selection in rice. *Molecular Breeding, 13,* 377–387.

Khalili, E., Sadravi, M., Naeimi, S., & Khosravi, V., (2012). Biological control of rice brown spot with native isolates of three Trichoderma species. *Braz. J. Microbiol., 43,* 297–305.

Koide, Y., Kawasaki, A., Telebanco-Yanoria, M. J., Hairmansis, N. N. T. M., Bigirimana, J., Fujita, D., Kobayashi, N., & Fukuta, Y., (2010). Development of pyramided lines with two resistance genes, Pish and Pib, for blast disease (*Magnaporthe oryzae* B. Couch) in rice (*Oryza sativa* L.) *Plant Breeding, 129,* 670–675.

Krishnam, R. S., Bhuvaneswari, V., & Madhusudhan, P., (2012). Quantification of yield loss caused by red stripe disease in rice. *Oryza, 49*(4), 313–315. ISSN: 0474-7615.

Kulkarni, S., Ramakrishnan, K., & Hegde, R. K., (1980b). Ecology, epidemiology, and supervised control of rice brown leaf spot. *Intern. Rice Res. Newsl., 5,* 13, 14.

Laha, G. S., Singh, K. Y. A., Hajira, S., Sundaram, R. M., Hari, P. S., & Viraktamath, A. S., (2013). Identification of new sources of resistance to bacterial blight of rice in India. *Proceedings of 4th International Conference on Bacterial Blight of Rice* (p. 70). Hyderabad, India.

Li, Y. G., Kang, B. J., Feng, Y. X., Huang, D. J., Wu, D. B., & Li, T. F., (1984). A brief report on the studies of rice stem rot. *Guangdong Agric. Sci., 5,* 35–37.

Madhavi, K. R., Prasad, M. S., Madhav, M. S., Laha, G. S., Mohan, M. K., Sundaram, R. M., Jahnavi, B., Vijitha, S., Rao, P. R., & Viraktamath, B. C., (2012). Introgression of Blast resistance gene Pi-kh into elite indica rice variety Improved Samba Mahsuri. *Indian Journal of Plant Protection, 40,* 52–56.

Malik, R., Vishnu, V., Reddy, P., Khusi, R., Dhillon, S., & Boora, K. S., (2013). Marker assisted selection for introgression of bacterial blight (BB) resistance genes in rice (*Oryza sativa* L.). *Proceedings of 4*[th] *International Conference on Bacterial Blight of Rice* (p. 59). Hyderabad, India.

Mew, T. W., (1987). Current status and future prospects of research on bacterial blight of rice. *Annual Review of Phytopathology, 25,* 359–382.

Mew, T. W., (2001). Planning workshop on red stripe. *Limited Proceedings* (pp. 1–56).

Mogi, S., Sugandhi, Z., & Baskaro, S. W., (1988). A newly discovered disease (bacterial red stripe) on rice in Indonesia, its symptoms and distribution. In: *Proceedings of the 5*[th] *International Congress of Plant Pathology* (p. 388). Kyoto Japan. Abstr.

Mohapatra, T., Rao, G. J. N., Reddy, J. N., Kar, M. K., & Das, K. M., (2013). Genetic improvement of rice for bacterial leaf blight resistance at CRRI. *Proceedings of 4*[th] *International Conference on Bacterial Blight of Rice* (pp. S–21). Hyderabad, India from 2nd–4th December, 2013.

Natarajkumar, P., Sujatha, K., Laha, G. S., Srinivasa, R. K., Mishra, B., Viraktamath, B. C., et al., (2012). Identification and fine mapping of Xa33, a novel gene for resistance to *Xanthomonas oryzae* pv. *oryzae. Indain Phytopathology, 102,* 222–228.

Nino-Liu, D. O., Ronald, P. C., & Bogdanove, A. J., (2006). *Xanthomonas oryzae* pathovars: Model pathogens of a model crop. *Molecular Plant Pathology, 7,* 303–324.

Ou, S. H., (1985). *Rice Diseases* (2nd edn., p. 380). Commonwealth Mycological Institute, Surrey.

POS (Production Oriented Survey), (1979 & 1980). DRR, ICAR, Rajendranagar, Hyderabad, India.

Priyadarisini, B. V., & Gnanamanickam, S. S., (1999). Occurrence of a subpopulation of *Xanthomonas oryzae* pv. *oryzae* with virulence to rice cv. IRBB21 (Xa21) in Southern India. *Plant Disease, 83,* 781.

Rajamannar, M., Krishnam, R. S., Vijay, K. K. K., Mohana, R. V., & Adinararyana, M., (2007). Red stripe disease on rice in East and West Godavari districts of Andhra Pradesh. *Oryza, 44*(1), 90–91.

Rajan, C. P. D., (1987). Estimation of yield losses due to sheath blight of rice. *Indian Phytopathology, 40,* 174–177.

Ranganathaiah, K. G., (1985). Incidence of grain discoloration of paddy in Karnataka. *Madras Agric. J., 72,* 468–469.

Rao, G. J. N., Prasad, D., & Das, K. M., (2013). Genetic enhancement of Lalat and Tapaswini through marker assisted gene pyramiding against bacterial blight. *Proceedings of 4*[th] *International Conference on Bacterial Blight of Rice* (p. 50). Hyderabad, India.

Rao, P. S., & Kauffman, H. E., (1971). *Current Science, 40,* 271–272.

Ratna, M. K., Sundaram, R. M., Laha, G. S., Rambabu, R., Vijay, K. S., Aruna, J., Abilash, K. V., Seshu, M. M., Viraktamath, B. C., & Srinivas, P. M., (2013). Combining blast and bacterial blight resistance into rice cultivar Improved Samba Mahsuri through marker assisted selection. *Proceedings of 4*[th] *International Conference on Bacterial Blight of Rice* (p. 62). Hyderabad, India.

Roy, A. K., (1993). Sheath blight of rice in India. *Indian Phytopathology, 46,* 97–205.

Saad, A., (2001). The occurrence of red stripe disease in Malaysia. *Paper Presented at the Planning Workshop on Red Stripe* (pp. 37–40). Ho Chi Minh City, Vietnam.

Saddler, G. S., (1998). Acidovorax avenae subsp.avenae. Descriptions of fungi and bacteria. *IMI-Descriptions of Fungi and Bacteria, 122,* 1211.

Savary, S., & Mew, T. W., (1996). Analyzing crop losses due to *Rhizoctonia solani:* Rice sheath blight, a case study. In: Sneh, B., Javaji-Hare, S., Neate, S., & Dijst, G., (eds.), *Rhizoctonia Species: Taxonomy, Molecular Biology, Ecology, Pathology And Disease Control* (pp. 237–244). Kluwer, Dordrecht.

Savary, S., Teng, P. S., Willocquet, L., & Nutter, F. W. Jr., (2006). Quantification and modeling of crop losses: A review of purposes. *Annual Review of Phytopathology, 44,* 89–112.

Savary, S., Willocquet, L., Elazegui, F. A., Castilla, N., & Teng, P. S., (2000). Rice pest constraints in tropical Asia: Quantification and yield loss due to rice pests in a range of production situations. *Plant Disease, 84,* 357–369.

Shim, J., Cabunagan, R., Choi, I., Gideon, T., & Ha, W., (2013). Breeding for RTSV resistant tropical Japonica rice using marker assisted selection with eco-background selection. In: *Proceedings of 7th International Rice Genetics Symposium* (p. 724). IRRI, Los Banos, Philippines: International Rice Research Institute.

Singh, R. P., (2013). *Plant Pathology* (2nd edn., p. 724). Kalyani Publishers, New Delhi.

Singh, S., Sidhu, J. S., Huang, N., Vikal, Y., Li, Z., Brar, D. S., Dhaliwal, H. S., & Khush, G. S., (2001). Pyramiding three bacterial blight resistance genes (xa5, xa13, and Xa21) using marker-assisted selection into indica rice cultivar PR-106. *Theoretical and Applied Genetics, 102,* 1011–1015.

Srivastava, A., Yamini, D., Ramya, D., Balachiranjeevi, C. H., Naik, S. B., Abhilash, K. V., Laha, G. S., Prasad, M. S., Sundaram, R. M., Viraktamath, B. C., & Ram, T., (2013). Marker-assisted introgression of bacterial blight (*Xa21, Xa33*) and blast resistance genes (*Pi-54, Pi2*) into the background of high yielding variety Sampada. *Proceedings of 4th International Conference on Bacterial Blight of Rice* (p. 46). Hyderabad, India.

Srivastava, D. N., Rao, Y. P., & Durgapal, J. C., (1966). Can Taichung Native-1 stand up to bacterial blight? *Indian Farming, 16*(2), 15.

Standard Evolution System for Rice, (1996). International Rice Research Institute, Philippines, p. 52.

Sujein, C., Jia-ling, Y., Chun-Wei, C., Dong-Hong, W., & Chang-Sheng, W., (2013). Study on breeding of resistance to bacterial blight (*Xanthomonas oryzae* pv. *oryzae*) in rice (*Oryza sativa* L.) in Taiwan. *Proceedings of 4th International Conference on Bacterial Blight of Rice* (p. 71). Hyderabad, India.

Sundaram, R. M., Laha, G. S., Viraktamath, B. C., Sujatha, K., Natarajkumar, P., Hari, Y., et al., (2011). Marker assisted breeding for development of Bacterial blight resistant rice. In: Muralidharam, K., & Siddiq, E. A., (eds.), *Genomics and Crop Improvement: Relevance and Reservations* (pp. 154–182). Institute of Biotechnology, Acharya NG Ranga Agricultural University, Hyderabad 500030 India.

Sundaram, R. M., Vishnupriya, M. R., Biradar, S. K., Laha, G. S., Reddy, A. G., Rani, N. S., Sharma, N. P., & Sonti, R. V., (2008). Marker assisted introgression of bacterial blight resistance in Samba Mahsuri, an elite indica rice variety. *Euphytica, 160,* 411–422.

Sunder, S., Singh, R., Dodan, D. S., & Mehla, D. S., (2005). Effect of different nitrogen levels on brown spot (*Drechslera oryzae*) of rice and its management through host resistance and fungicides. *Pl. Dis. Res., 20,* 111–114.

Swathi, G., Jamaloddin, M. D., Durga, R. C. V., Seshumadhav, M., Vanisree, S., Sundaram, R. M., Arunprem, K. N., & Sri Chandana, B., (2013). Introgression of bacterial blight resistance genes into rice variety Jagtial Sannalu using marker assisted selection. *Proceedings of 4*th *International Conference on Bacterial Blight of Rice* (p. 61). Hyderabad, India.

Teng, P. S., Torries, C. Q., Nuque, F. L., & Calvero, S. B., (1990). Current knowledge on crop losses in tropical rice. In: Teng, P. S., Torries, C. Q., Nuque, F. L., & Calvero, S. B., (ed.), *Crop Loss Assessment in Rice* (pp. 39–54). International Rice Research Institute, Los Banos.

Vinh, M. T., (1997). Etiological studies on the yellow leaf syndrome of rice (*Oryza sativa* L.). *M.S. Thesis* (p. 77). University of the Philippines Los Banos. College, Laguna.

Vinh, M. T., Mew, T. W., & Bien, P. V., (1999). Etiological studies on the yellow leaf syndrome of rice (*Oryza sativa* L.). *Paper presented at the International Workshop on Red Stripe*. Ho Chi Minh City, Vietnam.

Wakimoto, S., Kim, P. V., Thuy, T. T., Tsuno, K., Kardin, M. K., Hartini, R. H., et al., (1998). Micro-fungus closely associated with the lesions of red stripe disease of rice. *Int. Congr. Plant Pathol., Abstracts of Papers* (Vol. 3, pp. 3, 6–11). Edinburgh. Scotland.

Yazid, M. E., Saad, A., & Jatil, A. T., (1996). Historical profile & current rice disease management practices in Malaysia. *Paper Presented at the International Workshop on Rice Disease Management Technology in the Tropics*. Sungai Petani, Kedah, Malaysia.

Yohei, K., Nobuya, K., Xu, D., & Yoshimichi, F., (2009). Resistance genes and selection DNA markers for blast disease in Rice (*Oryza sativa* L.). *JARQ, 43*(4), 255–280.

Yugander, A., Sundaram, R. M., Ladhalakshmi, D., Hajira, M., Sheshu, M. M., Srinivas, P. M., Viraktamath, B. C., & Laha, G. S., (2013). Virulence analysis and identification of new pathotypes of *Xanthomonas oryzae* causing bacterial leaf blight of rice in India. *Proceedings of 4*th *International Conference on Bacterial Blight of Rice* (p. 25). Hyderabad, India.

CHAPTER 2

Disease Spectrum in Wheat and Barley Under Different Agro-Ecological Conditions in India and Management Strategies

D. P. SINGH, SUDHEER KUMAR, and PREM LAL KASHYAP

ICAR-Indian Institute of Wheat and Barley, Karnal–132001, Haryana, India, E-mail: dpkarnal@gmail.com

2.1 INTRODUCTION

Wheat and barley are the major Rabi cereal crops of India and contribute greatly to the food security of the country. The latest crop estimates during the 2016–17 crop season are given in Table 2.1. The crops have witnessed higher degrees of change in the cropping system, climatic conditions, tillage practices, and larger area under single variety, private sector varieties, and other crop management practices in the last one decade. As a result, the diseases which were of low importance are taking greater magnitude in its spread and losses which requires new thinking and approach in their proper management to realize the maximum yield and quality potential of cultivars. The area under crop cultivation is shrinking every year due to urbanization, transport high ways, airports, irrigation canals, etc.; thus, result in more pressure to increase the factor productivity in these crops. Rice-Wheat cropping system is still very popular in entire Indo-Gangetic plains (IGP) due to assured support prices and foolproof crop production technologies of rice and wheat. However, the burning of crop residue is adding to the depletion of essential nutrients like organic Carbon in soils of North-Western states in India besides the shortage of water. The growing demand of malt barley from malt industries for food products and brewing purposes demanding cultivation of barley under medium input conditions in the states like Punjab, U.P., and Haryana, unlike traditional barley cultivation under

Diseases of Field Crops: Diagnosis and Management, Volume I

low input and in marginal lands. Use of sprinkler irrigation in drier areas and water scare parts of Rajasthan, Haryana, and M. P. at boot leaf to heading stages is creating conditions favorable for higher infection of floral diseases like Karnal bunt in wheat. The country has witnessed increased cultivation of wheat in non-traditional areas like the northeastern plains zone (NEPZ) due to surplus land available after rice cultivation and cultivation of rice in the northwestern plains zone (NWPZ) under irrigated conditions in past three decades. The warm and humid climate in NEPZ is quite favorable for leaf rust, spot blotch, and head scab. The occurrence of wheat blast in Bangladesh in 2016 (Malaker et al., 2016) created threats to wheat cultivation in adjoining states like West Bengal, Assam, and other north-eastern states.

TABLE 2.1 Estimate of Production and Productivity of Wheat and Barley (2016–17 Crop Season)

State	Wheat	Barley	Wheat	Barley	Wheat	Barley
	Production: ('000 Tonnes)		Area: ('000 Hectares)		Yield: kg/Hectare	
Assam	34.0		25.0		1360	
Bihar	4718.5	14.0	2095.2	11.0	2252	1270
Chhattisgarh	200.7	1.7	150.1	2.2	1337	773
Gujarat	2879.0		976.0		2950	
Haryana	11480.2	150.0	2440.0	42.0	4705	3571
Himachal Pradesh	634.2	32.6	324.0	18.3	1958	1786
Jammu & Kashmir	463.3	4.4	282.9	6.7	1638	658
Jharkhand	400.7		193.9		2066	
Karnataka	110.0		155.0		710	
Madhya Pradesh	17778.4	239.8	5940.0	120.0	2993	1999
Maharashtra	1405.0	3.0	913.0	8.0	1539	375
Odisha	1.3		0.8		1671	
Punjab	16040.5	44.0	3500.0	12.0	4583	3667
Rajasthan	8704.8	865.0	2790.0	308.8	3120	2801
Telengana	5.0	0.0	3.0		1667	
Uttar Pradesh	29911.8	463.0	9726.0	176.0	3075	2631
Uttarakhand	799.0	25.0	348.0	21.0	2296	1190
West Bengal	970.0	3.0	340.0	2.0	2853	1500
Others	106.8	1.3	28.0	1.2	3810	1072
All India	96643.2	1846.8	30231.0	729.1	3197	2533

Source: Agricultural Statistics Division Directorate of Economics and Statistics New Delhi.

The important wheat and barley diseases occurring in India at different crop stages are given in Table 2.2.

TABLE 2.2 Important Diseases of Wheat and Barley in Different Agro-Ecological Zones in India

Diseases	Pathogen		Distribution	
	Wheat	Barley	Wheat	Barley
Leaf Rust (Brown rust)	*Puccinia triticina* Eriks	*Puccinia hordei* Otth	All the six agro ecological zones but more prevalent in NEPZ, CZ, and PZ whereas in NWPZ it appears late	All the four five agro-ecological zones
Stripe Rust (Yellow rust)	*P. striiformis* Westend._f. sp. *tritici*	*P. striiformis* Westend. f. sp. *hordei*	Northern hill zone and NWPZ	Northern hill zone and NWPZ
Stem Rust (Black rust)	*Puccinia graminis* Pers.: Pers.	*P. graminis* Pers.: Pers.	PZ, CZ	PZ, CZ
Karnal Bunt	*Tilletia indica* Mitra syn. *Neovossia indica* (M. Mitra) Mundk.	—	Major problem in NHZ, NWPZ, minor in NEPZ	Not present
Loose Smut	*Ustilago tritici* (Pers.) Rostr. Syn. *U. segetum tritici*	*Ustilago tritici* (Pers.) Rostr.	Major problem in NWPZ, NHZ, and NEPZ	NWPZ, NHZ, and NEPZ
Covered smut	—	*Ustilago hordei* (Pers.) Lagerh.	Not present in wheat	NWPZ, NHZ, and NEPZ
Spot blotch	*Bipolaris sorokiniana* (Sacc.) Shoemaker Syn. *Helminthosporium sativus*	*B. sorokiniana* (Sacc.) Shoemaker [anamorph]	Major problem in NEPZ, moderate in CZ, PZ, and NWPZ	Major problem in NEPZ, PZ and NWPZ, moderate in CZ
Net blotch	—	*Drechslera teres* (Sacc.) Shoemaker, *Pyrenophora teres* Drechs. [teleomorph]	—	NEPZ and NWPZ
Stripe disease	—	*Drechslera graminea* (Rabenh.) Shoemaker	—	NHZ, NWPZ
Powdery Mildew	*Erysiphe graminis* DC. f. sp. *tritici* Em. Marchal, *Blumeria graminis* (DC.) E. O. Speer	*Erysiphe graminis* DC. f. sp. *hordei* Em. Marchal	Generally, prevails in cooler and humid areas in NHZ and NWPZ but occurrence is erratic	NHZ and NWPZ as minor disease

TABLE 2.2 *(Continued)*

Diseases	Pathogen		Distribution	
	Wheat	**Barley**	**Wheat**	**Barley**
Head Scab	*Fusarium graminearum* Schwabe	—	NWPZ, NEPZ	Not reported
Flag smut	*Urocystis agropyri* (G. Preuss) Schrot.	—	NWPZ	Not present
Wheat blast like disease	Unknown	—	NEPZ	Not reported

Abbreviations: NEZ-northern hills zone, NWPZ-north western plains zone, NEPZ-north eastern plains zone, CZ-central zone, PZ-peninsular zone.

2.2 SYMPTOMS

The symptoms of different diseases (Figures 2.1–2.9) at different stages of crop growth are given in Table 2.3.

      (a)   (b)

FIGURE 2.1   Yellow rust (a) wheat, (b) barley.

FIGURE 2.2   Brown rust of wheat.

FIGURE 2.3   Stem rust of wheat.

(a)   (b)   (c)

FIGURE 2.4   Loose smut (a) wheat, (b) barley, (c) covered smut of barley.

FIGURE 2.5   Flag smut of wheat.

FIGURE 2.6   Karnal bunt of wheat.

(a)   (b)   (c)

FIGURE 2.7   (a) Wheat blast on leaf, (b) wheat blast on spike (Islam et al., 2016).

FIGURE 2.8   Powdery mildew of wheat.

(a)   (b)

FIGURE 2.9   (a) Spot blotch on leaves (b) spot blotch infected spikes.

TABLE 2.3   Diseases Occurring at Different Growth Stages of Wheat and Barley and Symptoms Produced

Crop Growth Stage	Disease	Symptoms
Seed	Karnal bunt	The grains some grains in the spike are partially or wholly converted into black powdery masses. The embryo tissue is generally not destroyed. Not all the grains in a spike are affected few grains and generally, five to six grains are infected per spike. The pericarp may be intact partially whereas grain looks hollow inside. The black powder give is foul smell due to the presence of trimethylamine (Figure 2.6)
	Black point and Seed discoloration	The black point infected seeds have brown discoloration only at embryo region whereas in seed discouration, one third to whole of the grain surface may look dark brown and seed may be shriveled.
	Head scab	The seeds are shriveled and discolored usually with a tint of pink or orange.
	Wheat blast	The infected seeds are dark brown to light black in color, shriveled, and some may have eye like symptoms.
	Loose smut	It is exclusively seed borne. While the dormant mycelium remains internal in scutellum portion of embryo it is not visible outside and infected seed looks healthy. The embryo when macerated and stained reveals presence of mycelial mat in scutellum portion under stereo binocular microscope.
Seedling stage	Seedling blight	The infected seedlings may develop poor roots and brown discoloration at the base. The leaves may have dark brown necrotic spots with yellow halo which increases with age and covers larger area of leaf resulting into blight.
	Foot rot	White silky growth of fungal mycelium may appear at collar region and small black dot like sclerotia may develop later. The infected tillers may die.

TABLE 2.3 *(Continued)*

Crop Growth Stage	Disease	Symptoms
	Yellow rust	The plumula and leaves may develop minute yellow-colored sori in rows along the veins on upper surface which later covers whole leaf thus green leaves turn yellow and powdery mass of uredospores droops on the ground.
Tillering to stem elongation stage	Yellow rust	The yellow streaks may be in continuation or in partial with length of leaves depending on level of resistance in cultivar. The leaves may dry premature and plant may be dwarf and not bear spike (Figure 2.1).
	Brown rust	The symptoms develop on upper leaf blades mostly and occasionally on sheaths, glumes, and awns as brown colored circular and scattered pustules which usually do not coalesce, and contain masses of orange to orange-brown uredospores. With increase in temperature in the months of mid-March to April, these changes into teliospores which look dark brown to black colored spots in place of uredospores (Figure 2.2).
	Black rust	Symptoms may be seen on all above-ground parts of the plant but are most common on leaf sheaths, upper and lower leaf surfaces and stem. Pustules (containing masses of urediospores) are dark reddish-brown and are usually separate and scattered initially. The black rust pustules are bigger than brown rust pustules and are having torn epidermis which gives the infection sites feel rough to the touch (Figure 2.3).
	Spot blotch	The initial symptoms develop as water-soaked spots after 48 h of inoculation which turn into necrotic small spots after a week. Later oval-shaped spots develop with light brown center and yellow halo on the margins. In severe cases, such spots coalesce to cause leaf blight (Figure 2.9).
	Powdery mildew	The disease appears as white, powdery patches on the upper surface of leaves and stems. The grayish-white powdery growth may cover the entire leaf, sheath, stem, and floral parts. The fruiting bodies later appear as black dots in powdery growth and infected leaves dry prematurely (Figure 2.8).
	Flag smut	The infected leaves and stem tend to elongate few times, turn into twisted, thick, and silver in color which changes into dark brown with time and black mass of smut teliospores (Figure 2.5).
	Wheat blast	Infected plants show the typical elliptical and eye-shaped greyish to tan necrotic lesions with dark borders on the leaf.

TABLE 2.3 *(Continued)*

Crop Growth Stage	Disease	Symptoms
Boot leaf, ear mergence, and grain filling stages	Karnal bunt	The infected spike may have some spreading of the glumes due to sorus production. Symptoms are most readily detected on seed after harvest. The infected seeds have black colored mostly at embryo region and may spread in whole grain thus leaving the outer seed coat only and black colored teliospores may fall from such sees thus leaving it hallow.
	Loose smut	The visible symptoms of diseases are black colored smutted spikes instead of normal green colored spikes. The entire inflorescence is commonly affected and appears as a mass of olive-black spores, initially covered by a thin silver colored membrane which ruptures thus leaving black powdery mass of teliospores. The spike may be half infected but infected portion is always the lower part of spike (Figure 2.4).
	Spot blotch	In warm and humid areas, it may produce spike blight and discolored, shriveled, and light weight seeds.
	Head scab	Fusarium head blight appears on spike as light brown lesions on spikelets with an orange fungal mass along the lower portion of the glume. Grains from plants infected spike are often shriveled and have a white chalky appearance. Some kernels may have a pink discoloration. It is not a major disease in India
	Wheat blast	The infected spike becomes partially or completely bleached with the blackening of the rachis at the point of infection. The spike may be partial or full blighted thus may or may not have seeds. If seeds develop these are light in weight, thin, and may have dark brown discoloration. It is not yet reported in India (Figure 2.7).

2.3 CASUAL ORGANISMS

The major diseases in wheat and barley are caused by fungi. Major diseases and their causal organisms are given in Table 2.2.

2.4 DISEASE CYCLE

2.4.1 YELLOW RUST

Wheat crop in certain parts of Punjab, Haryana, J&K, parts of Uttarakhand and bordering crop fields in Uttar Pradesh have been affected by stripe rust or yellow rust of wheat caused by a fungal pathogen, *Puccinia striiformis* (Dutta et al., 2016). The disease appears in the form of yellow stripes on wheat leaves. This disease appears if cold temperature with intermittent rains prevail in Punjab, Haryana, Jammu, and Kashmir and tarai regions of UP and Uttarakhand. Since, alternate hosts are not functional under Indian conditions (Mehta, 1940), the primary inoculum for the Gangetic plains of India has been said to be coming every year from the assortment of Himalayan hills, where it survives on volunteer plants or summer crop in the form of uredospores or some other grasses / plants in the catchment areas (Bhardwaj et al., 2016).

2.4.2 BROWN RUST

High relative humidity, free moisture, and temperatures ranging from 15–25°C are conducive for leaf rust to develop. The optimum temperature for urediniospore germination is 12–15°C (Junk et al., 2016). If these conditions exist, infection can occur within 8 h (Bolton et al., 2008). Dry, windy days, which disperse spores followed by cool nights with dew, also favor leaf rust epidemics. Urediniospores act as primary inoculum by virtue of long-distance dispersal by wind. The fungus is known to survive on alternate hosts like *Thalictrum* spp. (Jackson and Mains, 1921), *Isopyrum fumarioides* (Chester, 1946), *Clematis* spp. (Sibilia, 1960), and *Anchusa* spp. (de Oliveira and Samborski, 1966). However, under Indian conditions, none of the alternate hosts is functional, therefore, fungus perpetuates in the form of uredial stage only. An aerial stage observed on *Thalictrum*. The uredospores of this rust survive both on Northern hills and Southern hills (Nilgiris). The inoculum travels from both ways and results infection in wheat crops in plains.

2.4.3 BLACK RUST

Puccinia graminis Pers. f. sp. *tritici* Eriks. and Henn. is a macrocyclic, heteroecious rust. It survives on Southern hills and infects the wheat crop in the peninsular zone (PZ) and central zone (CZ). Its pycnial and aecial stages are produced on alternate hosts (*Berberis*, *Mahonia*, and *Mahoberberis*), whereas, uredial, and telial stages of the fungus occur on graminaceous host. Teleutospores germinate after a long dormancy period when exposed to freezing temperatures. On germination, a four celled promycelium is produced and each cell produces a sterigma which bears basidiospore which infect wheat to produce pycnia. Pycnia are flask-shaped and consist of spermatia (pycniospores) and receptive hyphae. Mating of opposite types in receptive hyhae and pycniospores results in aecia and aeciospores. Aeciospores infect wheat. Since, under Indian conditions alternate hosts are not functional (Mehta, 1940), the survival and perpetuation of the pathogen, therefore, occurs in the form of uredospores in the hills on self-sown plants, summer crop being grown there. The nonsynchronization of vulnerable tender barberry leaves when the basidiospores are available, drastically, curtails the role of alternate host under Indian conditions. In fact, alternate hosts are of no consequence in the recurrence of stem rust in India (Nagarajan and Joshi, 1985).

2.4.4   SPOT BLOTCH

Spot blotch development is influenced by high temperature and relative humidity accompanied by long duration of leaf wetness (Gurung et al., 2012; Duveiller et al., 2005). Despite the absence of rainfall, the high relative humidity occurring in the IGP of India as a result of high levels of soil residual moisture at the end of the monsoon and rice crop, along with foggy days that can last until late January, favor long hours of wetness on leaf blades, conditions that are ideal for the establishment and multiplication of pathogens (Duveiller et al., 2005). Usually, spot blotch development accelerated after flowering (starting from late February until early March) with the advancement of plant growth stage. Moreover, for spot blotch outbreaks to occur, leaves must remain wet for >18 h at a mean temperature of 18°C or higher (White and Rodriguez-Aguilar, 2001; de Lespinay, 2004).

2.4.5   POWDERY MILDEW

Powdery mildew is mainly present in the cooler areas and hilly region; foothills, and plains of North-Western India and the southern hills (Nilgiris). Now occasionally its occurrence in high severity is observed in NPEZ especially in Punjab, Haryana, and Uttaranchal (Singh et al., 2016). This disease becomes most susceptible during rapid growth periods of wheat plants, especially between the stem elongation and heading growth stages. The mildew fungus survives the winter on wheat straw in the form of cleistothecia or as mycelium on infected wheat. Spores germinate and infect plants under cool, humid conditions. Infection does not require free water on the plant surfaces, but high relative humidity (100%) favors infection. The disease is more severe in dense plants and application of heavy dose of nitrogenous fertilizer in wheat. Under optimum conditions, a new crop of conidia is produced every 7–10 days (Sharma et al., 2016; Singh, 2017). Dispersal of conidia is through wind to new leaves, causing the disease to spread up the plant and between plants. When temperatures increase late in the season and the plant and fungus aged, new cleistothecia are produced. Sharma et al. (2017) reported that disease development was positively correlated with temperatures and negatively correlated with evening relative humidity and rainfall under field conditions.

2.4.6 KARNAL BUNT

The environment plays a primary role in KB development. The teliospores germinate in response to free moisture and 65–185 primary sporidia (basidiospores) at the soil surface (Gill et al., 1993; Kumar et al., 2017) that are forcibly ejected and then dispersed by wind, splashing water, insects, etc. Sporidia have a short life-span, even at high relative humidity, and generally survive for only a few hours when airborne. Long-distance dispersal of secondary sporidia probably can occur only during moist periods and under cloud cover at night. Plants are most susceptible to infection when spikes emerge from the boot, but infection can occur throughout the flowering period (Singh and Krishna, 1982; Sharma et al., 2016). Sporidia infect the ovaries, directly penetrating the glumes and ovary wall. Infection can move across all florets of a spikelet and to the spikelets above and below the initial infection site. Diseased kernels may be partially or completely displaced by masses of teliospores in cool, humid, or wet weather (Kashyap et al., 2011). Sporulation starts at the embryo end and moves along the suture of the grain. The extent of kernel bunting depends on when infection occurs and how long favorable conditions last. Fresh teliospores require a dormancy period of up to 6 months, although a low level (6.7%) of germination of teliospores from freshly harvested kernels has been observed after 1 week of storage. They remain viable in the soil for up to 45 months. Teliospores survive longer when buried in dry soil than in moist, cropped soils or on the soil surface. Kaur et al. (2015) used REP-PCR fingerprinting for detecting the genetic variability in *Tilletia indica*.

The postharvest grain sample analysis of wheat for Karnal bunt Karnal bunt in May-June 2016 (Table 2.4) indicated that out of total 8732 grain samples, the highest incidence (53.3%) was recorded from Jammu region of J&K followed by Punjab (33.7%). Based on the overall Karnal bunt occurrence, it emerged that the KB incidence during 2016 was lower than two years. No sample from Maharashtra (Pune and Niphad) and Karnataka (Dharwad) was found infected with KB.

TABLE 2.4 Karnal Bunt Situation in the Country During 2016

State	Total Samples	Infected Samples	% Infected Samples	Range of Infection
Punjab	3074	944	30.71	0.01–0.249
Haryana	2078	334	16.07	0.05–1.15
Rajasthan	1312	402	30.64	0.1–12.50
Uttarakhand	72	05	15.85	0.01–0.5
Jammu	465	248	53.33	1.25–5.00
U.P.	291	44	15.12	0.01–1.6
M.P.	1023	225	21.99	0.01–2.05
Maharashtra	231	0	0	--
Karnataka	186	0	0	--
Total	**8732**	**2202**	**25.22**	0.01–12.50

2.4.7 LOOSE SMUT

The disease is internally seed borne, where pathogen infects the embryo in the seed. In India during the 1980s, loose smut incidence of up to 10% was reported from Northwestern areas (Joshi et al., 1980). Primary infection occurs during sowing when infected seeds are sown. Loose smut infection is most favored in cool, humid conditions during flowering period of the host plant (Kaur et al., 2014). The disease cycle of loose smut begins when teliospores are blown to open flowers and infect the ovary either through the stigma or directly through the ovary wall. After landing in an open floret, the teliospores give rise to basidiospores. Without dispersing to any alternate host plant, the basidiospores germinate right where they are. The hyphae of two compatible basidiospores then fuse to establish a dikarytic stage. After germination of spores inside the ovary, the mycelium of fungus invades the developing embryo in the seed. The fungus survives in the seed until the next growing season, when it is planted along with the seed. As the developing plant grows, the fungus grows with it. Once it's time for the flowers to form, teliospores are produced in place of the flowers and develop where the grain would be. Plants which are infected with fungus actually grow taller and flower earlier than their healthy counterparts. This gives the infected plants an advantage in that the flowers of uninfected plants are more physically and morphologically susceptible to infection. The teliospores in the smutted grain heads disperse to the open flowers of the healthy plants, and the cycle continues.

2.4.8 FLAG SMUT

The fungus produces teliospores in the infected leaves and stem, which may be wind dispersed or distributed through soils via farm machinery or animals. The dikaryotic teliospore germinates in soil (Purdy, 1965). The meiosis is followed by mitosis and finally four basidiospores are produces each containing a single nucleus. Basidiospores germinate on seedlings, and each hypha undergoes plasmogamy with a compatible hypha thus dikaryotic state of the fungus takes place. The hyphae form appressoria which penetrate the coleoptile of an emerging seeds' shoot through the epidermal tissue, and then hyphae grow between vascular bundles of the leaves. Some hyphal cells give rise to smut sori, bearing teliospores, which emerge through the leaf tissue for wind dispersal (Savchenko et al., 2017). Teliospores come to rest in soils, and when conditions are right, they give rise to more basidiospores, further spreading the infection. As a result, the disease may cause yield losses of up to 100% if environmental conditions are favorable (Purdy, 1965). Alternatively, teliospores can form in seeds when the mycelia grows throughout the plant, in which case they germinate within the seed to give rise to new infection, again via basidiospore production. Teliospores survive in the soil, senescent plant tissues, and in seeds (Ram and Singh, 2004). These spores maintain germination viability for 3–7 years.

2.5 DISEASES MANAGEMENT

Survey and surveillance of diseases is important in wheat and barley and regularly done during crop season and after grain harvest (for Karnal bunt and black point of wheat) using a network of All India Coordinated Research Project on Wheat and Barley, DAC, and FW, SAUs, KVKs, and state agriculture departments, and advisories are issued accordingly (Saharan, 2017).

The diseases are managed in an integrated manner using mainly host resistance in wheat and barley (Kumar et al., 2016; Singh et al., 2016a, b). Not all diseases are prevalent and causing losses uniformly in all agroecological zones. All the released varieties of wheat and barley are tested for at least 3–4 years against diseases under artificially created disease epiphytotic conditions at hot spot location in six agro-ecological zones of India. Disease resistance is a major criteria besides yield and quality in the release of wheat and barley varieties in India. Therefore, priorities for their management differ from one zone to another. Brown rust and foliar blights are present in all zones. However, brown rust pathotypes varies in North than in South accordingly to sources of inoculum which are Himalayan hills in North and Pulney and Nilgiri hills in South in Tamil Nadu. The fungicidal seed treatment and foliar spray is recommended for control of smuts and foliar diseases (Ram and Singh, 2004; Selvakumar et al., 2015; Sharma-Poudyala et al., 2016). The loose smut both in wheat and barley and covered smut of barley, the infected seeds transmit the disease from one season to another season and inoculum falling on soils dies during hot summer months. It is not enough to control one or two disease but for getting higher returns, an integrated management of all diseases is recommended. The management practices however, varies a bit with change of varieties due their varying degrees of resistance against a particular disease and it is more paying in old varieties than newly released varieties. However, certain diseases like rusts, wheat blast like disease and Karnal bunt, fungicidal foliar sprays are recommended especially when varieties grown are susceptible and occupies large areas (>5 million ha) in case of wheat. Survey and surveillance are of utmost importance in case of diseases like yellow rust since it appears sometimes in December and capable of causing heavy losses in both wheat and barley. Likewise, brown, and black rusts are monitored regularly in NEPZ, CZ, and PZ during crop season to take management practices at the right time and avoiding epidemics. The management of major wheat and barley diseases is given in subsections.

2.5.1   WHEAT

2.5.1.1   YELLOW RUST

- The initial disease control in susceptible varieties is done using foliar sprays of fungicides, viz., Propiconazole tebuconazole @ 0.1%, and tebuconazole+ trifloxystrobin @ 0.06% to avoid its further spread from initial infection foci. Usually, it is required in the first half of February till first half of March 2017 in Punjab, Haryana, Uttarakhand, and Western U. P.

- To avoid large scale spread growing diverse stripe rust tolerant varieties - avoiding single variety over large areas. For avoiding the losses due to stripe rust of wheat in NWPZ, varieties like WH 1105, HD 3086, and WB 02 for timely sown and DBW 71, WH 1021 and HD 3059 for late sown conditions may be preferred. In NHZ, varieties like HPW 349, HS 507, HS 365, HS 375, VL 616, VL 907, VL 829, VL 832, VL 892, HPW 155, SKW 196 etc. should be grown.

2.5.1.2   KARNAL BUNT

- Use of certified or disease free seed will help to check introduction disease in new areas.
- Grow resistant / tolerant varieties in disease prone areas viz. PBW 502 and PDW 223/ PDW291/ PDW314 (Durum) in Northern Plains Zones, HPW 251, HS 490, HS 507 in northern hills zone (NEZ) and GW366, HD 2864, MP 3336 and HI8498 (Durum) in CZ.
- Spray Propiconazole @ 0.1% at the time of 50% flowering and if rains the spray can be repeated after 15 days.

2.5.1.3   POWDERY MILDEW

Singh et al. (2016) identified number of wheat varieties and genotypes resistant to powdery mildew in India as well as genotypes showing differential host reaction at hot spot location.

2.5.1.4   SPOT BLOTCH

Foliar blight is the main problem in humid and warmer areas especially in NEPZ. It is managed using newly released and resistant varieties like, HD 2985, HI 1563, DBW 39, CBW 38, NW 1014, NW 2036, K 9107, HD 2733, DBW 14, HD 2888, K0307, DBW39, and HUW 468. The use of highly susceptible varieties may be avoided in NEPZ. A list showing status of Indian wheat varieties against spot blotch is given by Singh et al. (2015).

The disease in susceptible varieties may be managed using seed treatment with thiram, thiram+carboxin (1:1) @ 2.5 g/kg of seed as well as foliar sprays of propiconazole @0.1% at initiation of disease (Singh et al., 2008; Singh and Kumar, 2008).

2.5.1.5   LOOSE SMUT

Loose smut of wheat is an internally seed borne disease. In view of the horizontal distribution of the seed material among the farmers and the use of the carry over seed effective control measures for loose smut should be undertaken. The infected earheads emerge 15 days early then healthy, should be rouged out and burnt. For this, seed treatment with carboxin 75 WP @ 2.5 gm/kg seed or carbendazim 50 WP @ 2.5 gm/kg seed or tebuconazole 2DS @ 1.25 gm/kg seed or a combination of a reduced dosage of Carboxin (75 WP @ 1.25 gm/kg seed) and a bioagent *Trichoderma viride* (@ 4 gm/kg seed) is recommended. Integrated management of loose smut involving reduced the use of chemical fungicide and promote bioagent fungus which is more eco-friendly and equally effective as the chemical control measures and thus should be preferred. Use of bioagents also helps in improving the initial vigor of the crop. Seed treatment with fungicide should be done one or two days before sowing. In case of integrated management, the treatment with *T. viride* should be done 72 h before sowing, followed by the fungicide, 24 h before sowing.

Singh et al. (2017) identified number of wheat varieties and genotypes showing resistance against loose smut of wheat.

2.5.1.6   FLAG SMUT

Flag smut disease also poses problems in isolated fields in Punjab, Haryana, and Rajasthan.

- Use disease free seed.
- Seed treatment with carboxin 75 WP @ 2.5 g/kg seed or carbendazim 50 WP @ 2.5 g/kg seed or tebuconazole 2DS @ 1.25 g/kg.

2.5.1.7   IPM FOR WHEAT BLAST

Keeping in view of presence of wheat blast in nine districts of Bangladesh adjoining West Bengal during 2016 (Chowdhury et al., 2017; Saharan et al., 2016) and also in some of these during 2017 as well as mode of transmission of pathogen (Seed, air, and plant material) following IPM practices are suggested to prevent establishment and establishment of disease in West Bengal, Assam, and other states close to Bangladesh in India:

- Strict quarantine of wheat imported from affected countries, local quarantine of wheat grains in West Bengal.
- No wheat zone up to 5 km along the Indo Bangladesh borders.
- Replacement of wheat with nonhost crops like oilseeds and pulses.
- Wheat holiday in districts like Murshidabad and Nadia for the next three years.
- Monitoring of wheat blast like disease on wheat in West Bengal, Assam, and other adjoining states.
- Destruction of affected monocot weeds.
- *Complete* procurement of wheat grains in West Bengal and consumption locally after grinding.
- Use resistant varieties like HD 2967 and those with 2 NS translocations.
- Seed treatment with thiram+carboxin (1:1) @3 g /kg seed.
- Impose quarantine for seed /grain import from affected countries.
- Use foliar sprays of epoxiconazole + pyraclostrobin and tebuconazole+ trifloxystrobin @ 0.06% at flag leaf stage.

2.5.2 BARLEY

2.5.2.1 YELLOW RUST

- Spraying with propiconazole @ 0.1% followed by tebuconazole @ 0.1% and triademefon @ 0.1% gave the effective control of barley yellow rust (Devlash et al., 2015).

2.5.2.2 FOLIAR BLIGHT

- Seed treatment with carboxin @ 2 g/kg seed and spray with tebuconazole or propiconazole @ 0.1% to reduced disease severity.

2.5.2.3 SMUTS

- Seed treatment with carbendazim 50% WP or carboxin 75% WP @ 2 g/kg seed. However, in a study Kaur et al. (2014) carbendazim 50 WP @ 2.5 g, mancozeb 50%+ carbendazim 25% @ 3.0 g, tebuconazole 60 FS @ 1.0 g, carboxin 37.5% + Thiram 37.5% @ 1.5 g and tebuconazole 2DS 2% @ 1.5 g Kg^{-1} seed gave complete control of covered smut of barley.

2.5.2.4   INTEGRATED DISEASE MANAGEMENT

- Grow recommended resistant or tolerant varieties.
- Use disease free seed.
- Apply the balanced fertilizers as per recommendations.
- Rogue the loose and covered smut infected plant and burn.
- Seed treatment with carbendazim 50% WP or carboxin 75% WP @ 2 g/kg seed.
- Spraying with propiconazole @ 0.1% on the appearance of yellow rust (Singh, 2008; Singh et al., 2010).

KEYWORDS

- **central zone**
- **Indo-Gangetic plains**
- **northeastern plains zone**
- **northwestern plains zone**
- **northern hills zone**
- **peninsular zone**

REFERENCES

Bhardwaj, S. C., Prasad, P., Gangwar, O. P., Khan, H., & Kumar, S., (2016). Wheat rust research-then and now. *Indian J. Agric. Sci., 86*(10), 1231–1244.

Bolton, M. D., Kolmer, J. A., & Garvin, D. F., (2008). Wheat leaf rust caused by *Puccinia triticina. Mol. Plant Pathol., 9*(5), 563–575.

Chester, K. S., (1946). *The Nature and Prevention of the Cereal Rusts as Exemplified in the Leaf Rust of Wheat.* Chronica Botanica, Waltham, Massachusetts.

Chowdhury, A. K., & Saharan, M. S., (2017). Occurrence of wheat blast in Bangladesh and its implication for South Asian wheat production-a review. *Indian J. Genet. Pl. Br., 77*(1), 1–9.

De Lespinay, A., (2004). Selection for stable resistance to *Helminthosporium* leaf blights in non-traditional warm wheat areas. *M.S. Thesis.* Université Catholique de Louvain, Louvain-La-Neuve, Belgium.

De Oliveira, B., & Samborski, D. J., (1966). Aecial stage of *Puccinia recondite* on ranunculaceae and boraginaceae in Portugal. *Proceedings of Cereal Rusts Conference-1964* (pp. 133–150).

Devlash, R., Kishore, N., & Singh, G. D., (2015). Management of stripe rust of barley (*Hordeum vulgare* L.) using fungicides. *J. Appl. Natur. Sci., 7,* 170–174.

Dutta, S., Saharan, M. S., & Sharma, I., (2016). Wheat stripe rust detection of probable alarm zones in Rabi 2012–13 and 2013–14 seasons. *J. Wheat Res., 8*(2), 54–56.

Duveiller, E., Kandel, Y. R., Sharma, R. C., & Shrestha, S. M., (2005). Epidemiology of foliar blights (spot blotch and tan spot) of wheat in the plains bordering the Himalayas. *Phytopathology, 95,* 248–256.

Gill, K. S., Sharma, I., & Aujla, S. S., (1993). *Karnal Bunt and Wheat Production* (p. 153). Punjab Agricultural University Ludhiana.

Gurung, S., Sharma, R. C., Duveiller, E., & Shrestha, S. M., (2012). Comparative analyses of spot blotch and tan spot epidemics on wheat under optimum and late sowing period in South Asia. *Eur. J. Plant Pathol., 134,* 257–266.

Islam, M. T., et al., (2016). https://bmcbiol.biomedcentral.com/articles/10.1186/s12915-016-0309-7. *Under the terms of the Creative Commons Attribution 4.0 International License.* http://creativecommons.org/licenses/by/4.0))(c)Fusariumheadscab (accessed on 24 January 2020).

Jackson, H. S., & Mains, E. B., (1921). Aecial stage of the orange leaf rust of wheat, *Puccinia triticina* Erikss. *J. Agric. Res., 22,* 151–172.

Joshi, L. M., Srivastava, K. D., & Singh, D. V., (1980). Wheat disease newsletter. *Indian Agric. Res. Inst., 13,* 112–113.

Junk, J., Kouadio, L., Delfosse, P., & Jarroudi, M. E., (2016). Effects of regional climate change on brown rust disease in winter wheat. *Climatic Change, 135,* 439. doi: 10.1007/s10584-015-1587-8.

Kashyap, P. L., Kaur, S., Sanghera, G. S., Kang, S. S., & Pannu, P. P. S., (2011). Novel methods for quarantine detection of Karnal bunt (*Tilletia indica*) of wheat. *Elixir Agri., 31,* 1873–1876.

Kaur, G., Sharma, I., & Sharma, R. C., (2014). Characterization of *Ustilago segetum tritici* causing loose smut of wheat in northwestern India. *Can. J. Pl. Pathol., 36,* 360–366.

Kaur, M., Singh, R., Saharan, M. S., et al., (2015). REP-PCR fingerprinting based genetic variability in *Tilletia indica*. *Indian Phytopath., 68*(4), 380–385.

Kumar, A., Pandey, V., Singh, M., Pandey, D., Saharan, M. S., & Marla, S. S., (2017). Draft genome sequence of Karnal bunt pathogen (*Tilletia indica*) of wheat provides insights into the pathogenic mechanisms of quarantined fungus. *PLoS One, 12*(2), e0171323. doi: 10.1371/journal.pone.0171323.

Kumar, (2016). Evaluation of 19,460 wheat accessions conserved in the Indian National gene bank to identify new sources of resistance to rust and spot blotch diseases. *PLoS One, 11*(12), e0167702. doi: 10.1371/journal.pone.0167702.

Malaker, P. K., Barma, N. C. D., & Tiwari, T. P., (2016). First report of wheat blast caused by *Magnaporthe oryzae* Pathotype *triticum* in Bangladesh. *Plant Dis., 100*(11), 2330.

Mehta, K. C., (1940). *Further Studies on Cereal Rusts in India* (Vol. I, 14, p. 224). Imperial council agricultural research. New Delhi. Scientific monograph.

Nagarajan, S., & Joshi, L. M., (1985). Epidemiology in the Indian subcontinent. In: Roelfs, A. P., & Bushnell, W. R., (eds.), *The Cereal Rusts, Diseases, Distribution, Epidemiology and Control* (Vol. 2, pp. 371–402). Academic Press, Orlando, FL, USA.

Purdy, L. H., (1965). Flag smut of wheat. *Bot. Rev., 31,* 565–606.

Ram, B., & Singh, K. P., (2004). Smuts of wheat: A review. *Indian Phytopath., 57*(2), 125–134.

Saharan, M. S., (2017). Survey and surveillance of wheat biotic stresses: Indian scenario. In: Singh, D. P., (ed.), *Management of Wheat and Barley Diseases* (pp. 519–531). Apple Academic Press, USA.

Saharan, M. S., Bhardwaj, S. C., Chatrath, R., Sharma, P., Choudhary, A. K., & Gupta, R. K., (2016). Wheat blast disease - An overview. *J. Wheat Res., 8,* 1–5.

Savchenko, K. G., Carris, L. M., Demers, J., Manamgodab, D. S., & Castlebury, L. A., (2017). What causes flag smut of wheat? *Plant Pathol.,* doi: 10.1111/ppa.12657.

Selvakumar, R., (2015). Efficacy of fungicides as seed treatment and foliar application for managing leaf blight (*Bipolaris sorokiniana*) on wheat (*Triticum aestivum* L.). *J. Wheat Res., 7*(2), 14–18.

Sharma, P., (2016). Draft genome sequence of two monosporidial lines of the karnal bunt fungus *Tilletia indica* mitra (PSWKBGH-1 and PSWKBGH-2). *Published in Genome Announcements, American Society for Microbiology, 4*(5), e00928–16.

Sharma, V. K., Ram, N., Karwasara, S. S., & Saharan, M. S., (2017). Progression of powdery mildew on different varieties of wheat and triticale in relation to weather parameters. *J. Agromet., 19*(1), 84–87.

Sharma, V. K., Niwas, R., Karwasra, S. S., & Saharan, M. S., (2017). Progression of powdery mildew on different varieties of wheat and triticale in relation to environmental conditions. *J. Agromet., 19*(1), 84–87.

Sharma-Poudyal, D., Sharma, R. C., & Duveiller, E., (2016). Control of helminthosporium leaf blight of spring wheat using seed treatments and single foliar spray in Indo-gangetic plains of Nepal. *Crop Prot., 88,* 161–166.

Sibilia, C., (1960). La forma ecidica Della ruggine bruna delle foglie di grano puccinia recondita rob. ex desm. in Italia. Bull. Stn. Patol. *Veg. Rome 18*(3), 1–8.

Singh, D. P., et al., (2016a). Multiple rust resistances in *Triticum aestivum, T. durum, T. dicoccum* and triticale. *Intl. J. Curr. Res. Biosci. Pl. Biol., 3*(1), 46–52.

Singh, D. P., et al., (2016b). Assessment and impact of spot blotch resistance grain discoloration in wheat. *Indian Phytopath., 69,* 363–367.

Singh, D. P., (2008). Evaluation of barley genotypes against multiple diseases. *SAARC J. Agric., 6,* 117–120.

Singh, D. P., (2017). *Management of Wheat and Barley Diseases* (p. 643). Apple Academic Press, USA.

Singh, D. P., & Pankaj, K., (2008). Role of spot blotch (*Bipolaris sorokiniana*) in deteriorating seed quality in different wheat genotypes and its management using fungicidal seed treatment. *Indian Phytopath., 61,* 49–54.

Singh, D. P., et al., (2008). Management of leaf blight complex of wheat caused by *Bipolaris sorokiniana* and *Alternaria triticina* in different agroclimatic zones using an integrated approach. *Indian J. Agric. Sci., 78,* 513–517.

Singh, D. P., et al., (2017). Resistance in Indian wheat and triticale against loose smut caused by *Ustilago tritici. Indian Phytopathology, 70*(1), In Press.

Singh, D. P., Babu, K. S., Mann, S. K., Karwasra, S. S., Kalappanavar, I. K., Singh, R. N., Singh, A. K., & Singh, S. P., (2010). Integrated pest management in Barley (*Hordeum vulgare*). *Indian J. Agric. Sci., 80,* 437–442.

Singh, D. P., et al., (2016). Identification of resistance sources against powdery mildew (*Blumeria graminis*) of wheat. *Indian Phytopath., 69,* 413–415.

Singh, D. P., et al., (2015). Sources of resistance to leaf blight (*Bipolaris sorokiniana* and *Alternaria triticina*) in wheat (*Triticum aestivum, T. durum, T. dicoccum*) and triticale. *Indian Phytopath., 68,* 221–222.

Singh, R. A., & Krishna, A., (1982). Susceptible stage for inoculation and effect of Karnal bunt on viability of wheat seed. *Indian Phytopath., 35,* 54–56.

White, J. W., & Rodriguez-Aguilar, A., (2001). An agroclimatological characterization of Indo-Gangetic plains. In: Kataki, P. K., (ed.), *The Rice-Wheat Cropping Systems of South Asia: Trends, Constraints, Productivity and Policy* (pp. 53–65). Food Product Press, New York, USA.

CHAPTER 3

Ug99: Wheat Stem Rust Race: Exploring the World

SUDHA NANDNI, DEVANSHU DEV, and K. P. SINGH

Department of Plant Pathology, College of Agriculture, G.B. Pant University of Agriculture and Technology, Pantnagar, Uttarakhand–263145, India, E-mail: dev9105@gmail.com

3.1 WHEAT STEM RUST

Stem rust is one of the most shattering diseases of wheat, caused by the fungus *Puccinia graminis* Pers. f. sp. *tritici* Eriks. and E. Henn. The pathogen converts a healthy-looking plant to a mesh of black stems and few shriveled grains, just a few weeks prior to the harvesting. The disease is easily recognized by the appearance of black pustules on the stems and leaves of the plant. When these pustules burst, millions of spores burst out and search for a fresh host plant. The devastated plant then withers and dies off; its grains shrivel like pebbles which then remain no use. Stem rust can cause severe yield losses in susceptible cultivars of wheat in environments favorable for disease development. Worldwide, most of the wheat-growing areas provide a suitable environment for the development of wheat stem rust. For over three decades, wheat stem rust has largely been under control due to the extensive use of resistant cultivars. The last wheat stem rust epidemic was seen during 1993 and 1994 in Ethiopia where a popular wheat variety 'Enkoy' suffered serious losses but rest of the world remained safe for over three decades.

3.2 EMERGENCE OF NEW STEM RUST RACES

Stem rust of wheat is the most important threat to wheat cultivation all over wheat-growing areas of world due to the very high rate of evolution and mutation of the pathogen, *Puccinia graminis tritici*. This is the main reason due to which most of the widely grown wheat varieties remain at continuous risk of losing their resistance to it. Since, wheat is the next most important crop in the world after paddy, it is very important to safeguard it from pathogens and sustain global food security too. A new race of stem rust pathogen, Ug99 has evolved, universally known as 'shifting enemy of wheat' due to its rapid migration rate across the continents of the world. Researchers across the globe have already predicted Ug99 as one of the utmost threats ever confronted by wheat cultivation as it has the immense potential of breaking the strongest of resistant genes of stem rust, like *Sr*31.

Higher temperature induced by the climate change is answerable for the evolution of several new plant pathogenic races which are expected to reduce wheat production in developing countries by 29%. Black stem rust is most threatening disease which poses a serious risk to wheat cultivation due to rapid evolution of new races or variants of the stem rust pathogen.

3.3 SUPER RACE

Ug99 carries wide virulence range against many resistance genes including Sr31 and therefore called a 'Super Race.' Majority of wheat varieties growing across the globe show susceptibility against this race so, the emergence of race Ug99 is considered as highly significant issue and have consequences on global wheat production. According to an estimate, area under risk of Ug99 is about 25% of the world's wheat area. The race has projected towards South Asia where wheat is one of the main indispensable food crops for a gigantic population. As a result, stem rust has again become a menace to global wheat production, and hence to the food security too.

3.4 HISTORICAL MOVEMENT OF UG99

Wheat stem rust caused the serious infection to wheat nurseries planted at CIMMYT in Uganda in the year 1999. It was quite surprising that the planting material that was resistant to stem rust pathogen became susceptible. When uredospore samples were collected for race analysis, the presence of a new race with novel resistance was found against the resistance gene Sr31. The new wheat stem rust race was called as popularly as Ug99 after the country and year of discovery.

Ug99 is the first known race of *Puccinia graminis tritici* which have the virulence against stem rust resistant gene Sr31. It is also having virulence against many other resistance genes of wheat origin or from other sources. Hence, there is virulence combination in Ug99 which explains why the worldwide wheat varieties and breeding material are susceptible to it. The virulence profile exhibited by different races in the Ug99 race group is fully unique and that is why concern over these races is completely justified. All these factors make the Ug99 lineage an exclusive race group posing a strong threat to a huge population of currently cultivating commercial wheat varieties.

The East African highlands have been known for being the "hot-spot" for the evolution and survival of new rust races. The availability of favorable environmental conditions in addition to host plant round the year, favor accumulation of pathogen population. The pathogen has a higher rate of evolution and so is having thirteen known variants within Ug99 lineage.

1993: Race TTKSK may have been present in Kenya.

1998: Ug99 identified, characterized as having virulence on Sr31, and named.

2000: Race TTKSF identified in South Africa.

2001: TTKSK identified in Kenya.

2003: TTKSK identified in Ethiopia.

2006: TTKSK identified in Sudan and Yemen, TTKST a new variant of Ug99 with virulence to Sr24 identified in Kenya.

2007: TTTSK identified in Kenya, TTKSP identified in South Africa, PTKSK identified in Ethiopia, PTKST identified in Ethiopia.

2008: FAO declared the presence of Ug99 in Iran, PTKST identified in Kenya.

2009: TTKSK identified in Tanzania, TTKST identified in Tanzania, TTTSK identified in Tanzania, TTKSF identified in Zimbabwe, PTKSK identified in Kenya, PTKST identified in South Africa.

2010: TTKST identified in Eritrea, PTKST identified in Eritrea, PTKST identified in Mozambique, PTKST identified in Zimbabwe, TTKSF+ identified in South Africa, TTKSF+ identified in Zimbabwe.

2015: TTKTT identified in Kenya, TTKTK identified in Kenya, Egypt, Eritrea, Rwanda, and Uganda, TTHSK, PTKTK, and TTHST identified in Kenya.

3.5 EXTENSION OF RACE UG99

Blowing winds carried Ug99 and its variants have now breached the best defense barriers. It is acting as a shifting adversary. Its range is expanding unceasingly at a very high rate. Originally, it was reported from Uganda, later it was reported from eastern Kenya in 2001. The presence of Ug99 in Ethiopia was recorded in 2005. In the year 2006 (February/ March) it was reported from eastern Sudan. During the same year in the month of October-November, reports confirming presence of race Ug99 were found from western Yemen. In the year 2007, samples collected from two field sites in Iran also confirmed the presence of race Ug99. The observed expansion of Ug99 and its variant into new areas is in-accordance with previous predictions made on its likely movement and it fits the step-wise dispersal model following prevailing winds.

The recent recognition of Ug99 in Egypt is being painstaking mainly important, as its dispersal route gives resilient indication that Ug99 is spreading towards the important wheat-growing areas of the Middle East and South Asia. Ug99 is not only dispersing, but also mutating at a higher and alarming rate to defeat stem rust resistance genes. Unfortunately, if Ug99 spreads these wheat-growing regions, it would affect probably 1 billion people living in these parts of the world. To avoid such a catastrophic event one of the best strategies is to identify and deploy wheat rust-resistant varieties that are suitable for Ug99- prone areas. Widespread screening of elite genotypes across the Ug99 hotspots is prerequisite for identification of the area-specific wheat genotypes with resistance to Ug99.

3.6 UG99: AN EVER-EMERGING MONSTER: WHY?

To understand the ever-evolving nature of race Ug99 it is important to have a look on the modes of dispersal of rust pathogen. Rust pathogen possesses two very important features which affect their rapid large distance movement. First one is that, their spores are robust ensuring protection against environmental damage enabling them to travel long distance either by wind-assisted movement or sometimes accidental human transmission. Wind assisted long-distance dispersal makes quarantine a near impossible task in this context. It becomes very hard to keep an eye on the spread and movement of rust spores.

Second feature is extravagant ability of rust pathogen to change and evolve through mutation or sexual reproduction. The members of the Ug99 lineage are currently the principal stem rust pathotypes.

3.7 MEASURES TAKEN TO COMBAT UG99

Borlaug Global Rust Initiative (BGRI) earlier known as Global Rust Initiative was established on 9th September 2005 in response to Ug99's threat, at Nairobi, Kenya under the leadership of the late Nobel Peace prize winner, Dr. Norman Borlaug. The "BGRI" is using the following strategies to reduce the possibilities of major epidemics:

1. Monitoring the spread of race Ug99 beyond eastern Africa for early warning and potential chemical interventions.
2. Screening of released varieties and germplasm for resistance.
3. Distributing sources of resistance worldwide for either direct use as varieties or for breeding.
4. Breeding to incorporate diverse resistance genes and adult plant resistance into high-yielding adapted varieties and new germplasm.

World wheat community and some famous donor organization (such as Melinda and Bill Gates Foundation) recognized threat of Ug99 race on wheat production and responded positively, and numerous research and developmental projects across the globe are now ongoing to battle against Ug99 under the collaboration of the BGRI. Screening of genotypes carried out in Kenya and Ethiopia have identified a very low frequency of resistant wheat varieties and breeding materials against Ug99 race. Identification and relocation of new sources of race-specific resistance from several wheat relatives is underway to augment the diversity of resistance. Even though new Ug99-resistant varieties that yield more than existing popular varieties are being released and encouraged, foremost efforts are compulsory to relocate current Ug99 susceptible varieties with elite varieties with diverse race-specific or durable resistance for mitigation of the Ug99 threat.

One of the finest approaches to avoid catastrophe caused by Ug99 is to identify and deploy wheat rust-resistant genotypes that are suitable for Ug99-prone regions. Large scale genotypes screening across the Ug99 hotspots are compulsory for identification of the area-specific resistance to Ug99. Resistant genes are one of the most effective weapons against these fast mutating and evolving pathogens. Number of *Sr* genes have been identified and incorporated into wheat genome in last 50 years. Some of these genes including *Sr22, Sr25, Sr27, Sr32, Sr33, Sr35, Sr37, Sr39, Sr40, Sr44, Sr45, Sr46* and a few unnamed genes are still resistant to Ug99 and its derivatives. They are virtually unusable in their current form as most of these genes are derived from wild relatives of wheat and are located on chromosome translocations that include large donor segments that harbor genes possibly deleterious to agronomic and quality traits. Thus, to enhance the utility of genes in wheat breeding, currently there are several ongoing research efforts to eliminate the deleterious linkage drag and to produce lines with smaller chromosome segments containing the resistance genes. These genes have been introduced into wheat but have not been deployed in commercial cultivars.

According to the recent scenario of wheat, stem rust in most of the area of world may suffer from serious epidemics and losses due to decline in wheat production. Also, there is very less availability of resistant wheat cultivars and germplasm against Ug99 races. Detection of Ug99 in Egypt is of utmost importance because it gives indication that Ug99 may move towards major wheat-growing areas of the Middle East and South Asia. Effective resistance genes were identified that were subsequently deployed in resistant wheat cultivars in high risk areas such as in Kenya and Ethiopia.

Numerous vital initiatives have also been taken up by India in this course for combating Ug99. Various Indian Institutes are working under collaborative wheat program with CIMMYT to identify different resistant cultivars and genotypes and their deployment in different wheat zones before onset of Ug99. A set of 19 Indian wheat varieties and 3 genetic stocks were screened under natural outbreak of Ug99 at Njoro (Nakuru), Kenya in the summer nursery 2005. A variety HW 1085 developed by IARI regional station, Wellington, and three genetic stocks (FLW2, FLW6 and FLW8 at DWR Regional Station, Shimla, India) show resistance against race Ug99. The existing data suggests that varieties like HI 1531, HI 8627, HD 4672, MACS 2846, NI 5439, 9498, WH 147, GW 322, HI 8663, UAS 321, DL 788-2, MPO 1215, NIDW 295, Lok 1, HI, and UAS 431, which are under cultivation in Central and Peninsular India, possess substantial resistance to Ug99 variants.

Till now Ug99, pathotypes have not been confirmed from India and Pakistan. About only 25% of the total wheat-growing area in India are prone to stem rust. Therefore, it may not be a major threat in the main wheat-growing belt, but still we have to be vigilant and prepared for any unforeseen entry of these pathotypes into the country.

KEYWORDS

- **Borlaug global rust initiative**
- **genotypes**
- **germplasm**
- **pathogen**
- **pathotypes**
- **Ug99-resistant varieties**

REFERENCES

Braun, H. J., Atlin, G., & Payne, T., (2010). Multi-location testing as a tool to identify plant response to global climate change. In: Reynolds, M. P., (ed.), *Climate Change and Crop Production* (pp. 115–138). CABI London, UK.

Fetch, T., & Zegeye, T., (2016). Detection of wheat stem rust races TTHSK and PTKTK in the Ug99 race group in Kenya in 2014. *Plant Dis., 100,* 1495.

Nagarajan, S., (2012). Is *Puccinia graminis* f. sp. *tritici* virulence Ug99 a threat to wheat production in the North West plain zone of India? *Indian Phytopath., 65,* 219–226.

Nazari, K., Mafi, M., Yahyaoui, A., & Park, R. P., (2009). Detection of wheat stem rust (*Puccinia graminis* f. sp. *tritici*) race TTKSK (Ug99). *Iran Plant Dis., 93,* 317.

Prasad, P., Bhardwaj, S. C., Khan, H., Gangwar, O. P., Kumar, S., & Singh, S. B., (2016). Ug99: Saga, reality and status. *Curr. Sci., 110,* 1614–1616.

Pretorius, Z. A., Szabo, L. J., Boshoff, W. H. P., Herselman, L., & Visser, B., (2012). First report of a new TTKSF race of wheat stem rust (*Puccinia graminis* f. sp. *tritici*) in South Africa and Zimbabwe. *Plant Dis., 96,* 590.

Pumphrey, M. O., (2012). Stocking the breeder's toolbox: An update on the status of resistance to stem rust in wheat. In: McIntosh, R., (ed.), *Proceedings Borlaug Global Rust Initiative 2012 Technical Workshop* (pp. 23–29). Beijing, China.

Rautela, A., & Dwivedi, M., (2018). Wheat stem rust race Ug99: A shifting enemy. *Int. J. Curr. Microbiol. App. Sci., 7*(1), 1262–1266.

Roelfs, A. P., & Martens, J. W., (1988). An international system of nomenclature for *Puccinia graminis* f. sp. *tritici. Phytopathol., 78,* 526–533.

Santini, A., & Ghelardini, L., (2015). Plant pathogen evolution and climate change. In: *CAB Reviews: Perspectives in Agriculture, Veterinary Science, Nutrition and Natural Resources.*

Sharma, A. K., Saharan, M., Bhardwaj, S. C., Prashar, M., Chatrath, R., Tiwari, V., Singh, M., & Sharma, I., (2015). Evaluation of wheat (*Triticum aestivum*) germplasm and varieties against stem rust (*Puccinia graminis* f.sp. *tritici*) patho type Ug99 and its variants. *Indian Phtopath, 68,* 134–138.

Singh, R. P., David, P. H., Huerta-Espino, J., Jin, Y., Bhavani, S., Njau, P., Herrera-Foessel, S., Singh, P. K., Singh, S., & Govindan, V., (2011). The emergence of Ug99 races of the stem rust fungus is a threat to world wheat production. *Ann. Rev. Phytopath, 49,* 465–481.

Singh, R. P., Hodson, D. P., Jin, Y., Huerta-Espino, J., & Kinyua, M., (2006). Current status, likely migration, and strategies to mitigate the threat to wheat production from race Ug99 (TTKS) of stem rust pathogen. In: *CAB Reviews: Perspectives in Agriculture, Veterinary Science, Nutrition and Natural Resources, 1,* 1–13.

Wanyera, R., Kinyua, M. G., Jin, Y., & Singh, R. P., (2006). The spread of stem rust caused by *Puccinia graminis* f. sp. *tritici*, with virulence on *Sr31* in wheat in Eastern Africa. *Plant Dis., 90,* 113.

Yu, L. X., Lorenz, A., Rutkoski, J., Singh, R. P., Bhavani, S., Huerta-Espino, J., & Sorrells, M. E., (2011). Association mapping and gene–gene interaction for stem rust resistance in CIMMYT spring wheat germplasm. *Theor. Appl. Genet., 123,* 1257–1268.

CHAPTER 4

Current Status of Bajra / Pearl Millet Diseases and Their Management

RANGANATHSWAMY MATH,[1] KOTTRAMMA C. ADDANGADI,[2] and
DURGA PRASAD AWASTHI[3]

[1]Department of Plant Pathology, College of Agriculture, Jabugam,
Anand Agricultural University–391155, Gujarat, India
E-mail: rangu.math@gmail.com

[2]S. D. Agricultural University, Bhachau–370140, Gujarat, India

[3]Regional Research Station, Department of Plant Pathology,
College of Agriculture, Tripura, India

4.1 INTRODUCTION

Bajra/Pearl millet is an important cereal crop of tropical and subtropical countries. Many fungi, bacteria, viruses, nematodes, and parasitic plants are known to infect and cause substantial damage to crops, resulting in heavy economic losses. The present chapter highlights the important diseases affecting the crop and their management strategies.

4.2 DOWNY MILDEW OR GREEN EAR DISEASE

4.2.1 INTRODUCTION

Downy mildew of pearl millet also referred to as 'green ear' disease and this is most destructive. This disease is widely distributed and reported from many countries of Asia, Africa, and Europe. In India, the disease is present in most of the states cultivating pearl millet. The pearl millet production was dramatically reduced from 8 million tons in 1970–71 to 5.3 million tons in 1971–72 due to this disease.

4.2.2   SYMPTOMS

Symptoms can be observed in two phases, one is downy mildew phase appear on the leaves and other is green ear phase appear on the earhead. The affected leaves show patches of light green to light yellow color on the upper surface of leaves and on the lower surface white downy growth of the fungus can be seen. The downy growth seen on infected leaves composed of sporangiophores and sporangia of the pathogen. Foliage become pale and chlorotic streaks develop over entire length of the leaf extending from the base to the tip. The infected leaves become brown, distorted, and wrinkled. In severe cases, infected leaves tend to split along the veins due to formation of oospores. The infected plants become dwarf and produce excessive tillering. Badly infected plants fail to form ears or if formed they are malformed into green leafy structures. Mostly the entire ear is transformed into leafy structure, but sometimes only, a part of ear alone is affected. As the disease advance, the green leafy structures become brown and dry (Figure 4.1).

4.2.3   CAUSAL ORGANISM

Sclerospora graminicola (Sacc) schroet.
Kingdom: Chromista
Phylum: Oomycota
Class: Oomycetes
Order: Peronosporales
Family: Peronosporaceae
Genus: *Sclerospora*
Species: *Graminicola*

The fungus is an obligate parasite. Hyphae are coenocytic (nonseptate), intercellular with small bulbous haustoria. Sporangiophores emerge through stomata either single or in tufts which are long stout with many upright branches bearing sporangia at its tip. Sporangiophore is determinate in growth and bear sterigmata. The sporangia are borne on the sterigmata which are thin-walled, hyaline, elliptical, and bear prominent papilla. Sporangia geminate and produce biflagellated zoospores (3 to 12 zoospores). Maximum sporangial production occurs at 20 to 25°C with relative humidity more than 75%. The sex organs develop within the host tissues (leaves and malformed flowers) and sexual spore (oospore) is produced after fertilization. Oospores are thick-walled resting spores produced in infected leaves. Amature oospore is brownish yellow and spherical, and measures 32 μm (22 to 35 μm) in diameter. Oospores remain present in soil and act as primary inoculum in the next growing season.

FIGURE 4.1 (A) Chlorosis on leaves; (B) downy growth on lower surface of leaf; (C) partial infection of earhead; (D) complete infected and healthy ears.

4.2.4 DISEASE CYCLE

The disease is primarily soil-borne and to some extent, it is seed-borne also. The oospores survive on the infected plant parts or on the seeds produced on diseased ears. The oospores present in the soil germinate producing the germ tube which causes primary infection by direct penetration of root hairs and the coleoptiles. When the infected seeds are sown in the fields, as the seed germinate and grow, along with it mycelium also grows and causes infection. The secondary spread of the disease takes place through sporangio-spores and sporangia. These sporangia are disseminated by wind, water, and insects, land on the susceptible parts of the host, and germinate producing zoospores. At the end of the growing season, the pathogen produces oospores in the infected leaves and inflorescence, which help in the carryover of inoculum to next season.

4.2.5 EPIDEMIOLOGY

Environmental factors such as temperature, humidity, rainfall, and cloudiness play an important role in the development and spread of disease. The disease is more in low lying and poorly drained soil. Leaf wetness, High relative humidity (90%) and temperature between 22°C and 25°C are considered most favorable for disease initiation and spread. The minimum, optimum, and maximum temperature for infection are 11°, 20°, and 34°C, respectively. Maximum production of sporangia occurs at temperature range of 15–25°C. No sporangial stage develops at temperatures above 30°C. On the host, surface Sporangia germinate within 30 min at 20–25°C producing zoospores. Nitrogenous fertilizers influence incidence and severity of the disease.

4.2.6 MANAGEMENT

- Deep plowing in summer which reduces the primary source of inoculum.
- Avoid low laying and waterlogging fields.
- Selection of disease-free, healthy seeds.
- Sowing immediately after the onset of monsoon.
- Seed dressing with fungicide Ridomil @ 0.1%.
- Spray with fungicide metalaxy 18%+ mancozeb 64%WP@0.1%.

- Disease crop debries must be destroyed which reduces the disease buildup of inoculum.
- Use resistant varieties like, GHB-351, GHB-316, NH-179, MH-169, GHB-30, GHB-32. MH 1192, ICMH 451, Pusa 23, MBH 110, PHB 57, WC-C 75, ICTP 8203, and GHB 67.

4.2.7 INTEGRATED DISEASE MANAGEMENT PRACTICES

1. **Pre-Sowing:**
    - Deep plowing during summer, Field sanitation and timely sowing of the crop.
2. **Sowing:**
    - Sowing of disease-free healthy seeds, use tolerant/resistant varieties. Seed dressing with fungicide Apron S D @ 8 g/kg seed.
    - Long crop rotation with non-hosts.
3. **Post Sowing:**
    - Apply the recommended levels of Nitrogen fertilizers in split doses.
    - Spray with metalaxy 18%+ mancozeb 64%WP@ 0.1% immediately after appearance of initial symptoms.
    - Remove and burn severely infected plants.

4.3 ERGOT/SUGARY DISEASE

4.3.1 INTRODUCTION

Ergot disease is known to occur in most of the pearl millet growing countries. It is reported from India, Pakistan, and several countries in Africa. In India, it was first reported from Maharashtra in 1956 in epiphytotic form. Now it widely present in Delhi, Karnataka, Tamil Nadu, Maharashtra, and Uttar Pradesh. Severe epidemics of the disease are reported from Delhi, Rajasthan, Karnataka, Tamil Nadu, and Gujarat. During 1967–68, it has broken out in epiphytotic form affecting several newly introduced bajra hybrids. The disease cause considerable yield loss up to 70% in susceptible varieties. Apart from quantity, it also affects the grain quality due to the presence of sclerotia. These sclerotia contain harmful alkaloids which are poisonous to human and animals.

4.3.2 SYMPTOMS

The symptoms can be noticed at flowering stage. Small droplets of a light, honey-colored dew like substance exudes from diseased florets. This carbohydrate-rich droplet contains abundant conidia of the fungus. After sometimes, masses of hyphae filling in the ovary harden giving rise to sclerotia. These sclerotia replace the grain and are about 0.5–1.0 cm in length. Sclerotia are hard, dark brown to black structures. They get mixed with the seeds and carry over to next season (Figure 4.2).

(a)   (b)

FIGURE 4.2   (a) Honeydew symptom (b) Ergot symptom.

4.3.3 CAUSAL ORGANISM

Kingdom: Fungi
Phylum: Ascomycota
Subphylum: Pezizomycotina
Class: Sordariomycetes

Subclass: Hypocreomycetidae
Order: Hypocreales
Family: Clavicipitaceae
Genus: *Claviceps*
Species: *Fusiformis*

The fungus attacks the ovary and grows profusely producing masses of hyphae. The pathogen produces septate mycelium which produces conidiophores which are closely arranged. Microconidia and macroconidia are observed in honeydew on small conidiophores. Macroconidia are colorless and fusiform. Microconidia are hyaline, globular, and unicellular. Sclerotia vary in shape, size (3.6–6 1 × 1.3–1.8 mm), color, and compactness, depending upon the host genotype and environmental conditions prevailing during infection and sclerotial development. Sclerotia germinate form the perithecium at the periphery of capitulum. Perithecium is the sexual fruiting body of the fungus. It contains asci and ascospores. The asci are long and cylindrical having eight long, filiform hyaline ascospores.

4.3.4 DISEASE CYCLE

Disease cycle begins with the sclerotia present in the soil or admixed with seed. These sclerotia germinate producing stripes, which bear perithecial heads that contain many perithecia embedded in them. Perithecia represent the sexual fruiting body of the pathogen containing asci and ascospores. Asci contain filiform, hyaline, septate thin-walled ascospores which get disseminated after asci burst open. The disseminated ascospores, then fall on the spike and under suitable conditions germinate enter inside the ovary and incite primary infection. Pearl millet flowers are most susceptible to infection only after stigma emergence and before pollination-fertilization. After primary infection is established, the hypahe fill in the ovaries and produce honeydew symptoms within a week. Secondary disease cycle begins with the production of numerous conidia within the infected spikelet. The conidia are taken away by splashing rain, wind, and insects and spread the inoculum to other healthy plants in the field. The conidia germinate there and cause secondary infection. Late in the season, hard sclerotia develop in place of grain. Sclerotia perennate and act as primary source of inoculum in next year.

4.3.5   EPIDEMIOLOGY

The optimum conditions for infection and disease spread are 12 mm rainfall, average sunshine of 6 hours, higher relative humidity, and 20°C mean temperatures at flowering stage. Cloudy weather (75–100% sky covered with clouds), drizzling rain, temperature of 20–25°C, air movement during crop flowering, affect the infection and spread of disease. Cumbersomely hefty rainfall along with high relative sultriness (>85% RH), will maximize the disease incidence. Morning relative sultriness (85–95%) during flowering and withal in the evening (60–90%) as compared to mundane evening relative sultriness of 45–50% will result in higher disease incidence. Higher relative humidity of 85–95% in the morning and during flowering will enhance disease incidence.

4.3.6   MANAGEMENT

- Repeated plowing may reduce the viability of deep buried sclerotia in the soil.
- Balanced application of fertilizers.
- Growing of green gram as an intercrop in the bajra will reduce disease incidence.
- Use of clean seed.
- Seed soaking in 10–20% brine solution (NaCl) to separate out the sclerotia from seed.
- Timely sowing of the crop.
- Take two to three sprays with ziram 0.2% or mancozeb @ 0.2% at an interval of five to six days starting just prior to earhead emergence.
- Spraying panicles with 0.2% Tilt at flowering minimizes ergot disease.
- Rotation of bajra crop with non-hosts for 5–6 years.

4.3.7   INTEGRATED DISEASE MANAGEMENT

1. **Pre Sowing:** Deep plowing, field sanitation, selection of healthy seeds. Timely sowing of the crop, seeping the seed in 15–20% salt solution and use of disease resistant cultivars.
2. **After Sowing:** Balanced application of fertilizers. Apply the recommended N in three splits as 25:50:25% at sowing, 15 and 30 days after sowing and full dose of phosphorus and potassium at sowing.

Spraying panicles with 0.1% carbendazim or 0.2% propiconazole at flowering time.

4.4 SMUT OF BAJRA / PEARL MILLET

4.4.1 INTRODUCTION

The Smut of pearl millet is a widespread disease reported from Africa, India, Pakistan, and USA. In India, it is more prevalent in Tamil Nadu, Andhra Pradesh, Haryana, Punjab, Gujarat, and Rajasthan states. The yield loss may vary from 20–30% based on the severity of the disease.

4.4.2 SYMPTOMS

The initial symptom of the disease is observed when grain setting starts in the ear. Only individual grains in an ear are infected. The grains are replaced by sori, which is normally bigger than the grains. Sori are covered with membrane of host origin, which disintegrates at maturity releasing black spore mass (Figure 4.3).

FIGURE 4.3 Smut infection on Bajra earhead.

4.4.3 CAUSAL ORGANISM

Tolyposporium penicillariae
Phylum: Basidiomycota
Sub Phylum: Ustilaginomycotina
Class: Ustilaginomycetes

Order: Ustilaginales
Family: Ustilaginaceae
Genus: *Tolyposporium*
Species: *Penicillariae*

The fungus infects developing flowers and the mycelium aggravates in the ovary and rounds off into chlamydospores. The pathogen remains mostly confined to sori. The sori contain teleutospores which are held together in compact balls. No columella is present. The spore balls measure around 40–150 μ*m*. The spores round to irregular, light brown in color with a rough wall. Teleutospores germinate producing four celled promycelium. Each cell of promycelium separates and produces masses of sporidia by budding.

4.4.4   DISEASE CYCLE

Pathogen survives as spore balls that have fallen onto ground, remain in soil. These smut balls germinate at the time of ear formation in the next growing season and produce sporidia. Sporidia are carried by wind and fall on the florets of the plant. The sporidia germinate and cause infection. They profuse and produce chlamydospores which cause secondary infection. The smut ball falls on the ground or gets mixed with seeds during threshing.

4.4.5   EPIDEMIOLOGY

Disease incidence will be higher if humidity is higher. Monoculturing of the susceptible variety is favoring the disease. Temperature around 30°C is favorable for disease development.

4.4.6   MANAGEMENT

- Deep plowing helps to bury the teleutospores and thus prevent their germination and release of primary infective propagules.
- Timely sowing of the crop.
- Seed treatment with Vitavax power @ 2 g/kg seed.
- Spraying with Zineb @ 0.2% or carbendazim @ 0.1% on the panicle are effective in managing the disease.
- Field sanitation and crop rotation will help manage the disease.
- Use of Tolerant varieties like WC-C 75, CM 46, and MBH 110.

4.5 RUST

4.5.1 INTRODUCTION

Rust is an important disease present in most of the countries cultivating pearl millet. In India, disease is reported from Gujarat, Rajasthan, Karnataka, and Maharashtra. Early infection will result in greater yield reduction due to drying and premature defoliation of infected leaves. Rust disease not only affects the yield but also reduces the test weight and forage quality.

4.5.2 SYMPTOMS

Symptoms first appear mostly on lower leaves as minute, oval, or elongated raised reddish brown pustules. First distal half of the leaf is infected and later pustules spread over both the surfaces. The matured pustules break the epidermis and release brown rusty spores. Severely infected leaf tissues were wilt and show necrosis symptom from tip to base of the leaf. Infected leaves dry prematurely. Later in the season, the pustules spread to most of the aerial parts of the plant. At the end of the season, telial formation can be observed on the infected plant parts. Telia were black, elliptical, and subepidermal. In severe infection, plants remain stunted (Figure 4.4).

FIGURE 4.4   Rust pustules on infected leaf and stem severely infected leaf.

4.5.3 CAUSAL ORGANISM

Puccinia substriata var. *indica (*Syn.*, P. substriata* var. *penicillariae)*
Phylum: Basidiomycota
Sub Phylum: Pucciniomycotina
Class: Pucciniomycetes
Order: Pucciniales
Family: Pucciniaceae
Genus: *Puccinia*
Species: *Pennicillariae*

The pathogen produces five different types of spores *viz.,* basidiospores, pycniospores, aeciospores, urediniospores, and teliospores which appear in a definite succession. Pathogen is heteroecious forming uredia and telia on bajra and aecial and pycnial stages on members of *Solanum,* including brinjal. Uredospores are oval, elliptical or pyriform, yellowish to brown, sparsely echinulate with four equatorial germ pores, pedicel hyaline, up to 60 μm long. Teliospores are brown, two-celled, cylindrical to club shaped with apex flattered.

4.5.4 DISEASE CYCLE AND EPIDEMIOLOGY

Pathogen is heteroecious in nature. Uredospores act as a repeating spore. They are carried by wind and spread the disease. The basidiospores germinate on brinjal or other solanaceous species. Significance of alternate host in the Perennation and primary infection needs further investigations. Maximum germination of uredospores occurs at 25°C. Leaf wetness, closer spacing, presence of abundant brinjal plants and other Solanum species will help to initiate and spread the disease at faster rate.

4.5.5 MANAGEMENT

- Sowing at right time.
- Crop rotation with other cereals or pulses.
- Cultivate resistant varieties like IB 1203, ICML 11, and MH 1192.
- Two sprays of mancozeb @ 0.2% or hexaconazole @ 0.1% at 10–14 days gap starting from initial appearance of the disease.
- Uproot and burn infected plants early enough to avoid spread of the disease.
- Keep the field weed-free and avoid cultivation of brinjal near to the bajra field.

4.6 BLAST

4.6.1 INTRODUCTION

It was first time reported from Kanpur, Uttar Pradesh. Earlier it was considered a minor disease but in recent years, the disease incidence has increased alarmingly. It causes substantial yield losses.

4.6.2 SYMPTOMS

The symptoms can be observed on most of the aerial parts of the plant. Initially, grayish water-soaked elliptical lesions surrounded by a chlorotic halo appear on the leaves. These lesions soon turn brown. Later lesions become necrotic forming concentric circles. Extensive chlorosis causes shedding of leaves before maturity. On the infected parts, pathogen forms numerous spores under high humidity (Figure 4.5).

FIGURE 4.5   Spindle-shaped brown spots on infected leaf.

4.6.3 CAUSAL ORGANISM

Pyricularia grisea
Sub Division: Deuteromycotina
Class: Hyphomycetes
Order: Moniliales
Family: Moniliaceae
Genus: *Pyricularia*
Species: *Grisea*

Mycelium in cultures aerial or submerged, hyaline or olivaceous, 1.5–6.0 µm in width, septate branched. Conidiophores 1 to many, fasciculate, simple or rarely branched with 2–4 septate. Conidia are hyaline, pyriform, two setpate with 3-celled (Figure 4.6).

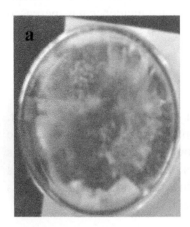 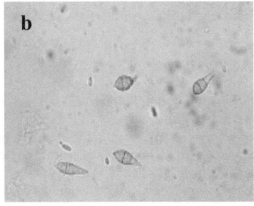

FIGURE 4.6   (a) Culture of *P. grisea*, (b) Conidia of *P. grisea* under microscope.

4.6.4 DISEASE CYCLE

- Primary infection caused by mycelium and conidia present in previously infected plant debries.
- Conidia produced as a result of primary infection serve as secondary source of inoculum which further spreads the disease.
- The mycelium survives in straw for 1–2 years under dry condition.

4.6.5 EPIDEMIOLOGY

- Disease is favored by high humidity, intermittent drizzles, overcast sky, and slow wind movement.

4.6.6 MANAGEMENT

- Selection of healthy seeds for sowing.
- Treat the seeds with either of the fungicide before sowing i.e., carbendazim or vitavax power@ 2 g/kg.

- Sowing the crop at proper time.
- Foliar spray with edifenphos 250 ml or iprobenphos 500 ml or carbendazim 250 g or tricyclazole 400 g/ha just after initiation of the disease.

KEYWORDS

- **chlorosis**
- **olivaceous**
- **pyriform**
- **Solanum species**
- **teliospores**
- **uredospores**

REFERENCES

Bhat, R. V., Roy, D. N., & Tulpule, P. G., (1976). The nature of alkaloids of ergoty pearl millet and its comparison with alkaloids of ergoty rye and ergoty wheat. *Toxicol. Appl. Pharmacol., 36*, 11–17.

Bhat, S., (1973). Investigations on the biology and control of *Sclerospora graminicola* on bajra. *PhD Thesis.* Department of Botany, University of Mysore, India.

Das, I. K., Nagaraja, A., & Vilas, A. T., (2016). Diseases of millets. *ICAR- Indian Institute of Millets Research* (pp. 1–71). Rajendranagar, Hyderabad.

Deepak, S. A., Chaluvaraju, G., Basavaraju, P., Amutesh, K. N., Shetty, H. S., & Oros, G., (2005). Response of pearl millet downy mildew (*Sclerospora graminicola*) to diverse fungicides. *In. J. Pest Management, 51*, 7–16.

Gupta, G. K., & Singh, D., (2000). Epidemiological studies on downy mildew of pearl millet (*Pennisetum glaucum*). *Int. J. Tro. Pl. Dis., 18*, 101–115.

Kumar, A., & Manga, V. K., (2011). Downy mildew of pearl millet. *Bioresearch Bulletin, 1*(4), 1–14.

Lukose, C., & Dave, H. R., (1995). Germination of oospores of *Sclerospora graminicola*. *Indian Phytopathol., 48*, 74–76.

Lukose, C. M., Kadvani, D. L., & Dangaria, C. J., (2007). Efficacy of fungicides in controlling blast disease of pearl millet. *Indian Phytopathol., 60*, 68–71.

Nene, Y. L., & Singh, S. D., (1976). Downy mildew and ergot of pearl millet. *Pest Articles and News Summary, 22*, 366–385.

Rangaswami, G., & Mahadevan, A., (2014). *Diseases of Crop Plants in India* (4th edn., pp. 227–234). PLI, Pvt. Ltd, New Delhi.

Rao, V. P., Thakur, R. P., Rai, K. N., & Sharma, Y. K., (2005). Downy mildew incidence on pearl millet cultivars and pathogenic variability among isolates of *Sclerospora graminicola* in Rajasthan. *International Sorghum and Millets Newsletter, 46,* 107–110.

Sharma, R., Upadhyaya, H. D., Manjunatha, S. V., Rai, K. N., Gupta, S. K., & Thakur, R. P., (2013). Pathogenic variation in the pearl millet blast pathogen *Magnaporthe grisea* and identification of resistance to diverse pathotypes. *Plant Dis., 97,* 189–195.

Shetty, H. S., & Kumar, V. U., (2000). Biological control of pearl millet downy mildew: Present status and future prospects. In: Upadhyay, R. K., Mukerji, K. G., & Chamola, B. P., (eds.), *Biocontrol Potential and its Exploitation in Sustainable Agriculture* (Vol. I, pp. 251–265). Springer, USA.

Singh, R. P., (2014). *Plant Pathology* (pp. 1–717.). Kalyani Publishers, New Delhi.

Singh, S. D., & Shetty, H. S., (1990). Efficacy of systemic fungicide metalaxyl for the control of downy mildew (*Sclerospora graminicola*) of pearl millet (*Pennisetum glaucum*). *Ind. J. of Agric. Sci., 60,* 575–581.

Singh, S. D., (1995). Downy mildew of Bajra. *Pl. Dis., 79*(6), 545–550.

Thakur, R. P., & King, S. B., (1988). *Smut Disease of Pearl Millet* (p. 20). Information Bulletin No.25, Patancheru, A.P. 502324, and India: International crops research institute for the semi-arid tropics.

Thakur, R. P., Rao, V. P., & Williams, R. J., (1984). The morphology and disease cycle of ergot caused by *Claviceps fusiformis* in pearl millet. *Phytopathol., 74,* 201–205.

Thakur, R. P., Sharma, R., & Rao, V. P., (2011). *Screening Techniques for Pearl Millet Diseases: Information Bulletin No. 89* (pp. 1–56). Patancheru 502324, Andhra Pradesh, India. International Crops Research Institute for the Semi-Arid Tropics.

Thakur, R. P., Shetty, H. S., & Khairwal, I. S., (2006). Pearl millet downy mildew research in India. *International Sorghum and Millets News Letter, 47,* 125–130.

Thakur, R. P., & King, S. B., (1988). *Ergot Disease of Pearl Millet: Information Bulletin No. 24.* Patancheru, A.P. 502324, India: International Crops Research Institute for the Semi-Arid Tropics.

The AESA Based IPM-Pearl Millet, (2014). *DPPQ&S and NIPHM, GOI, Dept. of Agriculture and Co-Operation, Ministry of Agriculture* (p. 51).

Tripathi, R. K., Kolte, S. J., & Nene, Y. L., (1981). Mycoparasite of *Claviceps fusiformis,* the causal fungus of ergot of pearl millet. *Ind. J. Mycol. Pl. Pathol., 11,* 114–115.

CHAPTER 5

Diseases of Maize Crops and Their Integrated Management

A. K. SINGH,[1] V. B. SINGH,[2] J. N. SRIVASTAVA,[3] S. K. SINGH,[1] and ANIL GUPTA[1]

[1]Division of Plant Pathology, Sher-E-Kashmir University of Agricultural Sciences and Technology (SKUAST-J), Chatha, Jammu–180009, Jammu & Kashmir, India

[2]Rainfed Research Sub-Station for Sub-Tropical Fruits, Raya, Technology (SKUAST-J), Jammu–180009, Jammu & Kashmir, India

[3]Department of Plant Pathology, Bihar Agricultural University, Sabour–813210, Bhagalpur, Bihar, India

5.1 INTRODUCTION

In India, maize is the third most important cereal crops after rice and wheat. According to advance estimate, its production is likely to be 21.80 m t (2015–16) mainly during Kharif season which covers 80% area. In India, its contributes nearly 9% in the national food basket. In addition to staple food for human being and quality feed for animals, maize also provides a basic raw material as an ingredient to many industrial products *viz.* starch, oil, protein, alcoholic beverages, food sweeteners, pharmaceutical, cosmetic, film, textile, gum, package, paper industries, etc.

In India, maize is cultivated throughout the year in Rabi, kharif, and zaid season in most states of the country for various purposes, i.e., grain, fodder, green cobs, sweet corn, baby corn, popcorn in peri-urban areas. The major maize growing states that contributes more than 80% of the total maize production are Andhra Pradesh (20.9%), Karnataka (16.5%), Rajasthan (9.9%), Maharashtra (9.1%), Bihar (8.9%), Uttar Pradesh (6.1%), Madhya Pradesh (5.7%), and Himachal Pradesh (4.4%). Apart from these major states, maize is successfully grown in Jammu and Kashmir and most of the

North-Eastern states. In recent years, the maize has emerged as a major crop in the non-traditional regions, i.e., peninsular India as the state like Andhra Pradesh which ranks 5th in area (0.79 m ha) has recorded the highest production (4.14 m t) with the productivity (5.26 t ha[-1]) in the country (Murdia et al., 2016).

As like another cereal crops, maize crops is also susceptible to many diseases caused by a fungi, bacteria, virus, and phytoplasma. Under favorable environmental conditions, a number of them are capable of causing severe losses and impair the quality of the produce or some time complete failure of the crop. Although 18 foliar diseases occur (common rust, 16 other fungal diseases and 1 bacterial disease), three are considered to be of major importance in terms of geographical distribution and potential to cause significant reductions. The total estimate of loss in the economic product per annum has been determined to be the order of 13.2% which is considerable more than the global disease loss estimates of 9.4% of total production.

5.2   FOLIAR DISEASES

5.2.1   *HELMINTHOSPORIUM LEAF BLIGHT*

5.2.1.1   *TURCICUM LEAF BLIGHT OR NORTHERN CORN LEAF BLIGHT OR LEAF STRIPE*

5.2.1.1.1   *Introduction*

The leaf blight occurs worldwide and particularly in areas where high humidity and moderate temperatures prevail during the growing season. When infection occurs prior to and at silking and conditions are optimum and causes significant damage to the maize crop. In India, the blights are prevalent in most of the maize growing areas but the species varied in different areas. The disease was first time reported from Italy in 1876. After that, it has also been recorded from other parts of the world *viz.* United States, South Africa, Japan, and the Philippines. While, in India this disease first time reported from Bihar in 1907. Turcicum leaf blight is quite common in many parts of India especially in hill region of India. The disease has been occurred as epiphytotic in Lal Mandi (Sri Nagar) India.

The disease appears as leaf necrosis and premature death of foliage which reduces the fodder yield and quality (Payak and Sharma, 1985). Rai 1987 reported the loss in grain yield from 27.6 to 97.5% and loss was directly proportional to intensity of the disease. Most of the composites and hybrids varieties, which are being cultivated commercially its susceptible to TLB. Patil et al. (2000) reported that, TLB in maize can cause yield loss in the range of 13.6 to 56.0% depending upon the genotype. The local cultivars recorded 66.0% reduction in yield due to TLB (Payak and Sharma, 1985).

5.2.1.1.2 Symptoms

The symptoms occurred as boat-shaped, grayish-green or tan color spots, measuring 4–20 cm in length and 1–5 cm in width produced firstly on lower leaves and further on the upper leaves. The spots gradually increase in size and number as the plant develops and can lead to complete burning of the foliage in susceptible plants (Figure 5.1). This disease may be appearing at any growth stage, but usually at or after a thesis. The surface of the leaves is covered with olive green velvety masses of conidia and conidiophores. Under high humidity when disease in sever conditions the entire leaf area becomes necrotic and plant appears as dead. Lesions may also be extended to husk.

Kernels are not infected even in the case of husk infection. In severely infected plants, the cobs are small and poorly filled. In severe infection causes a prematurely death and gray appearance that resembles frost or drought injury.

FIGURE 5.1 Turcicum leaf blight.

5.2.1.1.3   Casual Organism

Helminthosporium turcicum
Syn. *Exserohilum turcicum* (Pass.)
Sexual/Perfect Stage/Teleomorph:
Syn. *Setosphaeria turcica* (Luttrella) Leonard and Suggs.
Syn. *Drechslera turcica* (Pass) Subramanian and Jain
Syn. Bipolaris turcica (Pass), Shoemaker

The fungus belongs to division Eumycota, subdivision Deuteromycotina, order Moniliales, and class Dematiaceae. The teleomorph belongs to division Eumycota, subdivision Ascomycotina, order Pleosporales, and family Pleosporaceae.

The fungus develops conidia on the conidiophores which are geniculated and each conidium has 3–8 septa and is slightly curved. The conidia reach to the host through wind and cause infection through bipolar germination on free water on leaf and at a temperature from 18–27°C. They can cause leaf spot symptoms within 7–12 days.

5.2.1.1.4   Disease Cycle

Crop residues are the major source of survival and build-up of the primary inoculum. Once the infection takes place, it spreads through wind and splashing water. The crop may be severe in area where heavy dew and rainfall occurs. Spores germinate and penetrate the leaves within a few hours in the presence of free water and when the temperature range is 25–30°C. Losses are high if the disease occurs in the early stages of the crop in a severe form; blighted leaves are killed prematurely and lose their nutritive value especially for fodder purpose.

5.2.1.1.5   Epidemiology/Favorable Conditions

The disease is most common in areas where cooler condition prevails and maize is planted in high lands, winter planting in the plains as the cool/ moderate humid conditions (18–27°C) favors for disease developments. When infection occurs prior to and at silking stage and conditions are favorable for the development of the disease, it may cause significant economic damage.

5.2.1.1.6 Management

1. **Resistance Varieties:** Host resistance is the painless method of disease management not only safest but also economic. Two composite varieties, NAC-6002 (early maturity) and NAC-6004 (late maturity) were resistant against turcicum leaf blight of maize (Pandurange et al., 2001, 2002). However, Vivek 21, Vivek 23, Vivek 25, Pratap, Kanchan Rajendra Hybrid Makka-1, R.H.M, M-2, Deoki, Lakshmi, PRO345, JH10655, and MCH117 are tolerant varieties which can be recommended for cultivation in turcicum leaf blight endemic areas.

2. **Chemical Control:** Bowen and Pederson (1988) found that application of propiconazole was effective in reducing the rate of disease development when applied thrice at weekly intervals. Pandurange Gowda et al. (1993) evaluated eight fungicides on two maize hybrids Deccan and Ganga – 5 Maneb (0.25%) and mancozeb (0.25%) were significantly effective in reducing the disease severity and increasing grain yield. Harlapur (2005), observed that minimum PDI (25.38%) and maximum grain yield (64.41 g/ha) was recorded when seed treated with carboxin (power 2 g/kg) followed by three sprays of mancozeb (0.25%) at 30, 40 and 50 days after sowing of crop.

5.2.1.2 MAYDIS LEAF BLIGHT OR SOUTHERN LEAF BLIGHT

5.2.1.2.1 Introduction

Maydis leaf blight is an important disease in the India and in several other countries of the world also. The disease is generally distributed over the world in warm temperate and tropical (20–30°C) maize producing areas. Munjal and Kapoor (1960) for the first time reported its presence from India who isolated it from Maldah (West Bengal). In 1970, this disease reached epidemic proportions in the USA, causing a losses estimate at 1 billion dollars. The epidemic was caused by race "T" attacking corn with Texas male sterile cytoplasm which comprised of 85% of corn acreage at that time. Race 0 has never reached in epidemic proportions, but yield reductions up to 50% have been detected in susceptible cultivars (Payak and Sharma, 1978).

5.2.1.2.2 Symptoms

The symptoms "O" strain of the fungus appears as young minute and diamond shaped spots and elongate when they become mature. Growth is limited by adjacent veins, so final lesion shape is rectangular, and 2 to 3 cm long. Lesions may coalesce, producing a complete burning of large areas of the leaves, whereas "T" strain caused severe damage to maize cultivars in which the Texas source of male sterility had been incorporated. Lesions produced by the T strain are oval and larger than those produced by the O strain.

In Maydis leaf blight, individual spots are grayish, tan in color, up to 2.5 to 4.0 cm in length, oval-shaped with straight zonations. Disease appears on leaves as numerous, dead cinnamon buff or purplish lesions surrounded by a darker reddish brown margin often delicately variegated with brownish zonate bands. Lesions on the leaves are longitudinally elongated, limited to single vascular region and coalesce to form more extensive dead portions. Lesions may coalesce, producing a complete burning of large areas of the leaves (Figure 5.2). These symptoms appeared due to "O" strain of the fungus.

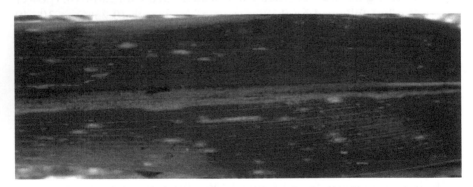

FIGURE 5.2 Maydis leaf blight.

In the early 1970s, the "T" strain caused severe damage to maize cultivars in which the Texas source of male sterility had been incorporated. Lesions produced by the T strain are oval and larger than those produced by the O strain. A major difference is that the T strain affects husks and leaf sheaths, while the O strain normally does not. Maydis leaf blight (or southern maize leaf blight) is prevalent in hot, humid maize-growing areas. The fungus requires slightly higher temperatures for infection than

Helminthosporium turcicum; however, both species are together found on the same plant.

5.2.1.2.3 Casual Organism

Helminthosporium maydis
Syn: *Bipolaris maydis*

5.2.1.2.4 Disease Cycle

The fungus survives as dormant mycelium and spores in maize debris in the field and on kernels in cribs, bins, and elevators. Conidia are carried by wind or splashing water to growing plants where primary infection occurs. Sporulation on the lesions produces additional primary or secondary inoculum. The disease cycle can be completed in about 60 to 72 hours (race T) under ideal conditions. Infected kernel may be potential means of overwintering and spread of *H. maydis*. Infected grains are not toxic to livestock.

5.2.1.2.5 Epidemiology/Favorable Conditions

This disease is commonly occurred in hot, humid, maize-growing areas where the temperature varies between 20–30°C during cropping period/during flowering period.

5.2.1.2.6 Management

1. **Cultural:** Considerable efforts have been made to manage Maydis leaf blight through altered tillage, planting date and mixed populations of maize containing *Tcms* and normal cytoplasm (Sumner and Littrell, 1974). The diseased crop debris should be buried deep or burnt as it helps in reducing the initial inoculurn and delays the appearance of the disease.
2. **Resistance:** Varieties like Ganga 5, Deccan 101, VL 42, Ganga 4, Jawahar, KH 101, KH 510, KH 517, KH 581, PSCL 3438, PRO 311, RIO 9637, Kiran, Seed Tech. 1240, are reported to be resistant from

India (Anonymous, 1999; Basandrai et al., 2000). Similarly, resistant sources have been reported from Pakistan (Khan et al., 1992), Korea (Park et al., 1991), China (Chin et al., 1989), and the USA (Coors and Mardans, 1989; Goodman et al., 1989). Inbred lines, CM 103, CM 105, CM 106, CM III, CM 113, CM 201, CM 600, Eto 28A, Eto 81, Eto I 82c, (Venz I × Venz 400), Ph DMR I, Ph DMR 5, Eva (MDR I, Eva (MDR 11)-76 are reported to have resistance from India (Sharma et al., 1993). Resistance against race is governed by several additive genes and single or two linked recessive genes. Recessive gene 'rhrni has been found to govern resistance to race '0' or 'T' (Sharma et al., 1993), Ceballos, and Gracen (1988) reported that resistance to *Bipolaris maydis* was controlled by two independent, complimentary, and highly dominant genes.

3. **Chemical:** Fungicides like mancozeb, thiram, and carboxin have been reported to be effective (Schenck and Stelter, 1974). Two to four foliar sprays of fungicides Dithane Z-78/Dithane M-45 @ 0.25% concentration are sufficient depending upon the disease severity.

5.2.2 BANDED LEAF AND SHEATH BLIGHT (BLSB)

5.2.2.1 INTRODUCTION

Banded leaf and sheath blight (BLSB) of maize is known as different names, viz; sclerotial disease, sharp eye spot, oriental leaf and sheath blight, Rhizoctonia ear rot, sheath rot, corn sheath blight, etc. (Rijal et al., 2007).

BLSB disease was considered as a minor disease until 1960s, however, in the subsequent years, it has become one of the major diseases in the countries of tropical Asia. In India, it is prevalent in hot humid foot-hill regions in Himalayas and in plains also.

BLSB disease of maize is the main production constraint in the several parts of the maize growing areas of the world. It was the first time reported from Sri Lanka (Bertus, 1927) under the name 'sclerotial' disease. In India, the disease was first reported as BLSB of maize caused by *Hypochonus sasakii* by Ullstrup in 1960 from Tarai region of Uttar Pradesh (Payak and Renfro, 1966). This disease mainly occurs in Jammu and Kashmir, Himachal Pradesh, Sikkim, Punjab, Haryana, Rajasthan, Madhya Pradesh, Delhi, Uttar Pradesh, and Bihar (Parihar et al., 2011). The menace shot into prominence in 1972 when it caused an unprecedented epidemic in foothills of Mandi district of Himachal Pradesh (Thakur et al., 1973). Rajput and Harlapur (2014) reported this disease severity a range of 12.57 to 52.45% from different districts of Karnataka state.

5.2.2.2 SYMPTOMS

The symptoms are characterized on lowermost leaf sheaths first show disease manifestation by the presence of irregular brownish patches with reddish brown margins. These spots increase in size and contaminate the upper leaf sheaths, eventually killing the blades. The sclerotia formed on the outer part of the dried leaf sheaths and blades are first white later becoming brown with varying sizes range from 1.4 to 2.7 mm and differentiate into rind and midula (Figure 5.4). The symptoms are more common on sheaths than on leaves. The invaded portion first show water-soaked appearance later becoming dirty white and appear as bands with length ranging from 3.25 to 4.65 cm across the leaf sheaths and leaf blades wherever initial infection. The disease appears on basal leaf sheaths as water-soaked, straw-colored, irregular to roundish spots on both the surfaces (Figure 5.3). A short of wave pattern of disease advancement can be seen not only on leaves but also on sheaths and husk leaves. In early stage, marginal chlorosis and rotting of laminae proceed inwardly. Later as the infection becomes older numerous sclerotial bodies are also seen (Saxena, 1997).

The ear rot phase may be differentiated into three types: (i) in case of infection develop prior to ear emergence, the development of ear is completely suppressed, (ii) if the pathogen reaches the ear shoot after ear emergence, the stalk fibers at the tip darkened become caked up and turn into hardened lump leading to poor filling, and (iii) if infection occurs after grain formation, the kernels become light in weight, chaffy, and lusterless. The developing ear is completely damaged and dries up pre-maturely with tacking of the husk leaves.

Profuse mycelia growth is seen on the affected areas of leaf sheath and in between the leaf sheath and stem. It is a destructive plant pathogen with an almost unlimited host range. It aggressively colonizes organic debris and thus has enough saprophytic survival ability. No sexual spores are formed and only sclerotia formed as soil-borne propagules. The disease after its occurrence causes direct loss due to premature death, stalk breakage, destruction of leaves, leaf sheaths, and ear rot.

FIGURE 5.3　Leaf sheath.

FIGURE 5.4　Sclerotia on cob.

5.2.2.3 CASUAL ORGANISM

Rhizoctonia solani f. sp. *sasakii* (Kuhn) Exner
Teleomorph: *Corticium sasakii,*
syn. *Thanatephorus cucumeris* (Frank) Donk

5.2.2.4 EPIDEMIOLOGY / FAVORABLE CONDITIONS

Warm and humid conditions temperature ranged 15–35°C along with water-logged and dense plant population situations are congenial for development of the disease.

5.2.2.5 MANAGEMENT

1. **Cultural:** Sharma and Hembram (1990) have observed that stripping of lower 2–3 leaves along with their sheaths considerably lower disease incidence and also does not affect grain yield. This method has been found to be effective in checking the disease spread significantly not only in India but in other countries of Asia also, where this disease causes considerable loss in maize crop.

2. **Chemical:** Several fungicides like carbendazim, bengard, thiophanate-methyl, iprobenfos, captan, quintozene, mancozeb + thiophenate-methyl, copperoxychloride, and thiram have been found effective in inhibiting the growth of pathogen in- *in vitro* conditions. Under field conditions also, all the fungicides except thiram were effective in reducing disease severity (Sharma and Rai, 1999; Puzari et al., 1998). Rani et al., (2013) recorded lowest percent (27.11%) disease incidence when seeds were treated with carbendazim further, Kumari (2012) also reported that, foliar spray of Tilt 25 EC was found superior over all other treatments giving lowest disease severity (19.75%), along with highest grain yield (48.41 q/ha) which was followed by Contaf (21.55%) and Bavistin (31.41%) along with (48.17 q/ha) and (47.17 q/ha) grain yield respectively. The formulation of Validamycin has shown good control against BLSB pathogen (Batsa, 2003).

3. **Bio-Control:** In recent years, there has been considerable popular and scientific resistance to use of pesticides/fungicides for the control of diseases and pests in food crop plant the world over. It, therefore, becomes all the getting importance that alternative environmental

friendly methods should be discovered to alleviate the hazardous effects of chemical control agents in the food chain. Biological control of diseases offers the preferred mode of environmental friendly control of diseases in maize including BLSB. Several antagonists such as species of *Trichoderma, Gliocladium virens, Bacillus subtelis,* and *Pseudomonas fluorescens* have been tried by different workers. However, the encouraging results have been obtained by the use of *P. fluorescens.* Shivakumar et al. (2000) demonstrated a very effective control by seed treatment of the peat-based formulation (@ 16 g/kg, or soil application 2.5 kg/ha, or spraying liquid formulation twice @ 5 'liter of water. Manisha et al. (2002) also observed that combined seed treatment and foliar spray with fluorescent pseudomonas from maize rhizoplane was most effective in giving 30.0% reduction in disease incidence of BLSB. Similarly, among several spp. of *Trichodertna* has shown good promise against this pathogen both in *in-vivo* and *in-vitro* conditions (Meena et al., 2003).

5.2.3   DOWNY MILDEWS OF MAIZE

There are several species of the genera *Peronosclerospora, Sclerospora,* and *Sclerophthora* are responsible for downy mildews:

1. **Crazy Top Downy Mildew:** *Sclerophthora macrospora.*
2. **Brown Stripe Downy Mildew (BSDM):** *Sclerophthora rayssiae* var. *zeae.*
3. **Java Downy Mildew:** *Peronosclerospora maydis.*
4. **Philippine Downy Mildew:** *Peronosclerospora philippinensis.*
5. **Sugarcane Downy Mildew:** *Peronosclerospora sacchari.*
6. **Sorghum Downy Mildew:** *Peronosclerospora sorghi.*

5.2.3.1   INTRODUCTION

Downy mildew diseases are of serious concern to maize producers in several countries of Asia, Africa, and throughout the Americas. The symptom expression is greatly affected by plant age, species of the pathogen, and environment conditions of the locality. Usually, there is chlorotic striping or partial symptoms in leaves and leaf sheaths, along with dwarfing. Downy mildew becomes conspicuous after development of a downy growth on or under leaf surfaces. This condition is the result of sporangia formation, which commonly occurs in the early morning.

The diseases are most prevalent in warm, humid regions. Some species causing downy mildew also induce tassel malformations, blocking pollen production and earhead formation. Leaves may become narrow, thick, and abnormally erect.

5.2.3.2 SYMPTOMS

Symptoms produced by most of the species on maize leaves. Long chlorotic streaks/strips and yellow-white color of streaks/strips appear on the leaves within a month and the plants exhibit a stunted and bushy appearance due to the shortening of the internodes. The disease becomes apparent before the plant has attained full height. White downy growth of the fungus can be seen on both surface of leaf lower and upper. Affected leaves often tear linearly causing leaf shredding. The downy growth also occurs on bracts of green unopened male flowers in the tassel. The important symptom of the disease is the partial or complete malformation of the tassel into a mass of narrow, twisted leafy structures. Proliferation of axillary buds on the stalk of tassel as well as the cobs is very common (Crazy top).

5.2.3.3 ETIOLOGY

The fungus grows as white downy growth on the lower and upper surface of the leaves, consisting of Sporangiophores and sporangia. Sporangiophores are short and stout, branch profusely into series of pointed sterigmata which bear hyaline, oblong, or ovoid sporangia (conidia). Sporangia germinate directly and infect the plants. In later stages, oospores are formed which are spherical, thick-walled and deep brown.

5.2.3.4 DISEASE CYCLE

The oospore present in soil and they plays a crucial role as a source of primary infection. Phillppine downy mildew *(Peronosclerospora philippinensis)* on maize plant also transmitted through collateral hosts (*Sorghum bicolor, Sorghum halapense*, etc), in Punjab, *Digitaria sanguinalis* serve as primary source of infection. Crazy top downy mildew (*Sclerophthora macrospora)* on maize plant also transmitted through seed and Further they form sporangia and the infection appear in the form of minute flecks (Characteristics of zoosporic infections) on the leaves which grow to brown streaks. Secondary infection is brought about sporangia formed on the lower surface of infected leaves. Secondary infection is through air-borne sporangia and sporangium germinates directly without production of zoospores.

5.2.3.5   EPIDEMIOLOGY / FAVORABLE CONDITIONS

High relative humidity (90%), water logging condition, light drizzles with a temperature of 20–25°C congenial the development of downy mildews. Young plants are more susceptible than older plants.

5.2.3.6   MANAGEMENT

Destruction of plant debris of maize and also removal and destruction of collateral hosts help the disease management. Deep summer ploughing and crop rotation with pulses can reduce the downy mildew infections. Varieties like (hybrids) Ganga Safed-2, Ganga Safed-101, DHM-1, DHM-103, DMR-5, and Ganaga II are reported to be resistant the disease from India. Seed treatment with Metalaxyl (Apron 35SD) at 1.75–2.00 g/kg reduce he disease incidence and severity. Spray the maize crop, 3–4 times, with Metalaxyl MZ (Ridomil MZ)@0.2% starting from 20th day after sowing control the disease. Spraying of Dithane M 45 @o.25% or any copper fungicides @0.3% is equally effective. First spraying should be given with the appearance of the disease and followed 2 to 3 times depending upon the severity of the disease.

5.2.4   COMMON RUST

5.2.4.1   INTRODUCTION

The rust disease was first recorded on tea mays L. and sorghum at Bathlehem, Pennsylvania, the USA by Schweinitz in 1832, and the pathogen was named as *Puccinia sorghi*. The disease is widely distributed throughout the world including the USA and India. It is more serious in Mexico and Central America and results in premature killing of plants. In India, it was first recorded from Mashobra (H.P.) in 1890 and appears mostly during the Rabi season in plain areas. Two more fungi, namely *Puccinia polysora* and *Physopellazeae* also cause rust symptoms on maize. Subedi, 2015 reported that, common rust, caused by *Puccinia sorghi* Schw., develops on maize plantings in subtropical conditions in Nepal, Bhutan, South China, and northern India. Losses ranging from 6.0–36.0% in yield have been recorded in different states of USA, whereas in cooler regions of tropics, 25.0–50.0% losses in yield have been recorded (Grath et al., 1992). In India, losses between 28–32% have been recorded on local varieties in Bihar (Sharma et al., 1982).

5.2.4.2 SYMPTOMS

The symptoms of rust appeared on both surfaces of leaves and are characterized by appearance of circular to elongate, cinnamon brown and powdery pustules. These pustules burst by breaking through the host epidermis to release reddish brown mass of uredospores which are repeating spores. In severe infection, the entire plants appear brownish and rusted. With the maturity of the crop, the uredosori change into teleutosori and produce a crop of black teleutospores (Figure 5.5).

FIGURE 5.5   Common rust.

5.2.4.3 CAUSAL ORGANISM

Puccinia sorghi.

The fungus belongs to division Eumycota, subdivision Basidiomycotina, class Teliomycetes, order Uredinales, and family Pucciniaceae. It is macrocyclic heterocious rust and produces uredial and telial stage on maize. Teleutospores produced on maize and produced promycelium which bear basidiospores. The basidiospores infect three species of *Oxalis*, i.e., *O. corniculata*, *O. europea*, and *O. stricta* or wood sorrel where pycnial and aecial stages occur. Uredospores are globose or sub-globose and measure 24–32 × 20–28 μm. Teleutospores are oblong to ovate, ellipsoid or sub-clavate, bright chestnut: brown, rounded or bluntly acuminate at the apex, slightly constricted at the septum, rounded or sub-alternate at the base, measuring 30–45 × 16–24 μm. Aeciospores are angular, globose, or ellipsoid, 18–26 × 14–20 μm, pale yellow and finely verrucose.

5.2.4.4   DISEASE CYCLE

The teleutospores remain viable on maize straw under natural conditions in areas above 1700 m, until mid-June and natural infection of *O. corniculata* occurs during April to June in high hills (Misra and Sharma, 1964). However, the role of telia in India is not well understood. Uredospores can survive by repeating life cycle on maize and other hosts, so perpetuation of disease through uredospores is continuous on maize plants. In some areas, the maize crop is grown the year round facilitating the perpetuation and survival of the pathogen. The role of *Oxalis* spp. as alternate host is not so important in India. In north India, primary inoculum of the disease comes from Nepal while in south India it comes from Nilgiri hills (Sharma et al., 1977).

5.2.4.5   EPIDEMIOLOGY / FAVORABLE CONDITIONS

The infection requires high humidity and low temperature. Optimum temperature for uredospore germination and infection are 17°C and 18°C, respectively. Cardinal temperature for this rust is 18–20°C (Kushalappa and Hegde, 1970) and optimum temperature for post infectional development of disease is 25°C. Mains (1934) identified two races based on differential reaction of *Oxalis europaea* to different collections of rust isolates. Subsequently, common rust races were identified in the USA and India (Groth et al., 1992; Sharma, 1998).

5.2.4.6   MANAGEMENT

Disease can be successfully managed by spray with Tebuconazole (0.1%) at 35 and 50 days after sowing or Tebuconazole (0.1%) at after 35 days sowing and Neemazole F 5% at 50 days after sowing (Dey et al., 2015). Crop rotation is not practically feasible because the spores are hardy and show long-delayed germination. Cultivation of resistant cultivars is a practically feasible method to manage the disease (Basandrai and Singh, 2002).

5.3   STALK ROTS

5.3.1   PYTHIUM STALK ROT

5.3.1.1   INTRODUCTION

Stalk rot due to Pythium was first noticed by M. T. Jenkins in Virginia in 1940 at Arlington Experimental Farm in Maryland, USA. In India. The disease was the first time reported from Delhi (Srivastava and Rao, 1964). This disease usually occurs at pre-tasseling stage. Renfro and Payak (1970) reported yield loss in Ganga-5 (7.9%) in VL-54 (37.3%), in Vijay (13.3%) and in Sona (42.4%). The yield reduction in susceptible genotypes has been estimated to the tune of 100% under favorable environmental condition (Payak and Sharma, 1978).

5.3.1.2   SYMPTOMS

Stalk rots are generally confined to a single internode near the soil line is soft and brown leading to destruction of pith parenchyma and subsequent weakening of the stalk. Stalk of infected plants are not completely broken off. The infected area becomes brown, water-soaked and soft. The stalk may be twisted distorted and collapsed, but they remain green and turgid for several weeks because the vascular bundles remain intact. Plants are most vulnerable when they are 6–7 weeks old. Roots, brass roots, and nodes are not affected by pathogen. The infection leads to sudden and rapid toppling of plants at pre-tasseling stage. Thus, the diseases become quite prominent and apparent. The internodes and leaf sheath have been found to be favorable sites of infection.

5.3.1.3   CAUSAL ORGANISM

Pythium aphanidermatum

Stalk rot is caused by *P. aphanidermatum* (Edson) Fitzp is a typical plant pathogen of warm region. *P. aphanidermatum* produces mycelium, which is cottony, white, and fluffy. It becomes matted on the surface of the medium. The hyphae are hyaline and non-septate except near fruiting structures. Sporangiophores are filamentous, with loose aerial mycelium without a special pattern. The hyphae in fresh cultures are coenocytic, hyaline, have granular cytoplasm, and are profusely branched in an irregular manner. The sporangia are terminal, intercalary, and palmate or delicate. At the time of zoospore production, a bladder like, thin-walled hyaline vesicle is formed at the apex of a lobule. The contents of the sporangium migrate into the vesicle and become segmented in reniform or plane-convex zoospores.

5.3.1.4 DISEASE CYCLE

P. aphanidermatum over winters in the soil in the form of dormant mycelium or oospores. Oospores are resting spore and can survive in the soil for longer period. They produce sporangia which release motile zoospores, they can move freely in water. Zoospores infect the plant directly at of just below the soil surface where free water is present, leading to the development of characteristic lesions. Mycelium within the infect host tissue gives rise to oogonium and antheridium to perform the sexual life cycle. The antheridium fertilizes the oogonium producing oospore, which is capable to overwinter in crop residue or in the soil.

5.3.1.5 EPIDEMIOLOGY / FAVORABLE CONDITIONS

The disease is most favor in poorly drained and low lying fields under conditions of high humidity, high temperature and also in crop with high plant population (>60,000) plant/ha and high dose of nitrogenous fertilizer (Diwakar and Payak, 1980).

5.3.1.6 MANAGEMENT

1. **Disease Resistant:** Planting of hybrid Ganga Safed-2, Hi-starch, and composite Suwan-1 is recommended for the disease prone areas.
2. **Chemical:** The disease can be effectively managed by spraying with Ridomil MZ 400 (Mancozeb + Metalaxyl) reported by Tu-CC et al. (1986). Sun et al. (1994) studied the mechanism of corn rot control by the application of potassic and siliceous fertilizers. Xu-Zuo Tin et

al. (1993) reported that genbao mixture at a concentration of 100 and 200 times gave 82.96% and 62.78% inhibition, respectively against *P. aphanidermatum*. In another study, the genbao mixture gave better results when maize seed was dressed with this mixture in pot trials.

3. **Biological Control:** Some biological control methods were also studied against this pathogen and found very effective. The pathogen showed high inhibitory effect when cultured with *Trichoderma* strain T_2 and bacterial strain P_6. They also confirmed that when seed dressed with either *Trichoderma* or bacteria and in another experiment, the application of *Trichoderma* combined with bacteria as seed dressing, all gave effective control against the pathogen in pot culture.

5.3.2 BACTERIAL STALK ROT

5.3.2.1 SYMPTOMS

This disease usually occurs at about mid-season in hot, humid weather as a tan to dark brown, water-soaked, soft or slimy disintegration of pith tissue at single and basal internodes, affected stalks suddenly collapse and are usually twisted, distorted, and suddenly collapse. The tip of uppermost leaves often wilt, followed by a slimy soft rot at the base of the whorl. The decay spreads rapidly downward until the affected plants collapse. Lodged plant usually has a foul odor (like vinegar). Stalks of infected plants are not completely broken off. With the advancement of the infection, the pith is completely destroyed leaving the bundles in a disorganized state and the stalks become extremely soft and pliable.

5.3.2.2 CAUSAL ORGANISM

Erwinia carotovora sp. *Zeae*

The bacterium was first named as *Erwinia carotovora* sp. *Zeae* by Sabet (1954) on the basis of its similarity to *E. carotovora* (Jones) Holland in cultural, morphological, and biochemical characteristics and its ability to attack maize and other graminaceous hosts in addition to other known hosts of *E. carotovora*. Rangarajan and Chakravarti (1971) did not consider establishment of forma special is as there was no host specialization at generic level. On the basis of morphological, cultural, and biochemical characters and host range, Rangarajan, and Chakravarti (1970, 1971) identified the bacterium as a strain of E. carotovora. Thind (1970) reported that the bacterium is closer to *E. chrysanthemi* Burkholder, McFadden, and Dimock than to *E. carotovora*. In the 8th Edition of Bergey's Manual (Buchanan and Gibbons, 1974) it was classified as *E. chrysanthemi* corn pathotype. In their list of pathovar, names and pathotype strains of phytopathogenic bacteria Dye et al. (1980) have listed it as *E. chrysanthemi* pv. *zea*.

5.3.2.3   DISEASE CYCLE

The causal bacterium lives as a saprophyte on plant debris in the soil. The organism also may be seed-borne. Infection occurs when the bacteria are blown or splashed on the plants followed by penetration through natural openings (stomates and hydathodes) or wounds made by hail or other injuries.

5.3.2.4   EPIDEMIOLOGY / FAVORABLE CONDITIONS

General infection may occur following flooding or where maize is sprinkler irrigated from a surface source of water, such as river, pond, or lake. The development of bacterial soft rot is favored by high temperatures (29–35°C) and poor air circulation.

5.3.2.5   MANAGEMENT

1.   **Chemical:** Many chemicals have been evaluated against this bacterium, both *in-vitro* and *in-vivo* (Randhawa and Thind, 1978). Streptocycline, Agrimycin-100, terramycin, streptomycin, synermycin, achromycin, ambistryn, Dithane M-22, Dithane M-45, Dithane Z-78, Fytolan, bisdithane, blitox, captan, farbam, ziram, potassium permanganate, and bleaching powder have been shown to be inhibitory *in-vitro*. Sharma et al. (1982) found that two applications of Klorocin (contains 22% chlorine) at the rate of 250 µg/m[1] chlorine resulted in significant disease control (48.18%). The effectiveness increased with the increase in concentration up to 1000 µg/ml. In a preliminary report Lal and Saxena (1978) stated that two applications of bleaching powder at the rate of 25 kg/ha, the first at flowering stage and the second 10 days later, considerably checked

the stalk rot incidence. Rangarajan and Chakravarti (1970) found that the application of streptocycline and streptomycin 48 hours before and at the time of inoculation reduced the disease incidence but the latter treatment was more effective. The evaluation of streptocycline (100 µg/sm¹), Agrimycin 100 (100 µg/m¹), stable bleaching powder (SBP) (100 µg/m¹ chlorine), Blitox-50 W (2000 µg/ml), potassium permanganate (100 µg/ml) and a combination of streptocycline + Blitox-50 WP in glasshouse and field experiments revealed that spray and soil drench applications of streptocycline alone and in combination with Blitox-60 WP proved most effective (Thind et al., 1984).

In spray, application Agrimycin-100 and SBP ranked second and third, respectively, whereas in soil drench application it was the reverse. Application of these chemicals 24 hours before inoculation yielded better results than after inoculation. Soil drench application was better than spray application. Disease incidence decreased from 64.4 to 28.8% with increases in the concentration of chlorine (SBP) from 100 to 1000 µg/ml. The persistence of chlorine in soil for longer periods and its uptake by the maize plants from soil, also favor the use of SBP over these antibiotics (Thind and Soni, 1983). Klorocin at 250 ppm provided significant control (48.26%) of ESR when applied by sprinklers along the basal internodes 24 h before inoculation with the pathogen (Sharma et al., 1982), however, Satnarayana and Begum (1996) manage the disease through seed treatment with captan @ 2 g/kg followed by 3 soil treatments with SBP (1000 ppm) at 45, 55, and 65 days after sowing.

5.3.3 POST FLOWERING STALK ROT (PFSR)

Post flowering stalk rot (PFSR) of maize is an important disease in India. It is a 'complex' where more than one pathogen is involved. The common pathogens are: *Fusarium verticillioides, Macrophomina phaseolina, and Harpophora maydis.* Yield loss is directly affected by premature plant death or by reduced kernel filling and lodging, resulting up to 100% loss.

PFSR is an important and destructive disease of maize. In India this disease occurring in Uttar Pradesh, Madhya Pradesh, Andhra Pradesh, Punjab Rajasthan, Bihar, and West Bengal. PFSR prevails namely charcoal rot, fusarium stalk rot, and late wilt caused by *Macrophomina phaseolina*, Fusarium verticilloides and Harpophora maydis. The disease incidence recorded in Karnataka ranged from 10 to 42% (Desai et al., 1992). Yield loss estimates are difficult to obtain because losses due to stalk rot may occur in several ways.

5.3.3.1   CHARCOAL ROT

5.3.3.1.1   Introduction

Charcoal rot is a common stalk rot disease in warm and dry areas of the world. It occurs in areas where drought conditions generally prevail for a longer period. This disease is economically important throughout the world, particularly in arid maize growing regions where extensive yield losses occur when the crop is infected early. Yield losses as high as 70% have been documented in Africa. The disease is particularly prevalent in drought years and in arid regions where maize is regularly cultivated in rotation with other host crops. The disease is heat and stress (drought) driven and is therefore rare in cooler climates and irrigated fields.

The disease in most common in Uttar Pradesh, Madhya Pradesh, Andhra Pradesh, Karnataka, Tamil Nadu, Punjab, Haryana, Rajasthan, Jammu & Kashmir, and the West Bengal state of India. Yield losses as high as 70% worldwide and In India reported 10–42%.

5.3.3.1.2   Symptoms

It is characterized by water-soaked, brown lesions, that later turn black, appear on the roots. As the plants mature, the fungus spreads into the lower internodes of the stalk. Infected plants appear to ripen prematurely and the interior of the lower internodes disintegrate. The disease is distinguished by the presence of numerous, tiny, round to irregular, pin-head like black sclerotia, which are present in large numbers along the vascular strands in the interior of shredded and rotted stalks giving them a charred appearance. Sclerotia also may be formed just beneath the stalk surface and on roots. The disease is usually confined to the first or second internode above the soil level.

5.3.3.1.3   Causal Organism

Macrophomina phaseolina (Tassil Goidanich = *M. phaseoli* (Maubl.) S. Ashby = *Botryodiplodia phaseoli* (Maubl.) Thirumalachar = *Sclerotium bataticola* Taubenhaus is the causal agent of the disease.

5.3.3.1.4   Disease Cycle and Epidemiology

The pathogen survives in the form of sclerotia and may penetrate roots and lower stems during growing season. This disease is also common in sorghum and soybean. Maize isolates of the pathogen. *Macrophomina phaseolina* are sterile and do not form conidia. However, some strains of the fungus produce conidia also. Sclerotia are generally smooth to irregular black and 0.05–0.22 mm in diameter.

The disease is favored by soil temperature ranging from 30 to 40°C, while either low soil temperatures or high soil moisture decrease severity (Basandrai and Singh, 2002).

5.3.3.2   FUSARIUM ROT

5.3.3.2.1   Introduction

Fusarium rot was first observed from the United States of America by Pammel (1914) as a serious root and stalk diseases. Later Valleau (1920) indicated that *Fusarium moniliforme* was a primary cause of root rot and stalk rot of maize. In India Fusarium, stalk rot was first reported from Mount Abu, Rajasthan (Arya and Jain, 1964). It has been infecting maize in the north and central Karnataka causing economic losses up to 85%. As per earlier records, the disease incidence ranged from 10 to 42% in Karnataka (Harlapur et al., 2002). An estimated loss due to Fusarium stalk rot has been reported as 38% in total yield (AICRP, 2014). Growers are not only worried of income but shortage of fodder which is starving the animals. Some are switching to alternate crops. Irrespective public and private bred hybrids all are succumbing to severe stalk rot.

The disease in most common in Uttar Pradesh, Andhra Pradesh, Rajasthan, Jammu & Kashmir, Bihar, and West Bengal states of India. Yield losses 100% under severe disease infestation.

5.3.3.2.2   Causal Organism

Fusarium moniliforme J. Sheld. = *Gibberella fujikuroi* (Sawada) Ito; *F. proliferatum* (T. Matsushima) Niternberg = *G. fujikuroi*; *F. subglutinans* (Wollenweb. and Reinking) P. E. Nelson, T. A. Toussom and Marasas is the of Fusarium stalk rot. However, some researchers have concluded that *F.*

moniliforme is not a stalk rot pathogen but usually remains associated with maize stalks whereas some scientists have proved that *F. moniliforme* is a more virulent pathogen even as compared to *F. graminearum*. Sometimes, both pathogens affect plants in the same fields.

5.3.3.2.3 Disease Cycle and Epidemiology

The fungus, *F. moniliforme* survives on crop residue in the soil or on the soil surface as conidia or chlamydospores. Under favorable conditions, it may infect roots or stalks. *F moniliforme* may be present throughout the life cycle of the plant, originating from infected seed. It causes comparatively more damage in tropical as compared to temperate regions.

The Fusarium stalk rot is very severe in the warm and dry regions and, more in sandy soil, sudden changes in weather condition at the time of flowering (Kumar and Shekhar, 2005).

5.3.3.3 LATE WILT

5.3.3.3.1 Introduction

Disease is occurring in India, Egypt, Hungary, Portugal, and Spain countries. *Harpophora maydis* is a soil-borne and apparently seed-borne fungus related to the root-infecting species in the genus Gaeumannomyces. It is known from only a few scattered countries, where it can cause significant losses, but may have been unobserved in others in which the primary host, maize, is grown. No dispersal by fungal propagules has been demonstrated, so that, other than in soil, its likely means of spread over borders would be in seed.

The disease in most common in Uttar Pradesh, Andhra Pradesh and the Rajasthan state of India. Yield losses in India reported up to 51%.

5.3.3.3.2 Symptoms

Leaves wilt moderately rapidly beginning in the tasseling (flowering) period or later. Progressing upward from the lower part of the plant, leaves become dry and dull green, rolling inward and eventually losing color. Vascular bundles in the stalk turn reddish-brown and then internodes also become discolored. Lower portions of the stalk are dry, shrunken, and hollow. Some plants develop yellowish to purple or dark brown streaks on the lower stem. Rotting of roots and lower internodes may involve secondary.

5.3.3.3.3   Causal Organism

Harpophora maydis

Mycelium appressed, felty, margin "rhizoidal," the outermost hyphae branching to resemble roots; hyphal "ropes" curving clockwise. Hyphae hyaline, septate, Conidiophores 60–250 µm or longer, mostly branched; conidia formed in phialides at apices, collecting in "heads." Conidia straight, mostly one-celled, hyaline, oblong, 3.5–14 × 3.3–3.6 µm. Sclerotia-like bodies, composed of several thick-walled pigmented cells, formed in old cultures.

5.3.3.3.4   Disease Cycle and Epidemiology

Soil-borne or seed-borne *H. maydis* infects the roots of young plants or the seedling mesocotyl, invades the vessels, and grows or is translocated from the roots up the stalk and into the ear stalks and grain. Initial superficial growth on the roots consists of short, brown, thick-walled swollen cells. In India, maximum disease occurred at a constant 24°C or when the temperature varied naturally between 20 and 32°C. Less disease was obtained at a constant temperature of 36°C. In infected stems kept inside at 20–35°C, the fungus survived and retained pathogenicity for up to 24 months. In the field, stem pieces on the surface of soil retained the pathogen for twelve months, but it could not be recovered after ten months from pieces buried at 10 cm (Singh and Sirhana, 1987b). The researchers suggested that survival with seed would be longer in cooler climates.

5.3.3.4   POST FLOWERING STALK ROT (PFSR) MANAGEMENT

Cultural practices such as use balanced fertilizers dose, maintain plant density and any other practice that lessens plant stress can help reduced amount of stalk rots. Control of stem borers is also helpful in management of the disease. Early harvest is highly desirable; as it is well known that longer standing maize exposed to weather more stalk breaking is likely to occur. This practice definitely reduces loss from stalk breakage and lodging. Bending over the stalk below the ear at about maturity also reduces stalk rot damage. This exposes the ear to drying conditions, accelerates the drying of ears, and thus retards the growth of the pathogen. This practice also prevents water from accumulating between the husk and kernels.

Based on the extensive research work done during the past at Directorate of Maize Research the sources of resistance against the post-flowering stalk rot of maize identified are — CM 103, CM 119, CM 125, CM 209, B-37, Cl 21 E, CML 31, CML 77, CML 79, CML 85, CML 90, CML 98, CML 101, CML 292 and CML 381., An efforts were made to test the efficacy of *T. harzianum* against rot of maize caused by *F. moniliforme* where *Th*3 strain showed 73.33% inhibition of disease (Bhandari and Vishunavat, 2013; Sharma et al., 2014). Singh et al. (2013) observed that application in furrows of *Drek* seed extract effective to manage the disease (56.25% control over control) whereas, seed treatment with carbendazim was found most effective (90.18% disease control).

5.4   SMUTS

5.4.1   INTRODUCTION

The diseases entitled as the smut, smut is the Greek word, and they represent the dark black colors in the reproductive bodies (Agarwal, 2017). Two smuts occur on maize crop in India, the Common smut (*Ustilago maydis*) and Head smut (*Sphacelotheca reiliana*). Head smut attacks on maize, sorghum, and other grasses and bas been found moderately destructive in the sub-temperate Himalayas and in the hilly areas of Rajasthan and Jammu and Kashmir.

5.4.2   SYMPTOMS

The most peculiar symptom of the smuts are-abnormal development of tassels, which become malformed and overgrown black masses of spores, which develop inside individual male florets; and masses of black spores which also grow instead of the normal ear, leaving the vascular bundles exposed and shredded.

Tassel infection may lead to its proliferation. Infection in ears produces leafy ears filled with smut sori. There is increased tillering and dwarfing of the plant. Symptoms appear as galls covered with white glistening membrane which ruptures to release black spore mass. On leaves, the galls remain small having few spores. Most prominent symptoms appear on ears which get swollen into black powdery mass (Figure 5.6).

FIGURE 5.6 Ear smut.

5.4.3 CAUSAL ORGANISM

Two smuts occur on maize in India, the Common smut (*Ustilago maydis*) and Head smut (*Sphacelotheca reiliana*). Head smut attacks maize, sorghum, and may cause significant economic damage in dry, hot maize growing areas. Aydogdu and Boyraz (2016) observed that disease incidence ranged from 16.6 to 74.1% depending upon variety.

5.4.4 DISEASE CYCLE

The smut pathogens survive is soil-borne in the form of teleutosori which germinate under favorable conditions to produce sporidia which further infect the crop at seedling stage. The infection is systemic because the fungus penetrates seedlings and grows within the plant without showing symptoms until plants reach tasseling and silking stages. Volunteer plants or grasses play a major role in transmission of spore in off season because fungus perpetuate on these plants.

5.4.5 EPIDEMIOLOGY / FAVORABLE CONDITIONS

Rain showers, high humidity, and warm temperatures are favorable for the development of smut diseases.

5.4.6 MANAGEMENT

For head smut, sanitation, rotation, and application of balanced dose of fertilizer are the measures to be adopted for management. Common smut is prevalent in humid tropical regions. It attacks all the above ground plant parts. The common smut is primarily soil-borne in the form of chlamydospores which produce sporidia though occasionally seed-borne infection may also occur. Sporidia infect through stomata or wounds. It may be controlled by avoiding injury to the plants during field operations, by maintaining balanced fertility and sanitation, e.g., rouging, and burning of smut galls before they rupture. Ivanova et al. (2017) found 15 hybrids were resistant against *U. maydis* and may be used as source of resistance, i.e., St20, Eks12, Eks10, Eks23, Eks16, Eks2, Eks3, Eks4, Eks6, Eks9, Eks13, Eks17, Eks19, Eks20, and 2Eks6.

5.5 EAR, COB, AND KERNEL ROTS

This group of diseases can be categorized into three types, i.e., those which affect cob, those which affect kernel and those which affect both cob and kernel. As many as 10 ear rot diseases (*F. moniliforme*) Aspergillus kernel rot (*Aspergillus flavus* and *A. niger*) and Botryodiplodia cob rot (*Botryodiplodia theobrotnae*) occur. Among the most, prevalent ones are Fusarium rot. Occasionally, Nigrospora, and Macrophomina cob rots have also been observed. Fusarium species may produce mycotoxins which cause acute toxicity in man and animals. Field observations suggest that their incidence is high in humid moist environments; Insect injuries also contribute to the increase in their incidence. In the breeding programme, emphasis should be given on developing cultivars with tight husk covers on the ears. Two sweet corn varieties Vega and Merit were found tolerant against smut disease (Aydogdu and Boyraz, 2016). Smut can be controlled by avoiding injury to the plants during field operations by maintaining balanced fertility and sanitation, e.g., rouging, and burning of smut galls before the rupture (Sharma, 2005).

5.6 VIRUS DISEASES

There are more than 40 total viral diseases of maize reported worldwide. Five of them have reportedly occurred in China. They are maize rough dwarf disease, maize dwarf mosaic disease, maize streak dwarf disease, maize crimson leaf disease, maize wallaby ear disease, and corn lethal necrosis disease (Cui et al., 2014). Three viruses occur on maize in India. These are: maize mosaic virus (MMV) I, maize mosaic, and vein enation (Sharma and Payak, 1993). Maize mosaic has been found to be prevalent in several states and the incidence ranges from 2.2 to 10.6%. Maize mosaic causes stunting of plants.

5.6.1 MAIZE CHLOROTIC DWARF

Maize chlorotic dwarf is caused by a virus that is spread by the leaf hopper *Graminella nigrifrons*. The most characteristic symptom of maize chlorotic dwarf is vein banding or vein clearing. These chlorotic stripes develop along the veins covering entire leaf blade. Leaves, internodes, and other plant parts including ear husks are also shortened. The stripes produced are continuous from base to the tip. Two strains of MMV can be differentiated, one which causes the stripes and other causing broader stripes. The disease is adapted to diverse agroclimatic conditions.

The maize dwarf chlorotic virus can overwinter in pearl millet, sorghum, Sudan grass Johnson grass, and wheat.

5.6.2 MAIZE DWARF MOSAIC

Maize mosaic has been found to be prevalent in several states and the incidence range from 2.2–10.6% (Sharma, 2009). The most common symptom is a light green to yellow mottling or mosaic pattern in the leaf tissue. This mosaic disease is most severe in seedlings stage of the plant, occasional shortening of the upper internodes gives a stunted, bushy appearance to the plants resulting ear size also reduced or plant may be barren. This virus disease of maize is transmitted by several species of aphids.

5.6.3 MANAGEMENT

Various cultural practices suggested for the control included the use of 'barriers' of bare ground between early and late planted maize fields to reduce leafhopper movement and subsequent mosaic streak virus spread (Bosque-Pérez, 2000). Chaboussou (2004) reported that in contrast to nitrate fertilizers, alkaline phosphate fertilizers have a beneficial effect against viral diseases, such that, by promoting maturity, they speed up the stage of resistance in the plant brought about by age. Thus, while P simultaneously stimulates plant growth and virus concentration, K on the other hand increases plant growth, and reduces viral concentration.

The potential of utilizing natural enemies (predators and parasitoids) and entomopathogenic microbes for the control of leafhoppers has been demonstrated in Asian countries (Mitsuhashi et al., 2002). Investigations on the chemical control showed that carbofuran granules applied to the planting furrow at 0.2 g a.i./m was significant in suppressing leafhopper populations (Drinkwater et al., 1979; Magenya et al., 2008).

5.7 DISEASE CAUSED BY PHYTOPLASMA

5.7.1 INTRODUCTION

Phytoplasm as belongs to class Mollicutes and are phytopathogenic prokaryotes without cell wall inhabiting the phloem of infected plants. Phytoplasma diseases are the major economic constraints in profitable cultivation of commercial crop plants production lowering its quantum and quality (Bertaccini et al., 2014).

5.7.2 SYMPTOMS

Affected plants show symptoms of reddening of the leaf midrib, leaves, and stalks; seed production and yields are also greatly reduced (Figure 5.7).

5.7.3 CAUSAL ORGANISM

Candidatus Phytoplasma asteris (16SrI-B subgroup)

In India, the 'Ca. P. asteris' group is reported as the major group of phytoplasm as infecting a wide host range of plant species including agricultural crops, ornamentals, tree, and weed species and recently reported on maize crop also from Siot, Rajouri, Jammu & Kashmir state of India in September 2016. Based on the 16S rDNA sequence similarity and virtual RFLP pattern, the phytoplasma strain associated with this disease was identified as a strain of 'Ca. P. asteris' and assigned to subgroup 16SrI-B (Rao et al., 2017a).

FIGURE 5.7   Redness of leaf in maize.

5.7.4   MANAGEMENT

An integrated approach include, together with the applied control measures based on clean propagating materials, vector control, and weed management, more ever a stimulation of plant defenses can become practically important. Since no effective control, measures are available for the phytoplasma associated diseases, effective management practices should be adopted (Rao et al., 2017b).

KEYWORDS

- **entomopathogenic microbes**
- **fusarium species**
- **maize chlorotic dwarf**
- **phytoplasm**
- **post flowering stalk rot**
- **stable bleaching powder**

REFERENCES

Agarwal, T., (2017). Smut of crops: A review. *Journal of Pharmacognosy and Phytochemistry,* *5*(1), 54–57.

Agrios, G. N., (2005). *Plant Pathology* (5ᵗʰ edn.) Burlington, MA. Elsevier Academic Press.

Anonymous, (1999). *42ⁿᵈ Annual Progress Report* (pp. 1–60). All India Coordinated Maize Improvement Project. Directorate of Maize Research IAR1, New Delhi.

Arya, H. C., & Jain, B. L., (1964). Fusarium seedling blight of maize in Rajasthan. *Indian Phytopathol., 17,* 51–57.

Aydogdu, M., & Boyraz, N., (2016). Responses of some maize cultivars to smut disease, *Ustilago maydis* (DC) Corda. *Mediterranean Agricultural Sciences, 29*(1), 1–4.

Basandrai, A. K., & Singh, A., (2002). Fungal diseases of maize. In: Gupta, V. K., & Paul, Y. S., (eds.), *Diseases of Field Crops* (pp. 103–125). Indus publishing company New Delhi.

Basandrai, A. K., Akhilesh, S., & Kalia, V., (2000). Multiple disease resistance in Indian maize hybirds and composites. *Ind. J. of Pl. Genetic Resource, 13,* 188–190.

Batsa, B. K., (2003). Integrated management of banded leaf and sheath blight of maize caused by *Rhizoctonia solani* f. Sp. *Sasakii. PhD Thesis, Division of Plant Pathology* (p. 48). IARI, New Delhi.

Bertaccini, A., Duduk, B., Paltrinieri, S., & Contaldo, N., (2014). Phytoplasmas and phytoplasma diseases: A severe threat to agriculture. *American Journal of Plant Sciences, 5,* 1763–1788.

Bertus, L. S., (1927). A sclerotial disease of maize (*Zea mays* L.) due to *Rhizoctonia solani* Kuhn. Yearbook. *Dept. Agric. Ceylon, 44,* 46.

Bhandari, P. C., & Vishunavat, K., (2013). Screening of different isolates of *Trichoderma harzianum* and *Pseudomonas fluorescens* against *Fusarium moniliforme. Pantnagar J. Res., 11*(2), 243–247.

Bosque-Perez, N. A., (2000). Eight decades of maize streak virus research. *Virus Res., 71*(1/2), 107–121.

Bowen, K. L., & Pederson, W. L., (1988). Effect of propiconazole on *Exserohilum turcicum* in laboratory and field studies. *Plant Disease, 72,* 847–850.

Bowen, K. L., & Pedersen, W. L., (1988). Effect of propiconazole on *Exserohilium turcicum* in laboratory in field studies. *Plant Disease, 72,* 847–850.

Ceballos, H., & Gracen, V. E., (1988). A new source of resistance to *Bipolaris maydis* race T in maize. *Maydica, 33,* 233–246.

Chaboussou, F., (2004). *Healthy Crops: A New Agricultural Revolution* (p. 244). Jon Carpenter for the Gaia Foundation.

Chin, T. C., Deng, D. X., Xu, M. L., & Liu, D. W., (1989). Characteristics and inheritance of new type of cytoplasmically male-sterile maize line. *J. Jiangsu Agric. College, 10,* 1–6.

Cui, Y. Z., Ai-hong, R., Ai-Jun, M., & Hong-Qin, (2014). Types of maize virus diseases and progress in virus identification techniques in China. *Journal of Northeast Agricultural University, 21*(1), 75–83.

Desai, S., Hegde, K. K., & Desai, S., (1992). Identification of suitable method and time for artificial inoculation of maize with stalk rotting fungi. *Indian Phytopathology,* *45*(3), 381–382.

Dey, U., Harlapur, S. I., Dhutraj, D. N., Suryawanshi, A. P., & Bhattacharjee, R., (2015). Integrated disease management strategy of common rust of maize incited by *Puccinia sorghi* Schw. *African Journal of Microbiology Research, 9*(20), 1345–1351.

Drinkwater, T. W., Walters, M. C., & Rensburg, J. B. J. V., (1979). The application of systemic insecticides to the soil for the control of the maize stalk borer, *Busseolafusca* (Fuller) (Lep.:

Noctuidae), and of *Cicadulinambila* (Naude) (Hem.: Cicadellidae), the vector of maize streak virus. *Phytophylactica, 11*(1), 5–11.

Dye, D. W., Bradbury, J. F., Goto, M., Hayward, A. C., Lelliott, R. A., & Schroth, M. N., (1980). International standards for naming pathovars of phytopathogenic bacteria and a list of pathovar names and pathotype strains. *Review of Plant Pathology, 59,* 153–168.

Elliott, C., (1943). A pythium stalk rot of corn. *Journal of Agriculture Research, 66,* 21–39.

Groth, J. V., Pataky, J. K., & Gingera, G. R., (1992). Virulence in eastern North American populations of *Puccinia sorghi* to Rp resistance genes in corn. *Plant Disease, 76,* 1140–1144.

Harlapur, S. I., (2005). Epidemiology and management of Turcicium leaf blight of maize caused by *Exserohilum turcicium* (pass.) Leonard and Suggs. PhD Thesis (p. 121). Department of Plant Pathology, College of Agriculture, University of Agricultural Sciences Dharwad-580005.

Ivanova, I., Ivanov, L., & Ivanova, G. A., (2017). Study on semi-late maize hybrids for resistance to smut of maize/*Ustilago maydis/*. *S M J Bioprocess Biotech., 1*(1), 1001–1004.

Khan, A., Ahmad, S., Khan, A., & Ahmad, S., (1992). Genotype assay of maize for resistance to Maydis leaf blight under artificial field epiphytotics of Peshawar region. *Sarhad Journal of Agriculture, 8,* 547–549.

Kommedahl, T., & Lang, D. S., (1973). Effect of temperature and fungicides on survival of corn grown from kernels infected with *Helminthosporium maydis*. *Phytopath, 63*(1), 138–140.

Kumar, S., & Shekhar, M., (2005). Post-flowering stalk rots of maize and their management. In: Ziadi, P. H., & Singh, N. N., (eds.), *Stress on Maize in Tropics* (pp. 172–194). Directorate of Maize Research, New Delhi, India.

Kumari, B., (2012). Studies on the management of banded leaf and sheath blight of maize caused by *Rhizoctonia solani* F. SP. Sasakii (KÜHN.) EXNER. *PhD Thesis* (pp. 83–89). Department of plant pathology, Govind Ballabh Pant University of Agriculture & Technology, Pantnagar.

Kushalappa, A. C., & Hegde, R. K., (1970). Studies on maize rust (*P sorghi*) in Mysore State III. Prevalence and severity on maize varieties and impact on yield. *Pl. Dis. Reps., 54,* 788–792.

Lal, S., & Sexena, S. C., (1978). Bacterial stalk rot of maize; loss assessment and possibility disease control by application of calcium hypochlorite (abstract). *Indian Phytopathology, 31,* 120.

Magenya, O. E. V., Mueke, J., & Omwega, C., (2008). Significance and transmission of maize streak virus disease in Africa and options for management: A review. *African Journal of Biotechnology, 7*(25), 4897–4910.

Mains, E. G., (1934). Host specialization of *P. sorghi*. *Phytopathology, 24,* 405–411.

Manisha, T., & Johri, B. N., (2002). In vitro antagonistic potential of fluorescent pseudomonads and control of sheath blight of maize caused by *Rhizoctonia solani*. *Indian Journal of Microbiology, 42,* 207–214.

Meena, R. L., Rathore, R. S., & Mathur, K., (2003). Evaluation of fungicides and plant extracts against banded leaf and sheath blight of maize. *Indian Journal of Plant Protection, 31,* 94–97.

Misra, D. P., & Sharma, S. K., (1964). Natural infection of *Oxalis corniculata* L. the alternate host of *P. sorghi* Schw. in India. *Indian Phytopathol., 17,* 138–141.

Mitsuhashi, W., Saiki, T., Wei, W., Kawakita, H., & Sato, M., (2002). Two novel strains of Wolbachia coexisting in both species of mulberry leaf hoppers. *Insect. Mol. Boil., 11*(6), 577–584.

Munjal, R. L., & Kapoor, J. N., (1960). Some unrecorded diseases of sorghum and maize from India. *Curr. Sci., 29*(11), 442–443.

Murdia, L. K., Wadhwani, R., Wadhwan, N., Bajpai, P., & Shekhawat, S., (2016). Maize utilization in India: An overview. *American Journal of Food and Nutrition, 6*(4), 169–176.

Pammel, L. H., (1914). Serious root and stalk diseases of corn. *IOWA Agriculturist, 15,* 156–158.

Pandurange, G. K. T., Puttarama, N., Sreerama, S. T. A., Hatappa, S., & Mallikarjuna, N., (2002). Performance of maize composite, NAC-6004, resistant to Turcicum leaf blight and downy mildew. *Environment and Ecolog., 20*(4), 920–923.

Pandurange, G. K. T., Shekara, S., Jayaramegowda, B., Prakash, H. S., & Sangamlal, (1993). Effect of foliar fungicides on *Exserohilium turcicum* leaf blight of maize, *Mysore J. Agri. Sci., 27,* 146–149.

Pandurange, G. K. T., Sreerama, S. T. A., Puttarama, N., Asadulla, M., & Mallikarjuna, N., (2001). Performance of maize composite, NAC-6002, resistant to Turcicum leaf blight and downy mildew. *Mysore J. Agri. Sci., 35,* 211–215.

Parihar, C. M., Jat, S. L., Singh, A. K., Kumar, R. S., Hooda, K. S., Chikkappa, G. K., & Singh, D. K., (2011). *Maize Production Technologies in India* (p. 30). DMR Technical Bulletin Directorate of Maize Research, Pusa Campus, New Delhi-110 012.

Park, K. Y., Moon, K. G., Park, S. U., Choi, K. J., Park, R. K., Sang, S. H., Lee, H. B., & Lee, H. D., (1991). A new single cross maize hybrid with leaf stay green for silage and grain type "Jungbuor. Research Reports of the Rural Development Administration. *Upland and Industrial Crops, 33,* 22–28.

Patil, S. J., Wali, M. C., Harlapur, S. I., & Prashanth, M., (2000). Maize Research in north Karnataka. *Bulletin,* (p. 54). University of Agricultural Sciences, Dharwad.

Payak, M. M., & Renfro, B. L., (1966). Diseases of maize new to India. *Indian Phytopath. Soc. Bull., 3,* 14–18.

Payak, M. M., & Sharma, R. C., (1978). Research on diseases of maize. In: *Project Final Technical Report* (p. 228). Indian Council of Agricultural Research, New Delhi.

Payak, M. M., & Sharma, R. C., (1985). Maize diseases and their approach to their management. *Tropical Pest Management, 31,* 302–310.

Puzari, K. C., Saikia, U. N., & Bhattacharya, A., (1998). Management of banded leaf and sheath blight of maize with chemicals. *Indian Phytopath., 51,* 78–80.

Rai, G. S., (1987). Status of breeding and management of Turcicum leaf blight and pre-mature drying. *Proc. 30th Annual Maize Workshop* (pp. 21–27). AICMPI, IARI, New Delhi.

Rains, S. S., & Dhaliwal, H. S., (1992). Downy mildews of maize. In: Singh, U. S., Mukhopadhyay, A. N., Kumar, J., & Chaube, H. S., (eds.), *Plant Diseases of International Importance, Diseases of Cereal and Pulses* (Vol. I, p. 488). Prentice-Hall, Englewood Cliffs, New Jersey.

Rajput, L. S., & Harlapur, S. I., (2014). Status of banded leaf and sheath blight of maize in North Karnataka. *Karnataka J. Agric. Sci., 27*(1), 82–84.

Rangarajan, M., & Chakravarti, B. P., (1970). Bacterial rot of maize in Rajasthan. Effect on seed germination and varietal susceptibility. *Indian Phytopathology, 230,* 470–477.

Rangarajan, M., & Chakravarti, B. P., (1971). *Erwinia carotovora* zones Holland. The inciting agent of corn stalk rot in India. *Phytopathologia Mediterranea, 10,* 41–45.

Rani, D. V., Reddy, N. P., & Devi, U. G., (2013). Management of maize banded leaf and sheath blight with fungicides and biocontrol agents, *Annals of Biological Research, 4*(7), 179–184.

Rao, G. P., Kumar, M., Madhupriya, & Singh, A. K., (2017a). First report of 'Candidatus Phytoplasma asteris' (16SrI-B subgroup) associated with maize leaf redness disease in India. *Phytopathogenic Mollicutes, 7*(1), 52–56.

Rao, G. P., Madhupriya, T. V., Manimekalai, R., Tiwari, A. K., & Yadav, A., (2017b). A century progress of research on phytoplasma diseases in India. *Phytopathogenic Mollicutes, 7*(1), 1–38.

Rijal, T. R., Paudel, D. C., & Tripathi, N., (2007). In: Gurung, D. B., Paudel, D. C., & G. K. C. S. R. (eds.), *Screening of Maize Inbreeds/Hybrids Against Banded Leaf and Sheath Blight Disease.*

Satyanarayana, E., & Begum, H., (1996). Relative efficacy of fungicides (seed dressers) and irrigation schedule for the control of late wilt in maize. *Current Res Uni. Agri. Sci. Bangalore, 25*(4), 59–60.

Schenck, N. C., & Stelter, T. J., (1974). Southern corn leaf blight development relative to temperature, moisture and fungicide application. *Phytopathology, 64,* 619–624.

Sharma, P., Sharma, M., Raja, M., & Shanmugam, V., (2014). Status of *Trichoderma* research in India: A review. *Indian Phytopath, 67*(1), 1–19.

Sharma, R. C., (2005). Disease of maize and their management. In: Thind, T. S., (ed.), *Disease of Field Crops and Their Management* (pp. 21–34). Daya Publishing House Delhi-110035.

Sharma, R. C., (2009). Maize disease management in India. In: Upadhyay, R. K., Mukerji, K. G., Chamola, B. P., & Dubey, O. P., (eds.), *Integrated Pest and Disease Management* (pp. 21–34). A. P. H. Publishing Corporation New Delhi.

Sharma, R. C., & Rai, S. N., (1999). Chemical control of banded leaf and sheath blight of maize. *Indian Phytopathol., 52,* 94–95.

Sharma, R. C., Carlos, D. L., & Payak, M. M., (1993). Diseases of maize in south and southeast: Problems and progress. *Crop Protection, 12,* 414–422.

Sharma, R. C., (1998). Diseases of maize and their management. In: Thind, T. S., (ed.), *Diseases of Field Crops and Their Management* (pp. 21–33).

Sharma, R. C., & Hembram, D., (1990). Leaf stripping: A new method to control banded leaf and sheath blight of maize. *Current Science, 59,* 745–746.

Sharma, R. C., Payak, M. M., Laxminarayan, C., Shankerlingam, S., & Lilaramani, J., (1977). Combating the common rust of maize. *Pesticides, II,* 37.

Sharma, R. C., Payak, M. M., Shankerlingam, S., & Laxminarayan, C., (1982). Comparison of two methods of estimating yield losses in maize caused by common rust of maize. *Indian Phytopathol., 35,* 18–20.

Sharma, S. C., Randhawa, P. S., Thind, B. S., & Khera, A. S., (1982). Use of kelrocin for the control of bacterial stalk rot and its absorption translocation and persistence in maize tissue. *Indian Journal of Mycology and Plant Pathology, 12,* 185–190.

Singh, A. K., Singh, V. K., & Singh, A. K., (2013). Evaluation of leaf extracts, fungicides and *Trichoderma viride* against fusarium stalk rot of maize (*Zea mays*). *Research on Crops, 14*(2), 455–458.

Sivakumar, G., Sharma, R. C., & Rai, S. N., (2000). Biocontrol of banded leaf and sheath blight of maize by peat based *Pseudomonas fluorescens* formulation. *Indian Phytopath, 53,* 190–192.

Srivastava, D. N., & Rao, V. R., (1964). Pythium stalk rot of corn in India. *Curr. Sci., 33,* 119–120.

Subedi, S., (2015). *A Review on Important Maize Diseases and Their Management in Nepal.*

Sumner, D. R., & Littrell, R. H., (1974). Influence of tillage, planting date, inoculums survival, and mixed population of epidemiology of southern corn leaf blight. *Phytopathology, 64,* 168–173.

Sun, X. H., Sun, Y. J., Zhang, C. S., Bai, J. K., Song, Z. H., & Chen, J., (1994). Study on the relationship among pathogens of corn stalk rot. *Journal of Shenyang Agricultural University, 23,* 93–96.

Thakur, M. S., Sharma, S. L., & Munjal, R. L., (1973). Correlation studies between incidence of banded sclerotial disease and ear yield in maize. *Indian J. Mycol. Pl. Path., 3*(2), 180–181.

Thind, B. S., (1970). Investigations on bacterial stalk rot of maize (*Erwinia carotovora* var. zeaSabet). *PhD Thesis* (p. 113). Indian Agricultural Research Institute, New Delhi.

Thind, B. S., & Soni, P. S., (1983). Persistence of chlorine in maize plants and soil in relation to control of bacterial stalk rot of maize. *Indian Phytopathology, 36,* 687–690.

Thind, B. S., Randhawa, P. S., & Soni, P. S., (1984). Chemical control of bacterial stalk rot (*Erwinia carotovora* var. *zea*) and leaf stripe (*Pseudomonas rubrilineans*) of maize. *Zeitsch rift fur Pflanzenkrankheiten und Pflanzenschutz, 91,* 424–430.

Tu-CC, Cheng, A. H., & Chen, G. M., (1986). Studies on control of stalk rot of corn. *Research-Bulletin-Tanian District Agriculture Improvement Station, 29*(38), 12.

Ullstrup, A. J., (1960). *An Abstract of a Report on Major Diseases in India* (p. 23). Progress Report of Coordinated Maize Breeding Scheme. I.C.A.R.

Upadhyay, &. Pokharel, B. B., (2007). *Proceedings of the 25*[th] *National Summer Crops Research Workshop on Maize Research and Production in Nepal, Held in June 21–23, 2007 at Nepal Agriculture Research Institute.* NARC, Khumaltar, Lalitpur, Nepal.

Valleau, W. D., (1920). Seed corn infection with *Fusarium moniliforme* and its relation to the root and stalk rots. *Ky. Agric. Expt. Statn., Res. Bull., 226,* 51.

Xu, Z. T., Sun, C. H., Zhang, C. M., Li, L., Wang, K. S., Han, Z. J., & Wang, D. C., (1993). Identification of maize inbred lines and hybrids for resistance to basal stalk rot at the seedling stage. *Crop Genetic Resources, 34,* 35.

CHAPTER 6

Brown Stripe Downy Mildew of Maize and Its Integrated Management

ASHWANI K. BASANDRAI and DAISY BASANDRAI

CSKHPKV, Rice, and Wheat Research Center, Malan District Kangra, Himachal Pradesh–176047, India, E-mails: ashwanispp@gmail.com, bunchy@rediffmail.com

6.1 INTRODUCTION

Downy mildews are important maize diseases in many tropical areas through the world. The disease has been reported to be very destructive in many areas of tropical Asia and causing more than 70% losses. Downy mildews are caused by up to ten various species of oomycete fungi comprising genera of *Peronosclerospora, Scleropthora, Sclerospora, and cause downy mildews.* Brown stripe downy mildew (BSDM) of maize is one of the most important downy mildew diseases of maize. In India, its prevalence and severity is very high in Uttar Pradesh, Uttarakhand, Himachal Pradesh, Rajasthan, and Punjab and Hills of West Bengal and Jammu and Kashmir. Losses in yield to the tune of 63% were recorded in Tarai region of Uttarakhand.

BSDM of maize was first observed in India during the year 1962 and was described in 1967 (Payak and Renfro, 1967). It is caused by the fungus *Sclerophthora rayssiei* pv *zeae*. It is a one of the common and the most destructive diseases of maize in the country causing losses of ranging from 20–90% (Payak, 1975). The maximum losses have been experienced in areas with high rainfall where, susceptible cultivars were grown. Its highest severity has been recorded in areas with 100–200 cm rainfall and it declines with the decrease in rainfall (Frederiksen and Renfro, 1977). In USA, McGregor (1978) listed '49 top-ranking exotic threatening pathogens,' out of 551 potential ones, based on expected economic impact (EEI), to agriculture. Out of which, *S. rayssiae* var. *zeae* was ranked 43rd with the EEI standing at US$ 53 million at 1978 prices and further it was calculated that even if only 20% of the crop were seriously affected, this could translate into a US $4 billion loss at 2005 prices (http://www.plantwise.org/KnowledgeBank/Datasheet.aspx?dsID=49244).

6.2 DISTRIBUTION AND ECONOMIC IMPORTANCE

This disease recoded in first time in 1962 in many maize growing areas of India (Putim, 2007), and further it has spread throughout India. It has also been recorded in Myanmar, Nepal, Pakistan, Bangladesh, and Thailand (CAB, International, 2006) also. BSDM diseases incidence was the highest in regions with high rainfall. In India, the most severe epidemics were reported in areas with annual rainfall of 100–200 cm, areas e 50–70 cm rainfall developed light incidence of the disease whereas, in areas with annual rainfall of 60–100 cm moderate disease development was observed (Payak and Renfro, 1967). In India, BSDM of maize disease has been reported from various states viz. Himachal Pradesh, Sikkim, West Bengal, Meghalaya, Punjab, Haryana, Rajasthan, Delhi, Uttarakhand, Bihar, Madhya Pradesh, and Gujarat. The BSDM of maize disease is favored by cool and moist conditions and severe infection (20–70% incidence) develops on susceptible cultivars in high rainfall areas. In some locations, disease incidence can be as high 100% (Singh, 1971; Basandrai, and Singh, 2002; Singh and Basandrai, 2012). Losses are the greatest in regions with abundant summer moisture and warm soils. Losses vary depending on time and severity the tissue affected. Disease severity of 75% or more, prior to flowering may result in total loss of the crop resulting in total suppression of ear formation or its marked attenuation. There may be 20–90% reduction in grain yield which may be more in highly susceptible cultivars under conditions conducive for disease development (Basandrai and Singh, 2002). Yield losses of up to 63% have been recorded in the *Tarai* area of Uttar Pradesh (Sharma et al., 1993). Maize cultigenes showed varied reaction in *S. rayssiae* var. *zeae* (Payak and Renfro, 1967).

6.3 SYMPTOMS

BSDM of maize manifest itself is the vein-limited striping of foliage especially leaves. Other parts of the plant including leaf sheaths, husk leaves, ears or tassels, do not develop symptoms. All the leaves including the flag leaf may be affected by the disease. The early zoosporic infections appear as vein-limited chlorotic flecks or blobs which enlarge lengthwise and coalesce. The merger of the flecks leads to the formation of rudimentary stripes in the inter-veinal areas. The stripes may vary 3–7 mm in breadth and may extend to the full length of the lamina on either side of the midrib. The stripes may turn yellowish-tan to purple ferruginous and necrotic with time. In some maize genotypes, the pathogen induces stripes with reddish to reddish-purple borders and with bleached centers. This process of necrosis of stripes coincides with the development of the teleomorphic stage (antheridia and oogonia) and indicates the cessation of sporangial production. The disease first appears on the lower most leaves proximate to the ground level which shows the highest level of striping leading to pale-brown, burnt appearance. The severely affected leaves may be shed prematurely. Leaves around the ear shoot show lesser amount of striping and it is least on leaves above it (near the tassel). Infections leading to severe striping result in blotching of extensive areas of leaf laminae. In the early stages, en masse zoosporic infections in large patches lead to rapid coalescence of spots and show blotching effect leading to more severe damage and premature defoliation. As the veins are not affected, laminar shredding is uncommon, however, under severe infection, leaves tear apart near the apices and hang in tatters. A grayish-white, downy growth of the fungus develops, which has a fine granular rather than fibrous appearance and it appears both on adaxial and abaxial surfaces of the stripes. However, it is more common on adaxial surface. Hence, it is amphigenous rather than hypophyllous. The downy growth has been observed in the afternoons and so it is not so evanescent (Payak and Renfro, 1967; Payak et al., 1970). As the stripes lose their chlorotic appearance and turn necrotic, the downy growth disappears. Oospores appear to be produced only in necrotic tissues. Striping of maize leaves can occur due to a variety of causes: genetic, nutritional, or pathological. However, the presence of granular downy growth the fungus on the undersides of vein-limited stripes without any malformation is a diagnostic feature of BSDM.

6.4   PHYTO-PATHOMETRY

6.4.1   INOCULATION TECHNIQUE FOR BROWN STRIPE DOWNY MILDEW (BSDM)

> ➢ **Method I:** Pathogen is an obligate parasite and cannot be cultured artificially. Hence, inoculum may be prepared by:

- Collecting infected maize leaves during the previous season and their storage in double-bagged containers in access-controlled cabinets or refrigerators.
- Collecting infected leaves full of oospores from early planting of susceptible maize grown during the same season and their drying.

The leaves are powdered and artificial epiphytotic is created by placing the powder in the furrows just before planting and placement of inoculums in proximity.

➢ **Method II:** Freshly infected leaf pieces (2–3 cm) containing fungal sporangia are put in the whorls of the seedlings during cloudy weather in the evening hours between 5–7 P.M, 4–5 weeks after sowing. In experimental plots where disease appears every year, this method is adequate for creating epidemics.

Disease rating of individual maize genotype may be scored (Table 6.1) by evaluation of all plants in the plot using 0–5 rating scale (slightly modified as was described previously) as described in Table 6.1.

TABLE 6.1 Disease Rating Scale of Brown Stripe Downy Mildew Disease of Maize

Sr. No.	Details	Score on Standard Rating Scale	Disease Reaction (R/MR/S/HS)
1	No infection	0	Free
2	Few scattered to moderate number of stripes on lower leaves. Infection (>5% disease severity).	1	Highly resistant
3	Abundant stripes on lower leaves and few on middle leaves. Light infection (Disease severity 5.1–10%).	2	Resistant
4	Abundant stripes on lower and middle leaves extending to upper leaves. Moderate infection (disease severity 10.1–20%).	3	Moderately resistant
5	Abundant stripes on lower and middle leaves extending to upper leaves. Heavy infection (Disease severity 20.1–40%),	4	Susceptible
6	Abundant stripes on all leaves, No cob formation. Plant may be killed prematurely. Very heavy infection (Disease severity >40%).	5	Highly resistant

*Score 1, 1.5, 2.0, 2.5, etc. on a scale of 1–5 and 1, 2, 3, 4, etc. on a scale of 1–9 must be given.

6.5 PATHOGEN BIOLOGY

6.5.1 CAUSAL ORGANISM

Sclerophthora rayssiae var. *zeae* R. G. Kenneth, Koltin, and I. Wahl Payak and Renfro.

The *sporangiophores* are short, determinate, and are produced from hyphae in the substomatal cavities. Sporangia are formed sympodially in groups of 2–6, arising in basipetal succession. Sporangia are hyaline; ovate, obclavate, elliptic or cylindrical and smooth-walled. These are papillate, possessing a projecting truncate, rounded, or tapering poroid apex. The sporangia are caducous, with a persistent, straight, or cuneate peduncle. Sporangia range in size from 18.5–26.0 µm × 29.0–66.5 µm; there may be lens shaped pores through which zoospores or cytoplasm may escape. Four-8 hyaline, spherical zoospores are formed in each sporangia which may vary 7.5–11.0 µm in diameter. Zoospores may encyst within or outside sporangia. Oogonia are hyaline to light colored, thin walled which may have one or two paragynous antheridia measuring 33.0–44.5 µm in diameter. Oospores are pleurotic, spherical or sub spherical, hyaline with one prominent oil globule. Cell walls are smooth, glistening, and uniform with 4 µm thickness and confluent with the oogonial wall which range from 29.5–37.0 µm in size (Basandrai and Singh, 2002; Singh, and Basandrai, 2012; Putim, 2007).

Sporangial production may be stimulated by placing chlorotic symptomatic tissue into a moist chamber and incubating at 22 to 25°C. Necrotic tissue will not produce sporangia. Sporangia are generally produced within 3–9 h.

Oogonia and oospores are produced in necrotic tissue. These may be visualized by clearing the leaf tissues in 2% sodium or potassium hydroxide solution at 45 to 50°C, washing in several changes of distilled water and staining with 0.1% cotton blue (= methyl blue) in 50% glycerin for 20 min at 45–50°C. Exact times for initial clearing and staining of tissues will vary depending on maturity and thickness of the tissues. Leaves should take less than 1 h to clear (Putim, 2007).

6.5.2 TAXONOMIC POSITION

As per Agrios (2005), the systematic of the fungi S. rayssie pv zeae is as follows:

1. Kingdom: Chromista (Stramenopiles);
2. Phylum: Oomycota;

3.   Class: Oomycetes;
4.   Order: Peronosporales;
5.   Family: Peronosporaceae;
6.   Genus: *Sclerophthora;*
7.   Species: *Rayssieae* var *zeae.*

6.6   DISEASE CYCLE AND FAVORABLE CONDITIONS RESPONSIBLE FOR DISEASE DEVELOPMENT

Oospores generally undergo indirect germination, producing sporangiophores that bear sporangia bearing 4–8 zoospores. Less frequently, the sporangium may germinate directly and produce a germ tube capable of penetrating maize leaves. Rapid spread of the pathogen in the field occurs with the production of sporangia (secondary inoculum), which are dispersed in wind and water splash, or from physical contact with an infected plant. Sporangia have been trapped 1.65 m above from an infected field, but the greatest numbers of sporangia were found to move less than 1 m, suggesting that there are less chances of long distance transport via wind (Basandrai and Singh, 2002; Singh, and Basandrai, 2012; Singh, and Renfro, 1971).

The disease cycle involves both sexual and asexual reproduction. Primary inoculum comes from oospores over seasoning in soil, plant debris or from mycelium in infected seed. Maize seed may be contaminated in two ways, i.e., seed surface may carry plant debris containing viable oospores (Lal and Parsad, 1989), and the seed may carry oospores or mycelium within the embryo (Lal and Parsad, 1989). Disease can become established on seedlings grown from infected seed (Singh et al., 1967), although seed transmission was found to occur at less than 1% (Lal and Parsad, 1989), and it is likely that infected leaf debris is more important than seed in initiating new infections (Singh, 1971). Oospores germinate to produce sporangia, which release zoospores which penetrate leaf tissue. Oospores in air-dried leaf tissue can remain viable for 3–5 years (Singh, 1971; Singh et al., 1970). Infected seed dried to 14% moisture or less and stored for 4 or more weeks will not be capable of transmitting the disease (White, 1999).

Lesions are initially inter-veinal and appear as chlorotic, brownish, or reddish stripes on the leaves. Asexual sporulation is favored by moderate temperatures (20–25°C) and high moisture. Sporangia are produced on non-necrotic leaf tissue and give the leaf a grayish-white appearance. Sporangia are dispersed to short distances via wind or rain splash which germinate to produce zoospores or, less commonly a germ tube to repeat the cycle. Oospores are produced in necrotic tissues and can survive for years in soil or in plant debris.

Warm soil (28 to 32.5°C) is required for disease development when seeds are inoculated with infected plant debris (Singh et al., 1970). There is no information on the ability of the pathogen to withstand winter temperatures in northern climates. Once the fungi has colonized host tissue, sporangiophores (conidiophores) emerge from stomata and produce sporangia (conidia) which are wind and rain splash disseminated and initiate secondary infections. Moisture is essential for infection by *S. rayssiae* var. *zeae* (Singh et al., 1967, 1970). Sporangia production, germination, and infection require a film of water. Twelve hours of leaf wetness were required for infection via zoospores with longer periods producing greater numbers of infected plants. Most sporangia are liberated at maturity during the day (Singh et al., 1970). Sporangial release occurs in the afternoon of sunny days when high moisture is present, rather than on cloudy or rainy days (Singh and Renfro, 1971). Generation time of secondary inoculum (sporangia) from primary inoculum (oospores) can be rapid. Under ideal conditions, sporangial production can occur as soon as 10 days post-inoculation. Sporangia are produced over a wide range of temperatures (18 to 30°C), but are most abundantly produced at 22 to 25°C. Infected leaves placed in a moist environment at 22 to 25°C can produce sporangia in as little as 3 h, with a second generation of sporangia arising 9 h later (Singh et al., 1970). Young plants are most susceptible to inoculation, with susceptibility decreasing as the plants age (Singh et al., 1970).

6.7 GENETICS OF RESISTANCE

Resistance to BSDM (*Sclerophthora rayssiae* var. *zeae*) was in the partial or complete dominance range and additive gene action appeared to play a role in disease resistance (Handoo et al., 1970; Asnani and Bhushan, 1970; Singh and Asnani, 1975). Resistance to BSDM in maize was reported to be inherited by a qualitative character poygenically controlled by partially dominant of dominant genes (Basandrai and Singh, 2002; Anonymous, 2004). Resistance in maize inbred lines DKI 9422, DKI 138, and CM 111 was controlled by single dominant gene whereas; it was controlled by single recessive gene DKI 3 (Anonymous, 2004).

6.8 INTEGRATED MANAGEMENT OF BROWN STRIPE DOWNY MILDEW (BSDM)

1. **Cultural:**
   - Sowing of non-host crops, viz. wheat, oats, cotton, and soybean which may induce oospore germination in infected soil.
   - Early planted maize escapes infection as by the time significant quantities of asexual spores are produced in collateral host, the plants develop resistance (Basandrai and Singh, 2002, 2012).
   - The eradication of collateral and wild hosts near maize field and rouging of infected maize plants has been recommended.
   - Destruction of plant debris by deep plowing and other methods.
2. **Chemical:**
   - The disease is effectively managed by seed treatment with Ridomil MZ 25WP (0.4%) and Ridomil MZ 75WP (0.3%) (Sharma et al., 1993; Basandrai, and Singh, 2002, 2012).
   - Seed treatment with Metalaxyl at 4 g/kg and foliar spray of Mancozeb 2.5 g/l or Metalaxyl MZ at 2 g/l is recommended.
   - Under middle Gujarat agro-climatic conditions, seed treatment with Apron 35 WS @ 7 gm/Kg seed followed by three foliar application of Ridomil MZ 72WP at 45, 55 and 65 DAS helped to increase the grain yield by 14.76% and 25.49% reduction in BSDM disease intensity over check.
   - In seed plots, 3–6 sprays of fungicide, mancozeb (0.3%) 10 days after sowing at 7 days interval were effective in reducing the disease.
   - Metalaxyl formulation Apron 35 WS and Apron 30 FW were more effective than Ridomil. Fungicides Furalaxyl (Fongarid WPG), Miliform (Patafol, Caltan WP), and Benalaxyl (Galben WPG) give effective control of downy mildew (Sharma et al., 1993; Singh and Basandrai, 2012).

6.8.1 HOST PLANT RESISTANCE

- Various hybrid and composite varieties resistant to BSDM viz. KH 150, X-1174 MY, PRO 312, PRO 316, JH 3189, MMH 69, EH 230792, X-1123G (Basandrai et al., 2000), DMR 1, DMR 5 and Ganga 11, EHVL 45, VLD 68, VLD 90, VL 54A, Ganga 5, Ganga 11, Nardi 16, Sneha Gold, Seed Tech 204 and Pro 348 were highly resistant to BSDM (Basandrai and Singh, 2000, 2002, 2012) were identified.

- Kumar et al. *(*2011) reported that hybrids PG 2465, P 3420, Hi-Shell, PSC 3322, PG 2408, 900 M Gold, F_1 Corn Hybrid, DMH 1107, KHB 63 and DKC 7074 were resistant to BSDM. Singh and Singh (2011) reported that ten stocks viz CMH 108–154, JH-12114, IDX-2901, PAC-T99, BIO-265, JKMH-31314, and KMH-3426 were highly resistant and 22 stocks were resistant to BSDM.
- Maize composite and hybrids PMH 2, Parkash, PAU 352, PMH 4, PMH 5, Gujrat Makkai 6, Gujrat Makkai 4, Partap Makka 3, Partap Makka 5, Aravali Makka 1 showed resistance to BSDM (Anony-mous, 2012).
- Among 2113 Indian maize inbred lines and other germplasm scored in the field, 58 were highly resistant, 667 resistant, 772 moderately resistant, 478 susceptible, and 138 highly susceptible (Putim, 2007).
- More than 600 Inbred lines were evaluated against *S. rayssiae* pv. *zeae* under artificial epiphytotic conditions during the year 2000 to 2003. In all, 80 lines were free from the disease and 42 lines were resistant. Inbred lines DKI 9726, DMR 89145 Sr 5, 8LT 32-1 and DKI 9204 showed multiple resistance against BSDM, Erwinia stalk rot (ESR) and banded leaf and sheath blight (BLSB). Among the inbred lines procured from CYMMIT, lines TL 99A-1206-31, TL 99A-1209-16 and TL 99A1220-24 showed combined resis-tance against BSDM, ESR, MLB, and BLSB whereas, lines TL 99A-1201-2, TL 99A-1201-6, TL 99A-1201-9, TL 99A-1201-22, TL 99A-1201-31, TL 99A-1201-44, TL 99A-1201-55, TL 99A-1208-86, TL 99A-1209-11 and TL 99A1220-25 were resistant to BSDM, MLB, and BLSB (Anonymour, 2004).

KEYWORDS

- **banded leaf and sheath blight**
- **brown stripe downy mildew**
- **epiphytotic conditions**
- **erwinia stalk rot**
- **expected economic impact**
- **oospore**

REFERENCES

Asnani, V. L., & Bhusan, B., (1970). Inheritance study on the brown stripe downy mildew of maize. *Indian Phytopathology, 23,* 220–230.

Handoo, M. I., Renfro, B. L., & Payak, M. M., (1970). On the inheritance of resistance to *Sclerophthora rayssiae* var. *zeae* in maize. *Indian Phytopathology, 23,* 231–249.

Anonymous, (2004). Final report of ICAR funded research scheme entitled, "Pathogenic variation in brown stripe downy mildew (*Sclerophthora rayssiae* var *zeae*) and its utilization for characterization of resistant sources in maize (*Zea mays.*)".

Anonymous, (2012). A compendium of hybrids and composites of maize (1993–2012). Directorate of maize research, New Delhi, 110112 (India). *Technical Bulletin No. 2012/5* (p. 172).

Asnani, V. L., & Bhusan, B., (1970). Inheritance study on the brown stripe downy mildew of maize. *Indian Phytopathology, 23,* 220–230.

Bains, S. S., Jhooty, J. S., Sokhi, S. S., & Rewal, H. S., (1978). Role of *Digitaria sanguinalis* in outbreaks of brown stripe downy mildew of maize. *Plant Dis. Rep., 62,* 143.

Ashwani, K. B., & Akhilesh, S., (2002). Fungal diseases of maize. In: Gupta, V. K., & Paul, Y. S., (eds.), *Diseases of Field Crops* (pp. 102–127). Indus Publishers, New Delhi.

CAB International, (2006). *Crop Protection Compendium.* Online. CAB International, Wallingford, UK.

Basandrai, A. K., Singh, A., & Kalia, V., (2000). Multiple disease resistance in Indian maize hybrids and composites. *Indian Journal of Plant Genetic Resources, 13*(2), 188–190.

Frederiksen, R. A., & Renfro, B. L., (1977). Global status of maize downy mildew. *A. Rev. Phytopathology, 15,* 249–275.

Handoo, M. I., Renfro, B. L., & Payak, M. M., (1970). On the inheritance of resistance to *Sclerophthora rayssiae* var. *zeae* in maize. *Indian Phytopathology, 23,* 231–249.

Kumar, A., Singh, A., Devlash, R., & Lata, S., (2011). Evaluation of maize hybrids against major diseases in Himachal Pradesh. *Plant Disease Research, 26,* 176.

Lal, S., & Prasad, T., (1989). Detection and management of seed-borne nature of downy mildew diseases of maize. *Seeds Farms, 15,* 35–40.

Lal, S., Saxena, S. C., & Upadhyay, R. N., (1980). Control of downy mildew of maize by maize hybrids and composites. *Ind. J. of Pl. Genetic Resource, 13,* 188–190.

Payak, M. M., & Renfro, B. L., (1967). A new downy mildew disease of maize. *Phytopathology, 57,* 394–397.

Payak, M. M., Renfro, B. L., & Lal, S., (1970). Downy mildew disease incited by *Sclerophthora. Indian Phytopathol, 23,* 183–193.

Putnam, M. L., (2007). Brown stripe downy mildew (*Sclerophthora rayssiae* var. *zeae*) of maize. Online. *Plant Health Progress.* doi: 10.1094/PHP-2007-1108-01-DG.

Sharma, R. C., De Leon, C., & Payak, M. M., (1993). Diseases of maize in South and South East Asia: Problems and progress. *Crop Protection, 12*(6), 414–422.

Singh, A., & Ashwani, K. B., (2012). Important diseases of maize and their eco friendly management. In: Vaibhav, K. S., Yogendra, S., & Akhilesh, S., (eds.), *Eco Friendly Innovative Approaches in Plant Disease Management* (pp. 357–387). International Book Distributors Dehradun.

Singh, I. S., & Asnani, V. L., (1975). Gene effects for resistance to brown stripe downy mildew in maize. *Indian Journal of Genetics and Plant Breeding, 35,* 123–127.

Singh, J. P., (1971). Infectivity and survival of oospores of *Sclerophthora rayssiae* var. *zeae*. *Indian J. Exp. Biol., 9,* 530–532.

Singh, J. P., & Renfro, B. L., (1971). Studies on spore dispersal in *Sclerophthora rayssiae* var. zeae. *Indian Phytopath, 24,* 457–461.

Singh, J. P., Renfro, B. L., & Payak, M. M., (1970). Studies on the epidemiology and control of brown stripe downy mildew of maize (*Sclerophthora rayssiae* var. *zeae*). *Indian Phytopath., 23,* 194–208.

Singh, R. S., Joshi, M. M., & Chaube, H. S., (1967). Further evidence of the seed borne nature of maize downy mildews and their possible control with chemicals. *Plant Dis. Rep., 52,* 446–449.

Singh, R. S., (1998). *Plant Diseases* (7th edn., p. 686). Oxford and IBH Publishing Co. Pvt. Ltd. New Delhi.

White, D. G., (1999). *Compendium of Maize Diseases* (3rd edn.). American Phytopathological Society, St. Paul, MN.

CHAPTER 7

Major Diseases of Sorghum and Their Management

YOGENDRA SINGH, DIVYA SHARMA, and
BHUPENDRA SINGH KHARAYAT

*Department of Plant Pathology, College of Agriculture,
GBPUA&T, Pantnagar, Uttarakhand, India*

7.1 INTRODUCTION

Sorghum (*Sorghum bicolor* (L.) Moench) is affected by many diseases caused by various plant pathogens. These diseases are classified as seedling diseases, foliar diseases, root, and stalk diseases, panicle diseases, and storage diseases. They can also be classified on the basis of symptoms produced on plants and are known as seedling blight, root, and stalk rots, leaf blight, leaf spot, rust, smut, ergot, wilt, downy mildew, grain mold, leaf streak, leaf mosaic, etc. Out of more than fifty diseases reported and described on sorghum, only some are economically important globally while several others are responsible for significant losses in particular agroecosystems. In most semi-arid tropical regions, economically important diseases are grain mold, anthracnose, leaf blight, downy mildew, zonate leaf spot, charcoal rot, ergot, smuts, stalk rot, and maize stripe and maize mosaic virus (MMV) diseases. These diseases, either individually or in combination causes substantial damage to crop resulting in heavy economic losses year after year. This communication deals with important diseases of sorghum and their management strategies.

7.2 FUNGAL DISEASES

7.2.1 GRAIN MOLD DISEASE

Grain mold is one of the major constraints in sorghum production. The sorghum cultivars with white grain pericarp are more susceptible in comparison to brown and red grain pericarp to grain mold. Grain mold results in reduced seed size and weight, discoloration of grains, decreased germination, decreased seedling vigor and mycotoxin contamination. Such molded grains are hazardous to human and animal health and fetch reduced market price to the farmers. The loss resulting from this disease range from 30% to 100% depending on cultivar type, flowering time and prevailing weather conditions (Singh and Bandyopadhyay, 2000).

7.2.1.1 SYMPTOMS

The initial symptom of the disease is discoloration of grains due to infection and colonization by mold fungi. Partially infected grains show whitish, grayish, orange, and pinkish to shiny black discoloration depending on infection by a particular fungal species. Fungal growth appears at the hilar end of the grain and thereafter extends to the pericarp surface. Often grains are colonized by many fungi. Heavily infected grains turn completely black.

7.2.1.2 CAUSAL ORGANISM AND ETIOLOGY

Fungi associated with grain mold complex include *Fusarium* spp., *Curvularialunata, Alternaria alternata,* and *Phomasorghina. Curvularialunata* appears as shiny black, fluffy growth on grain surface, *Fusarium* spp. generally produces pinkish white mycelium, powdery in appearance at first which later becomes pinkish fluffy. *Alternaria alternata* appears as dull with grayish black mycelium, often sparse and in stripes. *Phomasorghina* produces pin-like small, round, black pycnidia embedded in grain and produces a thick dirty black crust with rough surface on the pericarp.

7.2.1.3 DISEASE CYCLE AND EPIDEMIOLOGY

Most of the mold-causing fungi are weak parasite disseminated by seed-borne, soil-borne, and airborne spores. However, the role of seed-borne inoculum as a direct cause of grain mold is minimal. It has been observed that in sorghum infection and colonization of flowers occur prior to grain maturity. Mycelium penetrates the pericarp and ramifies within five to ten days. Fungus subsequently invades the endosperm and sometimes the embryo as well (Little, 2000). At the time of Anthesis sorghum flower is most susceptible to infection and colonization by grain mold fungi. Early infection on the apical portions of flower tissues occurs on glumes, lemma, and palea and subsequently grains are covered by fungal growth and sporulation. Most spores reproduce in plant debris and decaying organic matter on the soil surface. Abundant sporulation occurs during humid conditions on the lower leaves of the plants. Spores are readily disseminated by wind and rain splash and spread the disease.

Humid and warm conditions during flowering and grain development favor infection while dry conditions prevent it. Spore production increased in warm temperature (25–28°C) and high relative humidity (100%) and decreased with drop in temperature below 15°C and rise in temperature above 30°C. With sudden increase in the relative humidity following rainfall, the inoculum load also jumped several times (Indira and Muthusubramanian, 2004).

7.2.1.4 MANAGEMENT

- Grain mold losses can be avoided by adjusting planting dates so that plants do not face frequent rains during grain filling stage.
- Resistant accessions like IS 2815, IS 21599, IS 10288, IS 3436, IS 10646, IS10475 and IS 23585 may be useful in developing varieties and hybrids.
- Treatment with formulation of *Pseudomonas fluorescens* has been found effective in reducing *F. moniliforme* infection in sorghum seeds and also in increasing germination, vigor index and field emergence (Raju et al., 1999).

7.2.2 ANTHRACNOSE

Anthracnose is one of the most destructive diseases of sorghum throughout the world. The disease is particularly severe in warm and humid environments resulting in substantial economic losses. The pathogen causes seedling blight, leaf blight, stalk rot, head blight, and grain molding thus limiting both grain and forage production. Leaf blight phase is most devastating and may result in yield losses of 50% or more under severe conditions.

7.2.2.1   SYMPTOMS

Foliar symptoms usually appear 25–30 days after emergence (DAE). Symptoms are characterized by small, circular to elliptical spots, up to 5 mm in diameter but often smaller, which develop gray to straw-colored centers with wide purple or red margins depending on host cultivar. Under hot and humid conditions, the spots increase in number and coalesce to cover large leaf area. In the center of the spots, small, circular, black dot-like fruiting bodies (acervuli) develop. Midrib infection often occurs and is seen as elliptical to elongated red or purple lesions on which the black acervuli can be clearly seen. Leaf-sheath and panicles including grains and rachis are also infected (Figure 7.1).

FIGURE 7.1   Anthracnose symptoms on sorghum leaves.

7.2.2.2   CAUSAL ORGANISM AND ETIOLOGY

The disease is caused by *Colletotrichum graminicola* (Ces.) Wilson. Conidia are produced terminally on the erect, nonseptate and short conidiophores among the setae. They are hyaline, nonseptate, uninucleate, and cylindric to obclavate but become sickle-shaped with age measuring 4.9–5.2 × 25.1–38.8 µm. Acervuli produced on the infected host tissue are dark brown, oval to cylindrical and with or without setae.

7.2.2.3 DISEASE CYCLE AND EPIDEMIOLOGY

The pathogen survives on host residue, wild sorghum species, and weeds as conidia or mycelium on seed. It can persist for more than a year in diseased residues on the soil surface. Conidia from wild sorghum species or residues serve as the primary inoculum carried to sorghum leaves by wind or splashing rain. Conidia germinate and penetrate the epidermis directly or through stomata. Seedling infection occurs from the inoculums present in the crop residue. The disease becomes most severe during period of continuous rain, high humidity, and temperature of 28–30°C.

7.2.2.4 MANAGEMENT

- Several Sorghum lines identified with stable source of resistance include IS 3547, IS 6958, IS6928, IS 8283, IS 9146, IS 9249, IS 18758, M 35610, SPV 386 and ICSV 247. Some tolerant hybrid seed parents such as ICSA/B 295 are available at ICRISAT (Thakur et al., 2007).
- Clean cultivation and elimination of grasses have been used to manage the disease.
- *Trichoderma* spp. and fluorescent *Pseudomonads* have shown potential to be used as biocontrol agent for sorghum anthracnose. Application of *Pseudomonas marginalis* (C 21) reduced the disease severity by 28.23% under glass house conditions (Mischreff et al., 1994). *Trichoderma harzianum* (Th 43) applied as seed biopriming, colonized compost + seed biopriming, and foliar spray reduced the disease severity by 20, 22 and 21% respectively in addition to increasing plant height and green fodder yield significantly (Singh and Singh, 2008).
- Solarization of soil alone and when applied with *T. harzianum* (Th 43 and Th 39) and *P. fluorescens* (Psf 27) resulted in increased plant height, collar diameter and reduced disease severity (Singh, 2008).

7.2.3 LEAF BLIGHT

Leaf blight is prevalent in many humid regions of the world where sorghum is grown. Disease is favored by moderate temperature and heavy dews during the growing season. Pre-flowering infection of susceptible cultivars may result in up to 50%losses ingrain yield.

7.2.3.1   SYMPTOMS

Symptoms of the disease may appear from the seedling stage to the crop maturity stage. Small, reddish, or tan spots develop on the leaves of seedlings which later on enlarge and coalesce resulting in wilting of leaves. Long, elliptical, reddish purple or yellowish lesions develop first on lower leaves and then progresses to upper leaves in mature plants. These lesions may coalesce giving the crop a burnt appearance. Lesion size varies with resistance levels of host genotypes, pathogen virulence, and prevailing weather conditions. In humid weather, large numbers of grayish black spores are visible on the lesions in concentric zones (Figure 7.2).

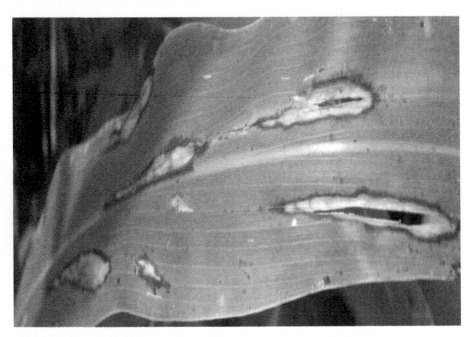

FIGURE 7.2   Leaf blight symptom on sorghum leaf.

7.2.3.2   CAUSAL ORGANISM AND ETIOLOGY

Leaf blight is caused by *Exserohilum turcicum* (Pass.) Leonard and Suggs (syns. *Helminthosporium turcicum* Pass., *Bipolaris turcica* (Pass.) Shoem., and *Drechslera turcica* (Pass.) Subram. and Jain). Conidia produced singly at the conidiophores tips are light gray, straight or spindle shaped or slightly curved with rounded ends. They are three to eight septate, have a protruding basal hilum, and measure $10–20 \times 28–153$ µm. They germinate by polar germ tubes. The pathogen can also infect other hosts like maize, Johnson grass, and Sudan grass.

7.2.3.3 DISEASE CYCLE AND EPIDEMIOLOGY

The pathogen survives as mycelia, conidia, and chlamydospores in infected crop residues on or in the soil. Secondary spread of the disease takes place mainly through wind borne conidia. A temperature of 20–22°C, high humidity, and high rainfall are most favorable for disease development.

7.2.3.4 MANAGEMENT

- Rotation with non-susceptible crops can minimize the losses occurring from this disease.
- Resistant sorghum germplasm accessions include IS 13868, IS 13869, IS 13870, IS13872, IS 18729, IS 18758, and IS 19670.
- Seed treatment with carbendazim (0.1%) combined with spray of mancozeb (0.2%) gave maximum efficacy of disease control (57.5%) and increased grain and fodder yield by 36.5% and 21.6% respectively (Bunker and Mathur, 2008).

7.2.4 ZONATE LEAF SPOT

Zonate leaf spot is fast emerging as a serious disease of sorghum. The pathogen may also infect other hosts such as maize, millet, sugarcane, and several weeds.

7.2.4.1 SYMPTOMS

The disease is characterized as circular, reddish purple bands alternating with straw colored or tan regions forming a concentric or zonate pattern with irregular margins on leaves. The Lesions are semicircular if they originate along the margins of leaves. These lesions range from 1–2 cm in diameter in early stages to 3–7 cm in advanced stages. Leaf sheaths are also encircled by dark red to blackish purple or brown lesions. Under warm humid conditions, the fungus produces large pinkish gelatinous spore masses on the necrotic areas of the lesions. In severe cases, lesions may coalesce and cover the entire leaf lamina. Sometimes small black sclerotia may form in mature lesions (Figure 7.3).

FIGURE 7.3   Zonate leaf spot symptoms on sorghum leaf.

7.2.4.2   CAUSAL ORGANISM AND ETIOLOGY

The disease is caused by *Gloeocercospora sorghi* Bain and Edg. The fruiting bodies of the pathogen (sporodochia) are formed on the surface of leaves. Conidia are hyaline, needle shaped, straight or slightly curved 4–14 septate and measure 20–195 × 1.4–3.2 μm. Sclerotia are black in color and measure 0.1–0.2 mm in diameter.

7.2.4.3   DISEASE CYCLE AND EPIDEMIOLOGY

The fungus survives in the form of sclerotia in the dead tissue of old leaf lesions. Sclerotia formed on the other grass hosts may also play a role in the survival of the pathogen. The sclerotia germinate sporogenically and form conidia, which infect the crop in the next season. During hot and humid weather, conidia produced on the new lesions cause further spread of the disease. The pathogen may also be carried on seed. The disease is more severe under moist condition.

7.2.4.4   MANAGEMENT

- Crop rotation and sanitation practices can reduce losses from this disease.
- Some grain sorghum genotypes viz., SPV 1659, IS 14332, SPV 1685, SPV 1713, SPV 1727, 7A and forage sorghum genotypes viz., UTMCH 1302, UTMC 532 and PC 5 have shown combined resistance against zonate leaf spot and anthracnose (Singh, 2009).

7.2.5   DOWNY MILDEW

Downy mildew is a widespread disease of sorghum in many tropical and subtropical regions of the world. The disease being systemic in nature is highly destructive and may result in death of plants in severe infections.

7.2.5.1   SYMPTOMS

Disease occurs as either systemic or localized infection. The systemic form is induced when the pathogen colonizes the meristematic foliar tissues. Infected seedlings become chlorotic and stunted and may die prematurely. Lower surfaces of chlorotic leaves are covered by a white, downy growth containing conidiophores and conidia of the pathogen. In later stages of the disease, leaves emerging from the whorl exhibit parallel stripes of green and white tissue. Later on, these interveinal tissues die and leaf shredding occurs. The shredded tissues contain a large number of thick walled oospores.

7.2.5.2   CAUSAL ORGANISM AND ETIOLOGY

The disease is caused by *Peronosclerospora sorghi* (Watson and Uppal) Shaw. Conidiophores are erect and dichotomously branched. Conidia produced on short sterigmata are hyaline, obovate, non-papillate, and measure 15–29 × 15–27 µm. They germinate by germ tubes. Oospores are spherical, hyaline to yellow and enclosed in irregular thick oogonial wall and germinate by producing germ tube.

7.2.5.3   DISEASE CYCLE AND EPIDEMIOLOGY

The primary infection is initiated by germinating oospores. Oospores formed in the diseased plants persist in the soil for several years. In the soil, these oospores germinate and infect the roots of sorghum seedlings. The pathogen moves upward systemically, colonizes the meristamatic tissues, and induces leaf chlorosis. Conidia produced on the chlorotic leaves are disseminated and cause the secondary infection. The pathogen can also survive in perennial weeds such as *Sorghum halepense*, which serve as source of conidial inoculum in the next cropping season. Low soil moisture and 10°C temperatures favor infection by oospores. Conidial production and infection are favored by cool environment (15–20°C) and high humidity (100%). The pathogen may also survive as oospores in the seed glumes and as plant debris mixed with the seed.

7.2.5.4   MANAGEMENT

- Cultural practices like deep plowing and crop rotation can effectively minimize the disease.
- Highly resistant sorghum lines and varieties against downy mildew include IS 3547, IS 8283, IS 5743, IS 8607, IS 8864, IS 7179, IS 8954, IS 18757, IS 22228 and IS 22230.

7.2.6   GRAY LEAF SPOT / CERCOSPORA LEAF SPOT

The disease is generally found in warm and wet sorghum growing regions and is responsible for extensive damage on susceptible varieties. Disease progress and losses are influenced by initial inoculums potential, duration of weather conditions favoring disease development and cultivar susceptibility.

7.2.6.1   SYMPTOMS

The symptoms begin as small, red spots on leaves which enlarge to form narrow, rectangular lesions measuring 2–5 × 5–15 mm in size. These lesions may coalesce to form longitudinal stripes. The lesions are typically dark red to purplish or straw colored in plants with tan reaction. Symptoms may also appear on leaf sheaths and upper stems. Sporulating lesions containing conidiophores and conidia of the pathogen give leaves a grayish appearance, from which the disease name is derived.

7.2.6.2   CAUSAL ORGANISM AND ETIOLOGY

The disease is caused by *Cercosporasorghi* Ellis and Everhart. The pathogen produces pale brown, fasciculate conidiophores which bear hyaline, 3–11 septate conidia (3.0–4.5×40–120 μm). The growth of the pathogen is slow on most culture media, producing a gray, non-sporulating amphigenous colony.

7.2.6.3   DISEASE CYCLE AND EPIDEMIOLOGY

The pathogen persists on crop debris, surviving host crop plants, infected wild sorghum, other grasses, and seed. Conidia serve as a source of primary and secondary inoculum, which is spread by wind and rain to leaves of susceptible plants. Moderate to high temperature and prolonged period of high relative humidity favor disease development.

7.2.6.4   MANAGEMENT

- Disease can be reduced by crop rotation, sanitation, and by killing surviving crop plants.
- Disease can be managed by using resistant cultivars.

7.2.7   TARGET LEAF SPOT

The disease has been reported from many sorghum growing countries including the United States, Sudan, Israel, India, Cyprus, the Philippines, and Taiwan.

7.2.7.1   SYMPTOMS

The symptoms appear as reddish or grayish elliptical or oval to cylindrical spots. Lesions vary from red to purple or grey depending on the variety. Sometimes the centers of lesions become brown or straw colored, surrounded by red or purple margins. Lesions may coalesce to form very long interveinal lesions.

7.2.7.2   CAUSAL ORGANISM AND ETIOLOGY

The disease is caused by *Bipolarissorghicola* (Lefebvre and Sherwin) Shoem. Conidia produced on the conidiophores are 3–6 septate with indistinct hilum and measure 72.5–75 × 10–12.5 µm. They are usually bipolar in germination. The germ tube may become a secondary conidiophore producing conidium.

7.2.7.3   DISEASE CYCLE AND EPIDEMIOLOGY

The pathogen attacks plants at all stages of development. It has been reported that disease developed readily on 14–49 days-old plants, but 28 days old plants were highly susceptible (Katewa et al., 2008). Conidia germinate readily in the presence of moisture. Typical lesions appear in three to four days after inoculation. Under humid conditions, these lesions produce numerous spores, which are then disseminated by wind. The pathogen may survive as dormant mycelium and spores on sorghum debris or on weed hosts such as Johnson grass. The development of disease is favored by high humidity and moist weather.

7.2.7.4   MANAGEMENT

- Target leaf spot can be managed successfully by using resistant/ tolerant varieties.

7.2.8   SOOTY STRIPE

First reported in the United States in 1903, the disease is currently widely distributed in most sorghum growing regions of the world. Sooty stripe is more prevalent in warm and humid conditions during the growing season.

7.2.8.1   SYMPTOMS

Elongated, elliptical or spindle shaped lesions develop on the leaves with straw colored centers and purplish to tan margins, depending on the host cultivar. Fully developed lesions can be 5–14 cm long and 1–2 cm wide. With the advance of the disease, the lesions may coalesce to produce large necrotic areas. In warm and humid conditions, center of mature lesions appear grayish in color due to abundant sporulation. Later on, these lesions become blackish or sooty with the production of numerous small black sclerotia. These sclerotia are superficial and can easily be rubbed off.

7.2.8.2 CAUSAL ORGANISM AND ETIOLOGY

Sooty Stripe is caused by *Ramulispora sorghi* (Ell. and Ev.) Olive and Lefebvre. The pathogen produces sporodochia. Conidia are borne singly at the tips of the conidiophores. Many conidia from each sporodochium aggregate into a gelatinous mass. Conidia are filiform, curved, tapering at the ends, hyaline, 3–8 septate, have one to three lateral, septate or nonseptate branches and measure 36–90 × 2–3 µm. Smooth, black sclerotia appear superficially on lesions. Often, the sclerotia germinate by producing sporodochia and conidia. The pathogen grows slowly in culture and form a circular, compact, and crumpled colony. Sporulation occurs after several weeks. Sclerotia are not produced in culture.

7.2.8.3 DISEASE CYCLE AND EPIDEMIOLOGY

The pathogen survives as sclerotia in leaf residue or as sclerotial sporodochia if infected leaves remain intact on the soil surface. In favorable conditions, sclerotia, and sclerotialsporodochia produce abundant conidia which are disseminated by wind or rain and cause fresh infection. The pathogen may also survive on some perennial sorghum hosts such as *S. bicolor* sub sp. *bicolor* and *S. halepense*. The hot and humid conditions favor the disease development.

7.2.8.4 MANAGEMENT

- Crop rotation and destruction of infected leaf debris is recommended for reducing initial inoculum of the pathogen.

7.2.9 ROUGH LEAF SPOT

The disease has been reported from almost all the sorghum growing countries of the world and is commonly found in humid areas.

7.2.9.1 SYMPTOMS

The disease is characterized by the presence of groups of round, hard, black pycnidia on chlorotic leaves. The pycnidia protrude above the leaf surface so that when rubbed they give the leaf a characteristically rough feel. Later on the infected tissues become necrotic and light-colored to reddish circular to oval lesions having darker margins and covered with the black pycnidia develop. Lesions coalesce to form large necrotic areas and whole leaves may be killed.

7.2.9.2 CAUSAL ORGANISM AND ETIOLOGY

The disease is caused by *Ascochyta sorghina* Sacc. Pycnidia of the pathogen are globose, papillate, and variable in diameter (140–300 μm). They contain numerous conidia which are hyaline, oblong elliptical, single septate, constricted at the septum, and measure 20×8 μm.

7.2.9.3 DISEASE CYCLE AND EPIDEMIOLOGY

Pycnidia are globose, papillate, and large in size. Pycnidia contain numerous hyaline, oblong to ellipsoid, two celled, slightly constricted conidia. The disease spreads mainly by air borne conidia during humid or wet weather.

7.2.9.4 MANAGEMENT

- Crop rotation and clean cultivation may help to reduce losses caused by the disease.
- Disease can be managed by using resistant cultivars.

7.2.10 RUST

Rust is widely distributed in most sorghum growing regions of the world. Rust epiphytotics have been reported to occur in cool, humid regions of central and South America, Southeast Asia and southern India, reducing forage quality, and grain yield. Under favorable environmental conditions, disease develops rapidly affecting panicle exertions and grain development resulting in poor grain yield.

7.2.10.1   SYMPTOMS

The characteristic symptoms of the disease appear as scattered purple or red flecks on both surfaces of leaves. In susceptible varieties, the flecks enlarge to form blister like, dark reddish brown pustules called uredinia. These uredinia are about 2.0 mm long which lie parallel to and between veins. Later on the dried epidermis over the pustules ruptures, releasing the powdery mass of urediniospores. As the disease advances, most uredinia are converted into telia and the color of the pustules changes from reddish brown to blackish brown. New telia are also produced on both leaf surfaces, especially on the lower surface. Brown to blackish brown uredinia and telia are also seen as long streaks or elongated lesions on peduncles.

7.2.10.2   CAUSAL ORGANISM AND ETIOLOGY

Rust is caused by *Puccinia purpurea* Cooke. In the rust pustules, urediniospores, and teliospores produced by the pathogen are interspersed with abundant paraphyses, which are short, clavate, hyaline, and bent inward. Urediniospores are unicellular, pedicellate, reddish yellow, oval to elliptical, finely echinulate and 30–42 × 22–30 μm in size. Teliospores are bicelled, oblong to ellipsoidal, dark brown, 25–32 × 40–60 μm in size and have a hyaline to yellowish-tinged, stout, and pedicel of variable length. Teliospores germinate by producing four celled promycelium. Each cell of the promycelium bears a single elliptical basidiospore on a sterigma.

7.2.10.3   DISEASE CYCLE AND EPIDEMIOLOGY

Urediniospores survive on sorghum plants and on several perennial and collateral hosts. Secondary spread of the disease occurs by means of airborne urediniospores. Infection and disease development is favored by light drizzles and morning dew. *Oxalis corniculata* is an alternate host of the pathogen on which the aecidial stage develops. However, the role of aeciospores in natural recurrence is not important. The disease is favored by warm humid weather and light rains.

7.2.10.4   MANAGEMENT

- In regions where rust is responsible for significant losses, resistant varieties should be grown.
- Early planting helps plant to escape disease.

7.2.11 ERGOT

Ergot also known as sugary disease is widespread in most parts of the world. The disease occurs due to infection of unfertilized ovaries in flowering sorghum panicles by the pathogen, thus preventing fertilization and seed set. Rapid pollination and fertilization of ovaries prevents ergot infection in most sorghum lines. Male sterile lines (A-lines) which are used as female parents in hybrid seed production, lack fertile pollen and thus are more vulnerable to ergot infection than male fertile lines (Futrell and Webster, 1965). Single cross hybrids where pollination is delayed also get infected. Ergot is thus a major problem in hybrid seed production where A-lines are used as female parents. The disease becomes highly damaging to hybrids when favorable weather conditions prevail at flowering. Ergot can also cause a widespread damage of male fertile cultivars at farmers' fields in favorable weather conditions to the pathogen at flowering (Kulkadia et al., 1982).

7.2.11.1 SYMPTOMS

The first symptom of the disease is exudation of sticky, dirty white to brownish liquid droplets (honeydew) from the infected florets on sorghum panicles. The name, 'sugary disease' for sorghum ergot originates from this sticky sweet fluid. Under favorable weather conditions, honeydew exudation is profuse and the droplets trickle down on the foliage and ground appearing as whitish patches. These honeydew droplets contain numerous conidia of the pathogen. Under favorable conditions, a saprophytic fungus *Cerebella volkensii* may overgrow on honeydew and convert it into a sticky, matted black mass. The infected florets do not set seed and develop sclerotia in some florets which are compact masses of whitish mycelium that later turn brown and become hardened (Figure 7.4).

7.2.11.2 CAUSAL ORGANISM AND ETIOLOGY

Ergot is caused by *Sphaceliasorghi* McRae. The perfect state of the fungus is *Clavicepssorghi* Kulk. *C. sorghi* has been described from Asia by Kulkarni et al., 1976 whereas, *C. africana,* which is most widespread, is found throughout Asia, Africa, America, and Australia. The pathogen (*C. Sorghi*) produces three types of conidia (macroconidia, microconidia, and secondary conidia). Macroconidia are hyaline, elliptical to oblong with rounded ends and measure 7.3 × 13.3 µm, microconidia are round to obovate (3–4 µ) and secondary conidia are pyriform (7.2 × 12.1 µm). Sclerotia which develop in infected ovaries may be cylindrical and slightly curved and either short or long depending on host genotype, environment, and nutritional factors. Sclerotia germinate to produce two or three stromata with stipes containing embedded flask shaped perithecia with ostiole. Asci are cylindrical; each ascus produces eight filiform, hyaline ascospores.

FIGURE 7.4 Symptom of ergot or sugary disease of sorghum.

7.2.11.3 DISEASE CYCLE AND EPIDEMIOLOGY

Sclerotia mixed with seed or in soil serve as source of primary inoculum. Ascospores released from germinating sclerotia or conidia contained on sclerotial grooves land on stigma to initiate infection under favorable conditions. Secondary spread of the disease in the field occurs by means of conidia in the honeydew, which are disseminated from flower to flower by insects, wind, and rain. Conidia produced on collateral hosts may also establish primary infection in the field. Weather condition that reduces pollen production and ovary fertilization increase the chance of ergot infection. Florets are vulnerable to infection from the time of panicle emergence through fertilization of the ovary. A period of high rainfall and high humidity during flowering season and cool night temperature and cloudy weather aggravate the disease.

7.2.11.4 MANAGEMENT

- Early sowing has been shown to avoid the disease (Singh, 1964; Anahosur and Patil, 1982).
- Clean seeds free from sclerotia should be planted. Sundaram (1968) suggested sowing pathogen-free seed by steeping seeds in 5% salt solution to remove sclerotia.
- Removal of collateral hosts and rouging of infected plants can minimize the disease.
- IS 8525, IS 14131 and IS 14257 resistant lines have been used in resistance breeding programme.
- Application of crude garlic extract gave complete inhibition of conidial germination in concentration up to 9% (Singh and Navi, 2000).

7.2.12 SMUTS

Four smuts viz. head smut, long smut, loose kernel smut, and covered kernel smut are known to affect sorghum.

7.2.12.1 HEAD SMUT

Head smut of sorghum has been reported from many countries of the world. In India, it occurs mainly in Maharashtra, Karnataka, Andhra Pradesh, and Tamil Nadu. The disease causes significant damage as it affects the entire ear transforming it into a smutted head.

7.2.12.1.1 Symptoms

The disease appears at the time of flowering. In the place of a normal inflorescence, a sorus fully covered with a grayish white membrane emerges from the boot leaf. The sorus is usually 3–4 inch long and 1–2 inch wide and maybe cylindrical in shape. When it has fully emerged the fungal membrane ruptures, exposing large mass of black, powdery spores. If the wind is blowing during sorus emergence, the spores resemble a smoky cloud around the head. When the spores are blown off, a network structure of filamentous vascular tissues of the host is exposed. Diseased plants remain stunted, tiller excessively and root system of the smutted plants is weakened.

7.2.12.1.2 Causal Organism and Etiology

The disease is caused by *Sporisorium reilianum* (Kuhn) Langdon and Fullerton (syn. *Sphacelotheca reiliana*). Teliospores or smut spores develop from a condensation of sporogenous hyphae into coils of dense cytoplasm, separated by partitioning hyphae. As the sorus matures, the dark, spherical, reticulate spores (9–14 μm in diameter) are liberated. Teliospores germinate and produce masses of sporidia.

7.2.12.1.3 Disease Cycle and Epidemiology

The young heads enclosed in the boot are completely replaced by a large smut galls covered by a thick whitish membrane. The membrane ruptures, often before the head emergence, exposing a mass of dark brown to black, powdery teliospores intermingled with a network of long, thin, dark, filaments of vascular tissue. The pathogen is soil borne in the form of teliospores which germinate and cause infection in the nodal region of the shoot apex. Mycelium grows both intracellularly and intercellularly. During flowering, mycelium grows vigorously and produces teliospores. Spores produced from infected plants reach the soil and survive until the next cropping season. Spores may also adhere to the seed surface and spread through seed. A dry soil with a temperature of approximately 24°C is favorable for infection. Spores are known to survive for long period in soil.

7.2.12.1.4 Management

- Collecting smutted heads in cloth bags and dipping in boiling water reduces the inoculum potential for the next crop.
- Head smut resistant sorghum lines Tx 2962 through Tx 2978 were developed and released by the Texas Agricultural Experiment Station in 2006. BTx 635 is one of the most popular lines and is believed to

posse's resistance against infection by all four races of the pathogen (Frederiksen, 2000).

- Seed should be treated with Captan or Thiram @ 4 g/kg seed.

7.2.12.2   LONG SMUT

Long smut, first reported from Egypt in 1887, is now prevalent in many countries in Africa and Asia.

7.2.12.2.1   Symptoms

Long smut appears as elongated, cylindrical, slightly curved sori, much longer than normal grains. The sori are covered by a whitish thin membrane that ruptures to release black powdery mass of spores. Usually a few smut sori are scattered sporadically throughout the panicle.

7.2.12.2.2   Causal Organism and Etiology

Long smut is caused by *Sporisorium ehrenbergii* Vanky (syn. *Tolyposporium ehrenbergii*). The pathogen produces characteristic sori filled with black masses of teliospores and also containing eight to ten dark filamentous structures of vascular tissues of the ovary arising from the basal portion of the sorus. Teliospores are firmly united into balls, light brown in color and measure 12–16 µm in diameter. They germinate in water droplets, potato agar, or nutrient agar and produce a three to six celled promycelium. The promycelium bears numerous sporidia at the septa and terminal ends. Sporidia are hyaline, spindle shaped, single celled and measure 8–24 µm in diameter.

7.2.12.2.3   Disease Cycle and Epidemiology

The fungus survives as teliospores on seed surface and in soil. The work of Vasudeva et al. (1950), Manzo (1976), and Omer et al. (1985) demonstrates that the pathogen is airborne and infects single florets. Under favorable conditions, the teliospores germinate and release numerous sporidia that may land in the boot leaf of plants and cause infection. Airborne spores may also produce sporidia on flag leaf sheath which can infect the florets in the panicle. A temperature of 30–35°C and >80% RH favor disease development.

7.2.12.2.4   Management

- Adjusting the sowing dates and destruction of infected heads and elimination of possible alternative hosts are effective in managing the disease.

7.2.12.3   COVERED KERNEL SMUT

This is most destructive of all the smuts, causing enormous losses throughout the country. Most of the varieties under cultivation are susceptible to the disease.

7.2.12.3.1   Symptoms

The disease appears at the time of grain formation in the ear, replacing most of the normal grains with smut sori. Generally, the sorus (spore sac) is larger than the normal grain. They are oval to cylindrical, 5–15 mm long and 3–5 mm broad and dirty gray in color.

7.2.12.3.2   Causal Organism and Etiology

Sporisorium sorghi (Ehrenberg) Link (syn. *Sphacelotheca sorghi*) causes covered Kernel Smut. The pathogen is present in the form of the sorus, which has a tough wall and a long, hard, central tissue, known as columellum. The columellum is made of host tissues, including parenchyma and vascular elements. A dense mass of brownish-black, minutely echinulate thick walled teliospores, measuring 4–7 µm in diameter are filled in the space between the columellum and sorus wall. These spores germinate and produce a four celled promycelium with a single sporidium from each cell. However, the germination of teliospores, the number of promycelia, and the number of sporidia produced from each promycelium may vary.

7.2.12.3.3 Disease Cycle and Epidemiology

The pathogen is externally seed borne. The teliospores germinate with the seed and infect the seedling, establish systemic infection that develops along in the meristamatic tissues and at the time of flowering, the spores are formed, replacing the ovary with the sori. Later on these sori are ruptured, releasing teliospores, which contaminate the healthy grains. The teliospores remain dormant on the seed until sown next season. The incidence of smut decreases when seed is planted during high temperature.

7.2.12.3.4 Management

- Seed treatment with sulfur and Thiram @ 4.0 g/ kg of seed has been found effective.

7.2.12.4 LOOSE KERNEL SMUT

The disease is reported from most sorghum growing regions of the world. However, it is less common than covered kernel smut.

7.2.12.4.1 Symptoms

The affected plants are shorter, produce thinner stalks, more tillers, and earlier flowering than the healthy plants. Generally, all the spikelets in an infected panicle are smutted. Individual infected kernels are replaced by smut sori. They also develop on glumes and pedicel. The sori (3–18 × 2–4 mm size) are covered by a thin grey membrane which ruptures, exposing powdery mass of dark colored spores, before the emergence of panicle from the boot.

7.2.12.4.2 Causal Organism and Etiology

Sporisorium cruentum (Kuhn) Vanky (syn. *Sphacelotheca cruenta*) is responsible for Loose kernel smut. As in Covered kernel smut, at the center of the sorus a columellum is present which is usually bigger and more curved than the columellum of covered smut. The spores are globose to sub globose, dark brown, minutely echinulate and measure 5–10 μm in diameter. They germinate by forming a four celled promycelium and sporidia.

7.2.12.4.3 Disease Cycle and Epidemiology

The pathogen is mainly externally seed borne. When the seed contaminated with teliospores are sown in the field, the spores germinate by producing hyphae, which infect young seedlings before emergence. Thereafter, the pathogen grows systemically and ultimately sporulates in the floral organs, resulting in smutted panicles. Seedling infection occurs at a varying soil moisture and temperature of 20–25°C. Since the disease is systemic, ratoon crops are also infected. Spores from a smutted panicle may infect late developing panicles of healthy plants thus causing secondary infection. This infection is localized, and no subsequent systemic spread of the pathogen occurs.

7.2.12.4.4 Management

- Seed treatment with fungicides as recommended for the management of covered kernel smut is also useful in protecting crop from the disease.
- Ratooning of the susceptible crop should be avoided.

7.2.13 CHARCOAL ROT

Charcoal stalk rot of sorghum is a major disease of sorghum in Asia, Africa Australia, and America. In India, the *Rabi* sorghum that is generally grown on residual soil moisture often get exposed to soil moisture stress during the grain filling stage in absence of rains. Dry weather condition during this period may further increase the moisture loss from the soil. Yield losses vary, depending on the cultivars, weather conditions, and disease severity. The disease reduces the grain yield and stover quality resulting from lodging and rotting and decaying of the stalk. Losses are more severe in hybrid sorghum. Grain yield losses have been reported to range from 15–55% (Anahosur and Patil, 1983).

7.2.13.1 SYMPTOMS

Infected roots of the plants show water-soaked lesions, which turn brown or black. Affected stalks become soft at the base and most often lodge even in moderate winds. When a lodged infected plant is split open and examined the pith of the stalk is found disintegrating across several nodes. The vascular bundles are separated from one another and are profusely marked by small, dark, charcoal, or black colored sclerotia of the pathogen. Occasionally, the seedlings may also get infected.

7.2.13.2   CAUSAL ORGANISM AND ETIOLOGY

Charcoal rot is caused by *Macrophomina phaseolina* (Tassi) Goid. Imperfect state of the pathogen is *Rhizoctonia bataticola* (Taub.) Butl. (syn. *Sclerotium bataticola*). The mycelium is aerial, hyaline to brown, septate, profusely branched. Older hyphae produce numerous sclerotia which are brown to black, hard when mature, shiny, irregular in shape and 100–400 µm in diameter. The pathogen produces numerous sclerotia on affected plant parts approaching senescence. Pycnidia rarely occur on sorghum.

7.2.13.3   DISEASE CYCLE AND EPIDEMIOLOGY

The pathogen survives as sclerotia which remain viable in soil for 2–3 years. Infection and disease development is favored by dry weather, high soil temperature, and low soil moisture.

7.2.13.4   MANAGEMENT

- Charcoal rot can be minimized by keeping the soil moist during the post flowering period.
- Sorghum genotypes showing stay-green trait are generally resistant to charcoal rot. SLB 7, SLB 8, SLR 17 and SLR 35 lines are tolerant to the disease.
- Soil amendments can minimize the disease. The combined treatment of straw mulch + neem cake+ seed treatment with *Trichoderma viride* resulted in significant reduction of charcoal rot incidence (Jamadar and Desai, 1996).

7.3   BACTERIAL DISEASES

7.3.1   BACTERIAL STALK ROT

Soft rot *Erwinias* are very important primary pathogens of both growing plants and the harvested crop (Perombelon and Kelman, 1980).

7.3.1.1 SYMPTOMS

The disease mainly affect sorghum stem showing water soaked symptoms that later turn reddish dark brown color. The infected stem pith is disintegrated and show slimy soft rot symptoms with foul smell and eventually the whole plant wilts. The rot may involve only one or two internodes, or the entire length of the stalk which finally dries up and its interior turns into a shredded mass of fibrous tissue. Lower leaves and leaf sheath covering the internodes are chlorotic and rind is pale straw instead of green in color (Figure 7.5).

FIGURE 7.5 Symptom of bacterial stalk rots of sorghum.

7.3.1.2 CAUSAL ORGANISM AND ETIOLOGY

Bacterial stalk rot is caused by *Dickeyadidantii* (*Erwinia chrysanthemi*). The pathogen is a motile, nonsporing, straight rod-shaped cell with rounded ends. Cells range in size from 0.8 to 3.2 μm by 0.5 to 0.8 μm and are surrounded by numerous peritrichous flagella.

7.3.1.3   DISEASE CYCLE AND EPIDEMIOLOGY

The pathogen survives in the soil. Infection occurs mainly through the wounds caused to the roots. Cloudy weather, high temperature (more than 30°C) and frequent rainfall are favorable conditions for the development of disease.

7.3.1.4   MANAGEMENT

- High nitrogen application should be avoided.
- Chlorination of irrigated water reduces the chances of disease.
- Soil drenching with stable bleaching powder (SBP) in standing crop is effective in minimizing the disease.
- Bioagents colonized vermicompost can be an alternative natural biological control of stalk rot diseases of sorghum. Maximum reduction of disease severity was recorded with vermicompost colonized by *Trichoderma harzianum* isolate Th-2 followed by Th-14, Th-R, and *Pseudomonas fluorescens* isolate Psf- 3 (Kharayat and Singh, 2016).

7.3.2   BACTERIA LEAF STRIPE

The disease is found in almost all the sorghum growing regions of the world. Host range is wide and includes Johnson grass, maize, clover, soybean, chickpea, velvet bean, teosinte among others.

7.3.2.1   SYMPTOMS

Symptoms of the disease include linear interveinal lesions on leaf and leaf sheath that are purple, red, tan, or yellow in color depending on the cultivar. Under favorable weather conditions, lesions may exceed 20 cm in length. Bacterial exudates appear on infected portions of the leaf. Lesions may also appear on the kernel, peduncle, and rachis branches.

7.3.2.2   CAUSAL ORGANISM AND ETIOLOGY

The disease is caused by *Pseudomonas andropogonis* (E. F. Smith) Stapp. Causal bacterium is an aerobic, gram negative, non-spore-forming rod, 0.5–0.7 × 1–2 μm in size, usually with one sheathed polar flagellum. Creamy, smooth, and round colonies are produced on yeast extract-glucose-calcium carbonate agar. The bacterium does not produce fluorescent pigment on King's B medium.

7.3.2.3   DISEASE CYCLE AND EPIDEMIOLOGY

Pathogen is disseminated mainly through wind and rain. Long distance dispersal may occur through infested seed or plant debris. The pathogen survives in infested debris, weed hosts, and volunteer plants.

7.3.2.4   MANAGEMENT

- Crop rotation and destruction of plan debris can minimize the disease.
- Resistant cultivars and hybrids should be planted.

7.4   VIRUS DISEASES

Important viruses affecting sorghum include sugarcane mosaic virus (SCMV), maize dwarf mosaic virus (MDMV), maize stripe virus (MStV), MMV, and sorghum mosaic virus (SrMV). Out of these, MStV, and MMV are reported to cause considerable losses in vegetative growth and grain yield of sorghum (Narayana and Muniyappa, 1995).

7.4.1   MAIZE STRIPE VIRUS (MSTV)

7.3.1.1   SYMPTOMS

Chlorotic stripes or bands appear between the veins on infected leaves. Infected plants are stunted with partial or no panicle emergence.

7.4.2   MAIZE MOSAIC VIRUS (MMV)

7.4.2.1   SYMPTOMS

Symptoms appear as continuous chlorotic or broken streaks between the veins that may become necrotic as the disease advances. Plants become severely stunted with shortened internodes and panicles having poor seed set.

7.4.3   *MAIZE DWARF MOSAIC VIRUS (MDMV)*

7.4.3.1   *SYMPTOMS*

On leaves, the mosaic pattern develops into narrow, light green or chlorotic streaks. Red leaf is the most characteristic symptom which appears as red discoloration on leaves, sheaths, and peduncles depending on the strain of the virus and the cultivar. As the red leaf symptom spreads, infected areas coalesce and become necrotic. Infected plants become dwarfed; tillering is reduced with reduced number and size of panicles. Seed setting is poor.

7.4.3.2   *DISEASE CYCLE AND EPIDEMIOLOGY*

MStV and MMV are transmitted by a plant hopper *Peregrinusmaidis*. Johnson grass, which generally is widespread where sorghum is grown, serves as a source of primary inoculum for both these viruses and MDMV. MDMV persists in maize, sorghum as well as in many grasses. Virus may also survive on ratoon sorghum. MDMV is transmitted mechanically and by species of aphids in a nonpersistent manner. Temperature above 24°C is favorable for buildup of vector population as well as the disease. MStV-S incidence is greatly reduced as showing dates are shifted from August to October in Maharashtra and incidence has been found to increase during higher rainfall season (Das and Raut, 2002).

7.4.3.3   *MANAGEMENT*

- The removal of grasses in and around sorghum fields reduces the source of virus.
- Early sowing of Rabi crop during September should be avoided to save the crop from exposure to high vector population and the resultant viral diseases.
- Insecticides should be used judiciously to manage the insect vectors. Spraying metasystox 35EC @ 5 ml/ 10 L water at 15 days interval starting from 20 DAE controls vector and spread of the virus.
- Resistant/tolerant varieties should be planted.

KEYWORDS

- maize dwarf mosaic virus
- maize mosaic virus
- maize stripe virus
- ratoon sorghum
- sorghum mosaic virus
- sugarcane mosaic virus

REFERENCES

Anahosur, K. H., & Patil, S. H., (1982). Effect of date of sowing on the incidence of ergot of sorghum. *Indian Phytopathology, 35*, 507–509.

Anahosur, K. H., & Patil, S. H., (1983). Assessment of losses in sorghum seed weight due to charcoal rot. *Indian Phytopathology, 36*, 85–88.

Anahosur, K. H., Gowd, B. T. S., & Patil, S. H., (1980). Inheritance of resistance to *Cercosporasorghi* causing gray leaf spot on sorghum. *Current Science, 49*, 637.

Bunker, R. N., & Mathur, K., (2008). Evaluation of neem based formulations and chemical fungicides for the management of sorghum leaf blight. *Indian Phytopathology, 61*(2), 192–196.

Das, I. K., & Raut, M. S., (2002). Effect of sowing dates and weather parameters on the occurrence of stripe disease in winter sorghum. *Indian Journal of Mycology and Plant Pathology, 55*, 313–314.

Frederiksen, R. A., (2000). Diseases and disease management in sorghum. In: Smith, C. W., & Fredericksen, R. A., (eds.), *Sorghum-Origin, History* (pp 497–533). Technology and Production New York: John Wiley & Sons, Inc.

Futrell, M. C., & Webster, O. J., (1965). Ergot infection and sterility in grain sorghum. *Plant Disease Reporter, 49*, 680–683.

Indira, S., & Muthusubramanian, V., (2004). Influence of weather parameters on spore production in major mold pathogens of sorghum in relation to mold severity in the field. *Indian Journal of Plant Protection, 32*, 75–79.

Jamadar, M. M., & Desai, S. A., (1996). Management of charcoal rot caused by *Macrophomina phaseolina* (Tassi.) Goid in *Sorghum bicolor* (L.) Moench. *Journal of Biological Control, 10*, 93–96.

Katewa, R., Mathur, K., & Bunker, R. N., (2008). Epidemiological studies on target leaf spot of sorghum incited by *Bipolarize sorghicola. Indian Phytopathology, 61*(2), 146–151.

Kharayat, B. S., & Singh, Y., (2016). Studies on interactions among bio agents colonized vermicompost, rhizospheric earthworm and stalk rot disease of sorghum caused by *Erwinia chrysanthemi. American Journal of Agriculture Research, 1*(5), 0015–0029.

Kukadia, M. U., Desai, K. B., Desai, M. S., Patel, R. H., & Raja, K. R. V., (1982). Natural screening of advanced sorghum varieties to sugary disease. *Sorghum Newsletter, 25,* 117.

Kulkarni, B. G. P., Seshadri, V. S., & Hegde, R. K., (1976). The perfect stage of *Sphaceliasorghi*. McRae. *Mysore Journal of Agricultural Science, 10,* 286–289.

Little, C. R., (2000). Plant responses to early infection events in sorghum grain mold interactions. In: Chandrashekar, A., Bandyopadhyay, R., & Hall, A. J., (eds.), *Technical and Institutional Options for Sorghum Grain Mold Management: Proceedings of an International Consultation* (pp. 169–182). ICRISAT, Patancheru, India. Patancheru 502324, Andhra Pradesh, India: International Crops Research Institute for the Semi-Arid Tropics.

Manzo, S. K., (1976). Studies on the mode of infection of sorghum by *Tolyposporium ehrenbergii*, the causal organism of long smut. *Plant Disease Reporter, 60,* 948–952.

Michereff, S. J., Silveira, N. S. S., & Mariano, R. L. R., (1994). Antagonism of bacteria to *Colletotrichum graminicola* and potential for biocontrol of sorghum anthracnose. *Fitopathologia-Brasileira, 19,* 541–545.

Narayana, Y. D., & Muniyappa, V., (1995). Effects of sorghum stripe virus on plant growth and grain yield of sorghum. *Indian Journal of Virology, 11,* 53–58.

Omer, M. E. H., Frederiksen, R. A., & Ejeta, G., (1985). A method for inoculating sorghum with *Tolyposporium ehrenbergii* and other observations on long smut in Sudan. *Sorghum Newsletter, 28,* 95–97.

Perombelon, M. C. M., & Kelman, A., (1980). Ecology of the soft rot *Erwinias*. *Annual Review of Phytopathology, 18,* 361–387.

Singh, S. D., & Navi, S. S., (2000). Garlic as a biocontrol agent for sorghum ergot. *Journal of Mycology and Plant Pathology, 30,* 350–354.

Singh, S. D., & Bandyopadhyay, R., (2000). Grain mold. In: Frederiksen, R. A., & Odvody, G. N., (eds.), *Compendium of Sorghum Diseases* (2nd edn., pp. 38–40). The American phytopathological society. St. Paul, MN, USA. APS Press.

Singh, Y., & Singh, U. S., (2008). Biocontrol agents of Anthracnose in Sorghum. *Journal of Mycology and Plant Pathology, 38*(3), 488–491.

Singh, Y., (2009). Combined disease resistance against anthracnose and zonate leaf spot in sorghum. *Indian Phytopathology, 62*(2), 263–265.

Singh, P., (1964). Sugary disease of sorghum. In: *Progress Report of the Accelerated Hybrid Sorghum Project and the Millet Improvement Programme* (p. 81). Indian council of agricultural research and cooperating agencies, New Delhi.

Sundaram, N. V., (1968). Sugary disease of jowar: How to recognize and control it. *Indian Farming,* pp. 21, 22.

Thakur, R. P., Reddy, B. V. S., & Mathur, K., (2007). *Screening Techniques for Sorghum Diseases: Information Bulletin No. 76,* (p. 92). Patencheru 502324, Andhra Pradesh, India: International Crop Research Institute for the semi arid tropics.

Vasudeva, R. S., Seshadri, M. R., & Iyengar, (1950). Mode of transmission of the long smut of Jowar (Sorghum). *Current Science, 19,* 123–124.

CHAPTER 8

Problem of Common Leaf Blight (*Exserohilum turcicum* (pass.) Leonard and Suggs) in Sorghum and Their Management

R. N. BUNKER

Department of Plant Pathology, Rajasthan College of Agriculture, Maharana Pratap University of Agriculture and Technology, Udaipur–313001, Rajasthan, India, E-mail: rnbunker@yahoo.co.in

8.1 INTRODUCTION

Sorghum [*Sorghum bicolor* (L.) Moench] belongs to family Poaceae. It is adapted to latitudes ranging from 40°S to 45°N. In India, it is mostly grown between 9°N and 21°N in tropical and sub-tropical climates. It is suitable for warm conditions at relatively less moisture and high temperatures. Sorghum stays 'dormant' during extreme drought stress and recovers with rains. In India, sorghum is mainly cultivated in the peninsular and central parts; more than 60% of the total crop is raised during the *kharif*. The rest is Rabi, except for some cultivation during late *kharif* (maghi) in some parts of Andhra Pradesh. The principal areas of sorghum cultivation are Maharashtra, Andhra Pradesh, Madhya Pradesh, and Karnataka. In Gujarat, Rajasthan, Punjab, Haryana, and Uttar Pradesh, sorghum is grown primarily for fodder under rainfed conditions.

Sorghum diseases are considered to be the major constraint in realizing proper yield potential (ICRISAT, 1980). Sorghum crop is attacked by a large number of diseases, infecting, grains, foliage, and roots. Prominent among the foliar diseases are Anthracnose [*Colletotrichum graminicola* (Ces.) Wils.]; Common leaf blight [*Exserohilum turcicum* (Pass.) Leonard and Suggs.], Rust (*Puccinia purpurea* Cooke), Downy mildew

[(*Peronosclerospora sorghi* (Weston and Uppal) Shaw], Zonate leaf spot [*Gloeocercospora sorghi* Bain and Edgerton ex Deighton], Gray leaf spot [*Cercosporasorghi* Ellis and Everh.] and Sooty stripe [*Ramulispora sorghi* (Ellis and Everh) Olive and Lefebvre]. Several other leaf spots are reported to occur sporadically on sorghum. These are rough leaf spot [*Ascochyta sorghina* Sacc.], Target leaf spot (*Bipolarissorghicola* (Lefebvre and Sherwin), Alcorn], Ladder leaf spot [*Cercospora fusimuculans* Atk.], Tar spot [*Phyllachora sacchari*, P. Henn.] and Oval leaf spot (*Ramulispora sorghicola E. Harris*] (Frederiksen and Odvody, 2000). Recently, two more diseases anthracnose caused by [(*Colletotrichum gloeosporioides* (Penz.) Sacc., (*Glomerclla cirgulata* (Stonem) Spauld and Schrenk.)] (Mathur et al., 1998) on yellow sorghum land races, and leaf blight caused by *Drechslera australiensis* (Bugnicourt) (Mathur and Bunker, 2002) have been reported for the first time on grain sorghum. Leaf blight of sorghum caused by *Exserohilum turcicum* (Pass.) Leonard and Suggs is a major foliar disease of sorghum, which substantially damages the foliage. In India, the disease was first reported by Butler (1918) on the leaves of cultivated sorghum and later by Mitra (1923) from Punjab.

8.2   DISTRIBUTION AND ECONOMIC IMPORTANCE

Exserohilum turcicum causing leaf blight exhibits good tolerance to varied ranges of agroclimatic conditions over the globe in several members of Poaceae causing considerable economic damage. The disease is widespread during humid and warm weather with heavy dew conditions (Indira et al., 2002). The pathogen is reported to cause severe damage in most of sorghum growing countries viz; United States, Argentina, Brazil Israel, Zimbabwe Southern Africa (Kenya, Rwanda, Sudan, and Uganda), and Japan, while it is moderately prevalent in Western Africa and Thailand In Central America and Caribbean basin (Tarumoto et al., 1977; Abadi et al., 1996; Frederiksen and Odvody, 2000).

In India, the disease is known to occur in all sorghum growing states (Ravindra Nath, 1980; Anahosur, 1992). It is commonly present but is of less importance in sub-tropical areas within 34° latitude and of moderate importance outside 34° latitude in temperate conditions and within 23.15° latitude in tropical high lands, and in summer season in tropical low lands. But it is a major yield deterrent in winter crops in tropical low lands within 23.15 latitude (Frederiksen, 2000).

Yield losses caused by leaf blight pathogens usually vary with sorghum growing region, prevalent climatic conditions and host pathotype combination (Anahosure, 1992). Until the eighties the economic importance of losses caused by leaf blight appeared minor (Olsen and Santos, 1976) but, it may cause direct loss of foliage due to premature drying and early seedlings blights, or loss in other forage components like fodder weight and reduced carbohydrate translocation to other parts of plants due to reduced photosynthesis and transpiration. The indirect losses are transmission of the disease in newer areas by the infected seed, and inferior quality of grain and fodder (Odvody and Hepperly, 1992). *E. turcicum* is known to cause plugging of nearby vessels in the infected leaves generating a "localized wilt" within leaf (Frederiksen, 1980); hence, the losses may be higher than envisaged. The disease occurs in both, *kharif*, and *Rabi* seasons, in India, yield may be reduced in susceptible cultivars as a result of seedling blight due to early infection and severe damage to foliage during favorable weather conditions (Ahmed and Ravinder Reddy, 1993). Incidence of leaf blight is generally low, but when infection occurs at flowering stage in susceptible cultivars, grain yield losses of up to 50% may occur. The epidemiological outbreaks of the disease have been reported during 1991–1994 on much valued *Rabi* sorghum cultivars M-35-I and Muguti in Karnataka, (Desai, 1998) and in Rajasthan in 2001 also on popular kharif hybrids CSH 14 and CSH 16 (Mathur and Bunker, 2001).

8.3 SYMPTOMS

The disease Symptoms are visible from the seedling stage to the crop maturity stage. Small reddish or ten spots develops on seedlings, later enlarge, and coalesce resulting in wilting of young leaves. On maturity, these spots become long, elliptical, reddish purple or yellowish lesions developing from lower leaves to upper leaves and stem. In humid weather, numerous black spores are produced in the lesions in concentric zones. Lesion size in leaf blight symptoms ranged from 5 to 155 mm in length and 2 to 20 mm in width (Bunkar and Mathur, 2010). The typical lesion size described (Bergquist, 2000) for *E. turcicum* on sorghum is 25 to 150 mm in length and 12 mm in width. The conidia of *E. turcicum* are also unique in function. These are known to thicken their walls and become conidiospores (chlamydospores) as the overwintering or over-seasoning spore (Figures 8.1 and 8.2).

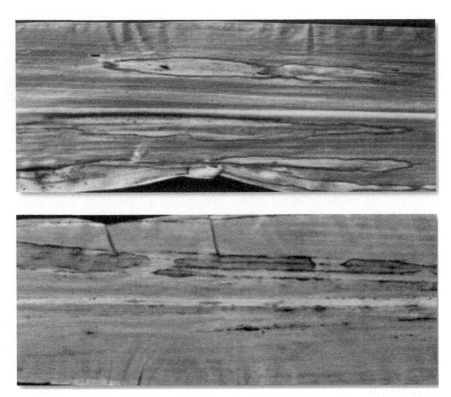

FIGURE 8.1	Typical symptoms of Turcicum leaf blight (*Exserohilum turcicum*) of sorghum.

## 8.4	ETIOLOGICAL AGENT (CASUAL ORGANISM)

The genus *Helminthosporium turcicum,* was first erected in the year 1876 by Passerini for the northern leaf blight of sorghum and by Butler et al. (1920) from India. In 1958, Luttrell described the perfect stage as *Trichometa sphaeria*. Regarding the asexual stage, Drechsler (1934) described the *Helminthosporium* conidial stage as those having the true *Helminthosporium* character including *H. turcicum* and the Cylindro-*Helminthosporium*. Shoemaker (1959) proposed a new genus for the graminicolus species with *Bipolaris turcica* as the type species Solane et al. (1975). Later on, Leonard, and Suggs (1974) proposed its perfect state, *Setosphaeria turcica* and removed the species having a protuberant conidial hylum from *Bipolaris* and established *Exserohilum turcicum* (Pass.) Leonard and Suggs as causal agent of sorghum leaf blight. The pathogen *Exserohilum turcicum* (telomorph *Setosphaeria turcica*; Dothidiales, Ascomycotina, Eumycota, Fungi) is characterized by typical conidia, measuring 50–144 × 18–33 μm; 3–8 distoseptate, pale to olivaceous brown, widest in the middle, fusoid or slightly curved and taper towards both ends. The conidia have a truncate and protuberant hilum in their basal cell. Pseudothecia have been observed under controlled conditions (Leonard and Suggs, 1974).

FIGURE 8.2 Symptoms of Turcicum leaf blight (*Exserohilum turcicum*) in the field on sorghum.

8.5 DISEASE CYCLE AND EPIDEMIOLOGY

The conidia germinate by the formation of a germ tube, which may or may not form appressorium on the surface of the leaf. An infection peg arises from appressorium that may penetrate through the cuticle and form hyphae with the host cells. The hyphae encounter a vascular bundle, enter a vessel begin to absorb nutrients and proliferate. Damage assigned results from the mycelial plugging of the vessel that causes a local or localized wilt (Jennings and Ullstrup, 1957). Toxins may be partially responsible for the death and collapse of host cells (Tuleen and Frederiksen, 1977). It was also suggested that toxic substances produced by *E. turcicum* were somewhat host specific and could differentiate between relative levels of resistance to *E. turcicum*. Vidhyasekaran et al. (1973) found the activities of exo-polygalacturonase, endo-polygalacturonase, and polygalacturonase transeleminase in leaf blight pathogen (*E. turcicum*).

The pathogen *E. turcicum* is seed-borne as well as soil-borne, in the infected plant debris in the tropics and as it forms chlamydospores within cells of the conidium in the temperate climate. These are air borne and can be transmitted to long distance and cause secondary infection. Infection and disease development is favored by moderate temperatures (18–25°C) and high humidity. Minimum temperatures between 14 and 16°C and mean temperatures between 20–22°C with high humidity are most favorable for disease development.

8.6   HOST RANGE

Pathogen *E. turcicum* also infects other hosts like maize, *sorghum halpencse*, teosinte, paspalum, *Echinoculoa, Triticum, Hordium, Avena, oryza,* and *Saccharum* under inoculation (Frederiksen, 1980). There are several reports about the possibility of existence of specialized races of *E. turcicum* infecting sorghum, Johnson's grass and maize have been suggested, but it is limited in sorghum (Ayala-Escobar et al., 1997). Physiological specialization in *E. turcicum* isolates with regard to virulence on sorghum hybrids and Johnson's grass have been studied by (Chiang et al., 1989). In India, leaf blight is also widely prevalent on maize and Johnson grass, but it is not known whether same isolate (s) can infect all the three hosts, or some physiological specialization exists, as has been reported by some workers (Bergquist and Masias, 1974; Bhowmik and Prasada, 1970; Hamid and Agragaki, 1975). Such information about host range of a particular pathogen is very useful for development and deployment of host plant resistance for dependable disease management.

8.7   PATHOGENIC VARIABILITY AND GENETICS OF RESISTANCE

Variability in fungal pathogens is studied through several characters, such as pathogenicity, fungicide resistance, culture characteristics, vegetative compatibility, isozymes ds RNAs and mycoviruses, Nuclear DNA, poly-morphism, mitochondrial DNA polymorphism and Karyotype polymor-phism (Caten, 1996). These characters may be independent or correlated in different fungi. While all those parameters have their own importance and applications, the use of virulence (pathogenicity) to access genetic vari-ability provides direct information on the effect of host selection. This varia-tion is detected on the basis of differential disease reaction among a set of host cultivars.

The genetic variability and pathogenicity of pathogen are the key factors for host-plant resistance and for the formulation of viable strategies for disease management. Variation in pathogenic properties of this pathogen has been reported in maize (Bigirwa et al., 1993) and more recently in sorghum (Mathur et al., 2011; Usha Sree et al., 2012). Three random amplified poly-morphic DNA (RAPD) markers closely linked with a locus for leaf blight resistance in sorghum has been identified (Boora et al., 2003). Little infor-mation is available on the variability at molecular level among the isolates of *E. turcicum* infecting corn (Abadi et al., 1996). However, there are no such reports on the existence of different races of *E. turcicum* on sorghum (Mathur et al., 2007). To determine the extent of genetic variability RAPD markers are useful in measuring genetic relatedness and for detecting variation within and among populations of *E. turcicum* thus helping one to understand the ecology and biology of fungus (Abadi et al., 1996). This technique also effectively applied for differentiating various races of patho-genic fungi, e.g., *Cochliobolus carbonum* and Macrophomina (Rajkumar et al., 2007; Das et al., 2008). Retrotransposons are mobile genetic elements and show variable insertional positions in genomes (Sabot et al., 2006), and the RAPD markers designed from long terminal repeats (LTR) sequences of the retrotransposon show more insertional polymorphism than conven-tional RAPD markers. Such type of research studies recently reported by Usha Sree et al. (2012) in India, taking 22 isolates of *E. turcicum* collected from five states of India on a set of seven sorghum differentials, and used retrotransposon based RAPD markers to study the molecular variability among these isolates.

The knowledge of pathogenic race/isolate variability at molecular level is a pre-requisite for resistance breeding. It helps in differentiation of various races of pathogen and to uncover pathogen population dynamics. Apart from this, molecular markers are useful in mapping of Vir and Avr genes of pathogen (Dioh et al., 2000).

8.8 CONTROL MEASURES

8.8.1 USE OF RESISTANT VARIETIES

Host résistance is the safest and most economical measure of disease management. Sorghum germplasm accessions like IS 13868. IS 13869, IS 13870, IS 13872, IS18729, IS 18758, IS 19669 and IS 19670 were developed in a trait-specific breeding program at ICRISAT. Two sorghum germplasm accessions IS 26866 and IS13996 are showed with high to moderate levels of resistance against eight isolates of *E. turcicum* of five states (Bunker and Mathur, 2010) which are now further used for stable source of breeding program (Mathur et al., 2011).

Some leaf blight tolerant hybrid seed parents, such as ICSA/B 296 TO ICSA/B328 were developed during 1989 to 1998 and available at ICRISAT, Patancheru.

8.8.2 CHEMICAL CONTROL

Foliar application of chemicals to manage foliar diseases of sorghum is usually considered impractical, because the cost may exceed benefits, especially with sorghum grown in stress environment with low cost inputs. However, in much valued *Rabi* crops, which are now again becoming popular and on which epidemics of *E. turcicum* have been observed, chemical control may be useful. Gangadharan et al. (1976) reported best control of leaf blight and rust in sorghum with spray of dithane-Z-78.

From Pakistan, Muhammad et al. (1991) reported that conidial germination and mycelial growth of *E. turcicum* were completely inhibited *in vitro* by antracol at 10 ppm and at 40 ppm by benlate, daconil, and cuprasan, Topsin-M reduced mycelial growth at 5 ppm, but had no effect on germination at any concentration tested.

Bunker and Mathur (2008) evaluated five fungicides and four neem based formulations in *vitro,* dithane M-45 and bavistin completely inhibited the mycelial growth and sporulation of *E. turcicum* at 1000 ppm concentration. Neem formulation, a chook, and neem seed extract were also found effective *in vitro*. In the field experiment, it was found that in sorghum cultivar CSH 14 suppression of leaf blight through application of bavistin seed treatment (0.1%) with dithane M-45 (0.2%) foliar spray fungicides resulted in 14.4 to 30.1% higher gains over the control. It was observed that under inoculated conditions, maximum reduction in disease severity (57.5%) and percent increase in grain (36.5), and fodder (21.6) yield was obtained with integration of seed treatment with bavistin (0.1%) and spray of dithane M-45 (0.2%) over control.

Besides direct control of the pathogen, these two fungicides are known to increased plant growth. Dithane M-45 provides zinc and manganese nutrition, while bavistin is known to have cytokinin like activity (Nene and Thapliyal, 1979).

KEYWORDS

- conidial germination
- long terminal repeats
- macrophomina
- pathogenicity
- random amplified polymorphic DNA
- retrotransposon

REFERENCES

Abadi, R., Perl-Treves, R., & levy, Y., (1996). Molecular variability among *Exserohilum turcicum* isolates using RAPD (random amplified polymorphic DNA). *Can. J. Plant Pathol., 18*, 29–34.

Ahmed, K. M., & Ravinder, R. C. H., (1993). A pictorial guide to the identification of seed borne fungi of sorghum, pearl millet, finger millet, chickpea, pigeon pea and groundnut. In: En. Summaries in Er. Es. & Ar., *Information Bulletin No. 34* (p. 200). ICRISAT, Patancheru, A.P., 502324, India.

Anahosur, K. H., (1992). Sorghum diseases in India: Knowledge and research needs. In: De Millano, W. A. J., Frederiksen, R. A., & Bengston, G. D., (eds.), *Sorghum and Millets Diseases: A Second World Review* (pp. 45–46). ICRISAT, Patancheru, A.P. 502324, India.

Ayala, E. V., Osada, K. S., Narro, S. J., Sandoval, & Islas, S., (1997). Pathogenic variability of *Exserohilum* [*Helminthosporium* Pass.] *turcicum* [*Setosphaeria turcica*] on sorghum in the Bajio, Mexico. *Agruciencia, 31*(2), 197–201.

Bergquist, R., (2000). In: Frederiksen, R. A., & Odvody, G. M., (eds.), *Leaf Blight in "Compendium of Sorghum Disease"* (2nd edn., pp. 9, 10). APS Press, St. Paul. Minn.

Bergquist, P. R., & Masias, O. R., (1974). Physiological specialization in *Trichometasphaeria turcica* f. sp. *Zeae* and *T. turcica* f. sp. *Sorghi* in Hawaii. *Phytopath., 64*, 645–649.

Bhowmik, R. P., & Prasada, R., (1970). Physiological specialization in *Helminthosporium turcicum* pass. from India. *Phytopathologische Zeitschrift, 68*, 84–87.

Bidari, V. B., Satyanarayan, H. V., Hegde, R. K., & Ponnappa, K. M., (1978). Effect of fungicides against the seed rot and seedling blight of hybrid sorghum CSH 5. Coll. Agric., Dharwad, India. *Mysore Journal of Agricultural Sciences, 12*(4), 587–593.

Bigirwa, G., Julian, A. M., & Adiala, E., (1993). Characterization of Ugandan races of *Exserohilum turcicum* from Maize. *African Crop Sci. J., 1,* 69–72.

Boora, K. S., Frederiksen, R. A., & Magill, C. W., (2003). A molecular marker that segregates with sorghum leaf blight resistance in one cross is maternally inherited in another. *Mol. Gen. Genet., 261,* 317–322.

Bunker, R. N., & Mathur, K., (2006). Host range of leaf blight pathogen (*Exserohilum turcicum)* of sorghum. *Indian Phytopathology, 59*(3), 370–372.

Bunker, R. N., & Kusum, M., (2008). Evaluation of neem based formulations and chemical fungicides for the management of sorghum leaf blight. *Indian Phytopath., 61*(2), 192–196.

Bunker, R. N., & Mathur, K., (2010). Pathogenic and morphological variability in *Exserohilum turcicum* isolates causing leaf blight of sorghum (*Sorghum bicolor*). *Indian Journal of Agricultural Sciences, 80*(10), 888–892.

Butler, E. J., (1918). *Fungi and Diseases in Plants* (Vol. 6, p. 206) Spink & Co., Calcutta.

Casela, C. R., Ferreira, A. S., & Schaffert, R. E., (1992). Sorghum disease in Brazil. In: De Milliano, W. A. J., Frederiksen, R. A., & Bengston, G. D., (eds.), *Sorghum and Millets Diseases* (pp. 57–62). A second world review. ICRISAT, Patancheru, A.P. 502324, India.

Caten, E. C., (1996). The mutable and treacherous tribe revisited. *Plant Pathology, 45,* 1–12.

Chiang, M. Y., Dyke, C. G., & Leonard, K. J. L., (1989). Evaluation of endemic foliar fungi for potential biological control of Johnson grass (*Sorghum halepense*): Screening and host range tests. *Pl. Dis., 73*(6), 459–464.

Das, I. K., Fakrudin, B., & Arora, D. K., (2008). RAPD cluster analysis and chlorate sensitivity of some Indian isolates of *Macrophomina phaseolina* from sorghum and their relationships with pathogenicity. *Microbiological Res., 163,* 215–224.

Desai, A., (1998). A note on the epiphytotic outbreak of leaf blight of sorghum in India. *Karnataka. J. Agri. Sci., 11,* 5–11.

Dioh, W., Tharreau, D., Notteghem, J. L., Orbach, M., & Lebrun, M. H., (2000). Mapping of a virulence genes in the rice blast fungus, Magnaporthegrisea, with RFLP and RAPD Markers. *Mol. Plant Microbe. Inter, 13*(2), 217–227

Drechsler, C., (1934). Phytopathological and taxonomic aspects of *Ophiabolus, Pyrenophora, Helminthosporium* and a new genus *Cochliobolus. Phytopath., 24,* 953–983.

Frederiksen, R. A., Rosenow, D. T., & Tullen, D. M., (1975). Resistance to *Exserohilum turcicum* in sorghum. *Pl. Dis. Reptr., 59,* 547–548.

Frederiksen, R. A., (1980). Sorghum leaf blight. In: *Proceedings of the International Workshop on Sorghum Diseases* (pp. 243–248). Hyderabad, ICRISAT, Patancheru, A.P. 502324, India.

Frederiksen, R. A., (1982). Disease problems in sorghum in the eighties. In: *Proceedings of the International Symposium on Sorghum* (pp. 263–271). ICRISAT, Patancheru, A.P. 502324, India.

Frederiksen, R. A., & Odvody, G. N., (2000). *Compendium of Sorghum Diseases* (2nd edn., p. 77). APS Press, St. Paul, Minnesota, USA.

Frederiksen, R. A., (2000). Diseases and disease management in sorghum. In: Wayne, S. C., (ed.), "*Sorghum: Origin, History, Technology and Production*" (pp. 497–533). John Wiley and Sons Inc.

Gangadharan, K., Subramanian, N., Mohanraj, D., Kandaswamy, T. K., & Sundaram, M. V., (1976). Efficacy of fungicides in the control of foliar diseases of sorghum. *Mad. Agril. J., 63*(5–7), 413–414.

Hamid, A. H., & Agragaki, M., (1975). Inheritance of pathogenicity in *Setosphaeria turica*. *Phytopathology*, *65*, 280–283.

ICRISAT (International Crops Research Institute for the semi-arid tropics), (1980). In: *Proceedings of the International Workshop on Sorghum Diseases* (pp. 465). Hyderabad, Patancheru, A.P.-502324, India.

Jones, M. J., & Dunkle, L. D., (1993). Analysis of *Cochliobolus carbonum* races by PCR amplification with arbitrary and gene-specific primers. *Phytopathology*, *83*, 366–370.

Leonard, K. J., Levy, Y., & Smith, D. R., (1989). Proposed nomenclature for pathogenic races of *Exserohilum turcicum* on corn. *Pl. Dis.*, *73*(9), 776–777.

Leonard, K. J., & Suggs, E. G., (1974). *Setosphaeria prolatum*. The ascigerous state of *Exserohilumprolatum*. *Mycologia*, *66*, 281–297.

Indira, S., Xu, X., Iamsupasit, N., Shetty, H. S., Vasanthi, N. S., Singh, S. D., & Bandyopadhyay, R., (2002). Diseases of Sorghum and Pearl millet in Asia. In: Leslie, J., (ed.), *Proc. of Global 2000 Sorghum and Pearl Millet Diseases* (pp. 393–402). Iowa State University Press, Ames, IA.

Jennings, P. R., & Ullstrup, A. G., (1957). A histologic study of three Helminthosporium leaf blights of corn. *Phytopath.*, *47*, 707–714.

Lutttrell, E. S., (1958). The perfect stage of Helminthosporium turcicum. *Mycologia*, *66*, 281–287.

Masias, O. R., & Bergquist, R. R., (1974). Host specific forms of *Trichometasphaeria turcica* in relation to homokaryons and heterokaryons in nature. *Phytopathology*, *64*, 436–438.

Mathur, K., & Bunker, R. N., (2001). *Annual Report of all India Coordinated Sorghum Improvement Project (AICSIP) Sorghum Pathology* (pp. 1–40). AICSIP Main Center Udaipur, Rajasthan College of Agriculture, Udaipur. 313 001.

Mathur, K., & Bunker, R. N., (2002). Leaf blight of sorghum caused by *Drechslera australiensis*. new report from India. *International Sorghum and Millets News Letter*, *43*, 60.

Mathur, K., Thakur, R. P., & Reddy, B. V. S., (2007). Leaf blight. In: *Screening Techniques for Sorghum Diseases* (pp. 24–30, 92). Information bulletin no. 76, Patancheru 502324, Andhra Pradesh, India: International Crops Research Institute for the Semi-Arid Tropics. ISBN: 978-92-9066-504-5.

Mathur, K., Thakur, R. P., Rao, V. P., Jadone, K., Rathore, S., & Velazhahan, R., (2011). Pathogenic variability in *Exserohilum turcicum* and resistance to leaf blight in sorghum. *Indian Phytopath.*, *64*, 32–36.

Mitra, M., (1923). *Helminthosporium* spp. In cereals and sugarcane in India: Part-1 (Diseases of *Zea mays* and *Sorghum vulgare*) caused by a spp. of (*Helminthosporium*) Memo. *Deptt. of Agric. India. Bot. Ser.*, *11*(10), 219–242.

Muhammad, S., Dogar, M. A., Khan, M. A., & Muhammad, S., (1991). Physiological studies and *in vitro* evaluation of fungicides against *Helminthosporium turcicum* Pass. *Sarhad. J. Agri.*, *7*(1), 95–100.

Narro, J., Betancourt, V. A., & Aguirre, J. L., (1992). Sorghum diseases in Mexico. In: De Milliano W. A. J., Frederiksen, R. A., & Bengston, G. D., (eds.), *Sorghum and Millet Diseases: A Second World Review* (pp. 75–84). ICRISAT, Pantancheru, Andhra Pradesh, India.

Odvody, G. N., & Hepperly, P. R., (1992). Foliar disease of sorghum. In: De Milliano, W. A. J., Frederiksen, R. A., & Bengston, G. D., (eds.), *Sorghum and Millets and Diseases: A Second World Review* (pp. 45, 46). ICRISAT, Patancheru, A.P. 502324, India.

Olsen, F. J., & Santos, G. L., (1976). Effect of nitrogen fertilization on the productivity of sorghum. Sudan grass cultivars and millet in Rio Graned do Sul, Brazil. *Trop. Agric.*, *53*(3), 211–216.

Raj, K., Fakrudin, B., & Kuruvinashetti, M. S., (2007). Genetic variability of sorghum charcoal rot pathogen (*Macrophomina phaseolina*) assessed by random DNA markers. *Plant Pathol. J.*, *23*(2), 45–50.

Ravindra, N. V., (1980). Sorghum disease in India. In: Williams, R. J., Frederiksen, R. A., & Mughogho, L. K., (eds.), *Proceedings of the International Workshop on Sorghum Diseases*, (pp. 57–66). ICRISAT, Patancheru, A.P. 502324, India.

Shoemaker, R. A., (1959). Nomenclature of *Drechslera* and *Bipolaris* grass parasites segregated from *Helminthosporium*. *Canad. Jr. Bot.*, *37*, 879–887.

Sabot, F., Kalendar, R., Jääskeläinen, M., Chang, W., Tanskanen, J., & Schulman, A. H., (2006). Retrotransposons: Metaparasites and agents of genome evolution. *Israel J. Ecol. Evol.*, *52*, 319–330.

Tarumoto, I., Isawa, K., & Watanabe, K., (1977). Inheritance of leaf blight resistance in sorghum-Sudan grass and sorghum sorghum hybrids. *Japan J. Breed, 27*, 216–222.

Tuleen, D. M., & Frederiksen, R. A., (1977). Characteristics of resistance to *Exserohilum turcicum* in *Sorghum Bicolar*. *Pl. Dis. Reptr., 61*, 657–661.

Usha, S. S., Nagarajareddy, Muralimohan, S., Madhusudhana, R., Mathur, K., Venkatesh, B., & Sanjay, R., (2012). Genetic diversity and pathogenic variation in isolates of *Exserohilum turcicum* causing common leaf blight of sorghum. *Indian Phytopath., 65*(4), 349–355.

Vidhyasekaran, P., Parambaramani, C., & Govindaswamy, C. V., (1973). Role of pectolytic enzymes in pathogenesis of obligate and facultative parasites causing sorghum diseases. *Indian Phytopath., 26*,(2), 197–204.

CHAPTER 9

Barnyard Millet / Japanese Millet or Sawan (*Echinochloa frumentacea* L.) Diseases and Their Management Strategies

BIJENDRA KUMAR[1] and J. N. SRIVASTAVA[2]

[1]*College of Agriculture, Department of Plant Pathology,*
G.B. Pant University of Agriculture and Technology, Pantnagar–263145,
Udham Singh Nagar, Uttarakhand, India

[2]*Department of Plant Pathology, Bihar Agricultural University,*
Sabour–813210, Bhagalpur, Bihar, India

Barnyard Millet/Japanese Millet or Sawan

(*Echinochloa frumentacea* L., Syn: *Echinochloa esculenta* L., and *Echinochloa colona* L.)

Vernacular names: *Jhangora, Sanwan, Madira, Udhalu, Oodhalu, Kuthiraivally, Shyama, Khira*

Barnyard (*Echinochloa frumentacea* (Roxb.) Link) also known as Japanese millet is one of the cultivated relative of barnyard grass. It is a short duration and fastest growing among all the millets and matures in about six weeks.

The crop is cultivated for dual purpose feed and fodder, grown in many countries like India, China, Japan, Malaysia, East Indies, Africa, and the United States of America. In China, Japan, and India it is often used as a substitute of rice. However, it is cultivated mainly for forage in the U.S. A., where up to eight harvests a year is produced. In India, it is cultivated in several states including Madhya Pradesh, Andhra Pradesh, Karnataka, Uttarakhand, Uttar Pradesh, Tamil Nadu, Maharashtra, and Bihar (Kumar and Prasad, 2009; Kumar, 2012).

It is highly nutritious having protein and fat content comparable to prosomillet, but the actual quality of the protein, but have the poorest amino acid values like that of little millet. It is very rich in fiber. It has antioxidant compounds and highly nutritious. Three antioxidant phenolic compounds, one serotonin derivative [N-(p-coumaroyl) serotonin] and two flavanoids (Luteolin and Tricin) have been isolated and identified. Serotinin is recognized as a messenger in the brain which also acts as a hormone in the intestine, is a metabolite from tryptophan.

9.1 HEAD SMUT

9.1.1 INTRODUCTION

Head smut disease of barnyard millet is caused by *Ustilago crusgalli*. *Echinocloa colonum* is also infected by the pathogen in Karnataka and Tamil Nadu. Head smut has been reported from the United States and India. In India, it has been recorded from Madhya Pradesh (Mundkur, 1943), Uttarakhand (Kumar et al., 2008) and Karnataka (Nagaraja and Reddy, 2010b). According to Kulkarni (1922), about 90% of the plants become infected when seeds mixed with smut spores are sown.

9.1.2 SYMPTOMS

The infected earheads are completely deformed and destroyed. Besides, gall-like swellings (smut sori) are also produced on the nodes of young shoots, the stem and in the leaf axils of older leaves. In some cases, twisted, and deformed clusters of leafy shoots with aborted ears may also be produced. The smut sori are enclosed by a hairy rough membrane of host origin which may be up to 12 mm in diameter (Mundkur, 1943). Only ovaries are infected by the pathogen. The infected ovaries are converted into round, hairy, grey sac, but do not increase in size than the normal grain (Figure 9.1).

FIGURE 9.1   Head smut of Barnyard millet or Japanese millet or sawan.

9.1.3   CAUSAL ORGANISM

Ustilago crus-galli Tracy and Earle

The spores are brown, spherical, and 9–12 μm in diameter. The epispore is densely echinulate. The spores germinate with a promycelium bearing lateral and terminal sporidia. The fungus becomes systemic. Head smut is externally seed borne in nature. It occurs late in the season when the crop is maturing and the temperature ranges within 20–25°C.

9.1.4   MANAGEMENT

- As the disease is seed borne in nature; it can be managed by seed treatment either with Carbendazim or Thiram @ 2 g/kg seed (Nagaraja et al., 2007b) and growing resistant variety PRJ 1 (Yadav et al., 2010).
- Roguing of infected plants in the early stages is also helpful in reducing the incidence and spread of disease.

9.2   GRAIN SMUT

9.2.1   INTRODUCTION

The disease is known to occur in Eastern Europe (Pall et al., 1980) and in several states of India including Karnataka, Bihar, Uttarakhand, and Maharashtra (Sharma, 1963; Vasudeva, 1954; Kumar et al., 2007a; Kumar 2016a). In different species of barnyard millet, up to 75% smut incidence has been recorded under natural conditions (Pawar et al., 1982).

9.2.2   SYMPTOMS

All the grains are not infected and only few grains in an ear smutted. The infected seeds enlarge in their size usually 2–3 times of their normal size with hairy seed surface. The disease becomes evident during grain formation when the temperature around 20–25°C (Figure 9.2).

9.2.3   CAUSAL ORGANISM

Ustilagopanici-frumentacei Bref.

The spore mass is powdery and black. The spores are brown, subglobose 6–10 μm in diameter and minutely echinulate. Spores germinate producing a septate promycelium and lateral and terminal sporidia. Secondary sporidia may be formed in chains.

9.2.4   MANAGEMENT

- Seed treatment with Ceresan dry, Dithane M45, MBC, Carbendazim and Thiram @ 2 g kg^{-1} (Pall and Nema, 1978) and vitavax @ 2.5 g kg⁻¹ or carbendazim @ 2 g kg⁻¹ (Kumar, 2013) has been reported to manage the disease.

• Four genotypes namely PRB 901, PRB 903, TNAU 141 and TNAU 155 were reported to be highly resistant to grain smut disease.

FIGURE 9.2   Grain smut of Barnyard millet or Japanese millet or sawan.

9.3   KERNEL SMUT

9.3.1   *INTRODUCTION*

Kernel smut was recorded for the first time in Italy. However, in India, the disease has been observed in Maharashtra, Bihar, and Tamil Nadu (Viswanath and Seetharam, 1989).

9.3.2 SYMPTOMS

In an earhead, only few grains are smutted. The affected grains become swollen greenish bodies/ smut sori (measuring 1.5–4 to 1–2 mm in size) which are scattered all over the earhead. Up to 25 infected grains may occur in an earhead.

9.3.3 CAUSAL ORGANISM

Ustilago paradoxa Syd., P. Syd. and E. J. Butler

The smut chlamydospores are smooth, olive brown, round, and measure 7–11 μm in diameter.

9.3.4 MANAGEMENT

- Since, the disease is seed borne; treating seeds with sulfur @4 g/kg or systemic fungicides like vitavax @ 2 g/kg can help in reducing the disease incidence.

9.4 LEAF SPOT OR BLIGHT

9.4.1 INTRODUCTION

Leaf spot or blight was first reported from the USA by Drechsler in 1923. The disease is known to occur in China, Japan, and India. Often, the disease occurs in humid conditions.

9.4.2 SYMPTOMS

The symptoms of the disease appear as isolated, dark brown, spindle shaped spots, fairly scattered on flag leaves. The spots measure 0.5 to 3.0 mm × 0.25 to 1.5 mm. as the disease advances, several nearby coalesce to cover the large area of the leaf and may cause blightening of the entire leaf. At this stage, the leaves become grey and finally dry up. Initially, the spots are grey to dark brown in color and a yellow halo surrounds them. Careful examination reveals dark points in the center of these lesions. If the conditions are moist, fungal growth may also be visible on these spots. In severe cases, the leaves may be blighted. Similar spots are also formed on the leaf sheath.

9.4.3   CAUSAL ORGANISM

Helminthosporium monoceros Drec.
[Syn. *Helminthosporium crusgalli* Nisikado and Miyake]

Conidiophores emerge either singly or in clusters of 2–5 through stomata or outer wall of the epidermis. Conidiophores are septate, dark olivaceous brown, slightly swollen at the base and produce 4–6 conidia. The condia are thin walled, fusoid, straight or slightly curved, yellow in color, 3–11 septate, measuring 49–158 × 10–15.5 μm in size.

9.4.4   MANAGEMENT

*   Infected seeds are the primary source of the inoculum; therefore seed treatment with systemic fungicides before sowing helps in disease control.
*   Foliar sprays of copper fungicides @ 0.3% are also effective in reducing the disease severity (Nagaraja et al., 2007b).

9.5   SHEATH BLIGHT

9.5.1   INTRODUCTION

Sheath blight disease of barnyard millet was reported for the first time by Kumar and Prasad (2009) from Uttarakhand, India.

9.5.2   SYMPTOMS

The disease manifests itself as light grey to dark brown, oval to irregular lesions on leaf sheath. The lesions enlarge in their size and central portions of the lesions become white to straw colored with reddish brown narrow borders, which appear as series of brown and copper color bands all across the leaf sheaths giving these lesions a characteristic banded appearance. The symptoms initially appear on leaf sheaths near the soil but rapidly enlarge in their size. Several lesions coalesce with one another and cover the upper sheaths and may extend up to leaves causing blighting of the foliage (Kumar 2016b; Nagaraja et al., 2016) (Figure 9.3).

FIGURE 9.3   Sheath blight symptoms of Barnyard millet or Japanese millet or sawan.

9.5.3   *CAUSAL ORGANISM*

Rhizoctonia solani Kuhn.

Temperatures between 28 and 30°C and high RH (~70%) are favor the development of banded leaf and sheath blight (BLSB) disease (Nagaraja et al., 2007b).

9.5.4   *MANAGEMENT*

- Not much information is available on the management aspects of this disease. The disease is known to occur in crops like; rice, maize, etc. The chemicals like; Carbendazim, hexaconazole, Propiconazole (@ 0.1%) and Validamycin @ 3 liters/ha, recommended for controlling the disease in rice and maize crops (Tewari, 2000) can help in managing the disease.

9.6 MINOR DISEASES

In addition to the above mentioned fungal diseases brown spot or sharp eye spot (*Pellicularia filamentosa*), leaf sheath rot (*Sclerotium hydrophyllum*), leaf spots (*Curvularialunata, Helminthosporium nodulosum, H. Oryzae, Brachysporium senegalense, Cercospora echinochloe, C. Sorghi*) (Pall et al., 1980), wheat streak virus (Sill and Agusiobo, 1955), sugarcane mosaic virus (SCMV), Eleusine virus 2 (Jaganathan et al., 1973), *Heterodera morioni* are also reported to infect this crop and are of very minor nature.

KEYWORDS

- **carbendazim**
- **chlamydospores**
- **conidiophores**
- **grain smut**
- **septate promycelium**
- **sporidia**

REFERENCES

Jaganathan, T., Padmanabhan, C., & Kandaswamy, T. K., (1973). A mosaic disease of Ragi in Tamil Nadu. *Madras Agric. J.*, *60*(3), 214–217.

Kulkarni, G. S., (1922). The smut of Nachani or Ragi (*Eleusinecoracana*). *Ann. Appl. Biol.*, *9*, 184–186.

Kumar, B., & Prasad, D., (2009). First record of banded sheath blight disease of barnyard millet caused by *Rhizoctonia solani*. *J. Mycol. Pl. Pathol.*, *39*(2), 352–354.

Kumar, B., Kumar, J., & Srinivas, P., (2008). Head smut: A new disease of barnyard millet in mid hills of Uttarakhand. *J. Mycol. Pl. Pathol.*, *38*(1), 142, 143.

Kumar, B., (2012). Management of important diseases of barnyard millet (*Echinochloa frumentacea*) in mid-western Himalayas. *Indian Phytopath.*, *65*(3), 300–302.

Kumar, B., (2013). Management of grain smut disease of barnyard millet (*Echinochloa frumentacea* (L) Beauv.). *Indian Phytopath.*, *66*(4), 403–405.

Kumar, B., (2016b). New addition to sheath blight of Barnyard millet caused by *Rhizoctonia solani*. *Internat. J. Plant Sci.*, *11*(2), 383–385.

Kumar, B., (2016b). Status of small millets diseases in Uttarakhand. *International Journal of Plant Protection*, *9*(1), 256–263.

Kumar, J., Kumar, B., & Yadav, V. K., (2007a). *Small Millets Research at G.B. Pant University* (p. 58). G.B. Pant University of Agriculture and Technology, Hill Campus, Ranichauri, Uttarakhand.

Mundkur, B. B., (1943). Studies in Indian cereal smuts. VI. The smuts on sawan, *Echinochloa frumentacea*. *Indian J. Agric. Sci.*, *13*, 631–633.

Nagaraja, A., & Anjaneya, R. B., (2010b). Head smut of barnyard millet: A new disease in Karnataka. *J. Mycopathol., Res.*, *48*(1), 167, 168.

Nagaraja, A., Kumar, B., Jain, A. K., & Sabalpara, A. N., (2016). Emerging diseases: Need for focused research in small millets. *J. Mycopathol. Res.*, *54*(1), 1–9.

Nagaraja, A., Kumar, J., Jain, A. K., Narasimhudu, Y., Raghuchander, T., Kumar, B., & Hanumanthe G. B., (2007b). *Compendium of Small Millets Diseases* (p. 80). Project coordination cell, All India Coordinated Small Millets Improvement Project, UAS, GKVK Campus, Bangalore.

Pall, B. S., & Nema, A. G., (1978). Chemical control of sawan (*Echinochloa frumentacea*) smut. *Food Farming and Agriculture*, *10*(4), 123.

Pall, B. S., Jain, A. C., & Singh, S. P., (1980). *Diseases of Lesser Millet* (p. 69). JNKVV, Jabalpur (MP), India.

Pawar, C. P., Rathod, R. K., Nivale, P. A., & Harinarayana, G., (1982). Natural incidence of grain smut in Sawan. *Curr. Sci.*, *51*, 480.

Sharma, B. B., (1963). *Proc. Natl. Aca. Sci. India*, *23*(Sect. B), 618–621.

Sill, W. H., & Agusiobo, P. C., (1955). Host range studies of the wheat streak-mosaic virus. *Plant Dis. Reptr.*, *39*, 633–642.

Vasudeva, R. S., (1954). The fungi of India. *Sci. Monogra, ICAR*, p. 12.

Viswanath, S., & Seetharam, A., (1989). Diseases of small millets and their management in India. In: Seetharam, A., Riley, K. W., & Harinarayana, G., (eds.), *Small Millets in Global Agriculture* (pp. 237–253). Oxford and IBH Publishing Co. Pvt. Ltd., New Delhi, India.

Yadav, V. K., Yadav, R., Kumar, B., & Malik, N., (2010). *Success Story of PRJ 1, Disease Resistant, High Yielding Variety of Barnyard Millet* (p. 24). GB Pant University of Agriculture and Technology, Hill Campus, Ranichauri, Tehri Garhwal.

CHAPTER 10

Common Millet or Proso Millet or Cheena Millet or French Millet (*Panicum miliaceum* L.) Diseases and Their Management Strategies

BIJENDRA KUMAR[1] and J. N. SRIVASTAVA[2]

[1]*College of Agriculture, Department of Plant Pathology, G.B. Pant University of Agriculture and Technology, Pantnagar–263145, Udham Singh Nagar, Uttarakhand, India*

[2]*Department of Plant Pathology, Bihar Agricultural University, Sabour–813210, Bhagalpur, Bihar, India*

Common millet or Proso millet or Cheena or French Millet

(*Panicum miliaceum* L.)

Hindi: *Chena, Variga, Baragu, Pani Varagu, Vari, Cheno, Cheena, Bachari Bagmu*

Common millet or Proso millet or Cheena or French Millet (*Panicum miliaceum* L.) is also called hog millet, broom corn millet or harshey millet. It is warm season grass capable of producing seeds 60–90 days after sowing. It is believed that proso millet was cultivated initially in eastern Asia, later spreading to India, Russia, the Middle East, and Europe (Baltensperger, 1996). Presently, it is cultivated in India, China, Russia, the Middle East (Baltensperger, 1996). Proso millet was domesticated in Manchuria and later introduced in Europe about 3000 years ago, then introduced in the Near East and India.

Common millet is one of the most delicious and highly nutritious crops. It can be cultivated in many soil types and climate conditions. Compared to other millets Common millet/proso millet is a short duration crop, maturs in about 60–75 days after sowing. The crop is often grown as a late summer crop. The plant is about three to four feet tall. Its compact panicle droops at the top like an old broom, hence the name broom corn. Seeds are round and about 1/8 inch wide and enclosed by a smooth, glossy hull. Grains contain relatively high percentage of indigestible fiber because the seeds are covered by the hulls which are difficult to remove by traditional milling processes.

Although, it is a self-pollinated crop; but occasionally natural cross-pollination may exceed 10%. It grows successfully further north than the other millets and also acclimatizes well to plateau conditions and high elevations, can be grown high in mountains; 1200 m in the former USSR up to 3500 m in India. It is a hardy crop and seems to be better adapted to traditional agricultural practices, requires less water, possibly the lowest of cereals, and most efficiently converts water to dry matter/grain. This is not due to its drought resistance but because of its short duration. Proso millet can be used in different ways. Grains are used as bird and livestock feed in the United States and for human consumption and livestock feed in other countries. The area under proso is in patches in Bihar, Uttar Pradesh, Jharkhand, Uttarakhand, and Tamil Nadu.

10.1   LEAF SPOT

10.1.1   INTRODUCTION

Leaf spot has been reported as one of the serious diseases of proso millet. Seed infection results in rotting, spot on coleoptile and seedling blight (Lee-DuHyung, 1997).

10.1.2   SYMPTOMS

On leaf lamina small oval brownish spots measuring 2.5 mm in length are seen which later turns black. The spots may coalesce under rapid disease progression stage and cause blighting. Symptoms can also be observed on culm, leaf sheath, and leaf blade. The disease does not cause much damage in the early stages but at times could be severe.

10.1.3 CAUSAL ORGANISM

Bipolarispanici-miliacei (Y. Nisik.) Shoemaker
[Syn. *Helminthosporium panici-miliacei* Y. Nisik.;
Drechslera panici-miliacei (Y. Nisik.) Subram. and B. L. Jain]

The conidiophores are olive in color; 2–8 septate and 131–296 × 3–6 µm (mean 186.3 × 4.2 µm). Conidia are slightly curved with round end cells, 3–6 septate and 34–87 × 9–12 µm (mean 54.2 × 11.3 µm) (Misra, 1973),

10.1.4 DISEASE CYCLE

It is an externally seed-borne disease; however, the pathogen can survive in soil along with the infected crop debris. In addition, the pathogen also perpetuation some grasses which serve as primary source of inoculum for the main crop. Secondary spread of the disease occurs through air borne conidia.

10.1.5 MANAGEMENT

- Since, the disease is seed-borne; seed treatment with suitable fungicides can manage the disease.
- Spraying of carbendazim @ 0.05% at the time of flowering has been recommended to prevent the secondary infection (Sinha and Upadhyay, 1997).
- Variety RAUM-7 is reported highly resistant.

10.2 LEAF STRIPE

10.2.1 INTRODUCTION

In India, this disease has been reported for the first time in 1970 from Dholi, Muzaffarpur (Misra, 1973).

10.2.2   SYMPTOMS

The disease manifests as water-soaked lesion on the leaves measuring 2–7 mm × 0.5 mm. Later they convert into brown stripes.

10.2.3   CAUSAL ORGANISM

Helminthosporium oryzae Breda de Haan
[Syn. *Bipolarisoryzae* (Breda de Haan) Shoemaker; *Drechslera oryzae* (Breda de Haan) Subram. and B. L. Jain]

Conidiophores are 7–14 septate, swollen at the base, geniculate, and 95–200 × 3–6 µm (mean 140 × 3.7 µm) in size. Conidia are slightly broader in the middle portion but narrow towards the end cells. Size of the conidia is 24–74 × 6–9 µm (mean 44.3 × 8.4 µm).

10.2.4   MANAGEMENT

- Spray Fytolan or Blitox-50 (3 g/liter of water) or Dithan Z-78 or Dithan M-45 (2 g/liter of water).

10.3   SHEATH ROT

10.3.1   INTRODUCTION

The disease was noticed for the first time by Rajagopalan et al. (1992) at the College of agriculture, Vellayani, Trivandrum during Jun-Aug 1991. Recently, the disease has been reported from Uttarakhand (Kumar and Prasad 2010a; Nagaraja et al., 2016).

10.3.2   SYMPTOMS

The symptoms of the disease appear first at the tillering stage of the crop and continue in the succeeding stages. The first symptoms appear as grayish green lesions on the leaf sheath. The lesions are ellipsoid or ovoid, 2–3 cm long which becomes grayish white with reddish-brown margins. The disease spreads further upwards causing blighting of leaf sheath and leaf blades.

10.3.3 CAUSAL ORGANISM

Rhizoctonia solani Kuhn

10.3.4 MANAGEMENT

Details of management practices are not available

10.4 HEAD SMUT

10.4.1 INTRODUCTION

This disease is widely prevalent in millet growing areas in Europe and Asia. It has been reported from Bulgaria, Kazakhstan, Japan, and Poland (Sinha and Upadhyay, 1997).

10.4.2 SYMPTOMS

The affected plants are much taller and green compared to healthy plants. Smut sori first become evident at the time of panicles emergence. The entire inflorescence is transformed into a sorus covered by a bright grayish-white false membrane. At crop maturity, the membrane ruptures, revealing the dark-brown spore mass.

10.4.3 CAUSAL ORGANISM

Sporisoriumdestruens (Schltdl.) Vánky.

10.4.4 DISEASE CYCLE

It is a loose smut and is externally seed-borne. The chlamydospores present on the seed serve as the primary source of inoculum. The chlamy-cospores germinate and infect the host at seedling stage. The mycelium becomes systemic and produces the symptoms only during the emergence of panicle.

The optimum temperature for spore germination is 22–30°C; however, they can germinate at 10–35°C (Lovik and Dahlstrem, 1936). Two races and biotypes of the pathogen have been reported by Shirokov and Maslenkova (1983).

10.4.5　MANAGEMENT

- Since head smut is externally seed-borne as teliospores, only fungicidal seed treatment is reported effective (Kovacs et al., 1997).
- Sharma and Sugha (1991) reported Carboxin and Benomyl to give good control of smut reducing the disease incidence by 99 and 95% and increasing the yield by 136 and 119%.

10.5　GRAIN SMUT

10.5.1　INTRODUCTION

Grain smut is also known as covered or kernel smut.

10.5.2　SYMPTOMS

Majority of the grains are converted into white grayish sacs (smut sori), which are slightly pointed to oval and filled with black powdery mass of spores (chlamydospores).

10.5.3　CAUSAL ORGANISM

Sphacelotheca sorghi (Ehrenb. ex Link) G. P. Clinton

[Syn. *Sporisorium sorghi* Ehrenb. ex Link.; *Tilletiasorghi* (Ehrenb. ex Link) Tul. and C. Tul.; *Ustilagosorghi* (Ehrenb. ex Link) Pass.; *Sphacelotheca sorghi* (Ehrenb. ex Link) Speg.]

The pathogen survives through contaminated seed.

10.5.4　MANAGEMENT

- The disease can be managed by collection and destruction of diseased ears at early stages on appearance and following crop rotation for 2–3 years.
- Seed treatment with thiram or any copper seed dressing fungicide reduces the disease incidence in the main crops.

10.6 RUST

10.6.1 SYMPTOMS

The disease is characterized by the formation of numerous minute brown uredosori or pustules on both the surfaces of leaf. Rust pustules are oblong, brown in color and often arranged linearly in rows. They may also be formed on other plant parts viz; leaf sheaths, culms, and stems. In case of severe infection, leaves dry prematurely and poor grain set was also observed. If the infection occurs before flowering the damage will be more.

10.6.2 CAUSAL ORGANISM

Uromyceslinearis Berk. and Broome

[Syn. *Caeomurus linearis* (Berk. and Broome) Kuntze; *Ustilagolinearis* (Berk. and Broome) Petch].

10.6.3 MANAGEMENT

Details of management practices are not available

10.7 DOWNY MILDEW

10.7.1 INTRODUCTION

Downy mildew disease of proso millet is reported to occur mainly in America, Europe, and Asia. In India, it is known to occur in Tamil Nadu and Maharashtra states (Sinha and Upadhyay, 1997).

10.7.2 SYMPTOMS

The disease first becomes visible on the leaves as narrow stripes running parallel to the leaf veins. Under favorable environment, white mycelial downy growth of the pathogen becomes evident on these stripes. Mycelial growth comprises of large number of conidiophores and conidia of the fungus. Such leaves dry-up in few days and sometimes they may split in their length. In severe form of the disease heads are not formed and if formed either grains do not develop or the entire plant dries before the emergence of inflorescence. In partially infected plants, ears develop only in few shoots and grains develop in them.

10.7.3   CAUSAL ORGANISM

Sclerospora graminicola (Sacc.) J. Schröt.

The size of conidiophores, conidia, and oospores of the pathogen are similar to the pathogen that causes downy mildew of pearl millet or bajra.

10.7.4   DISEASE CYCLE

The pathogen perpetuates mainly in soil through oospores and infects the plants at young stage in the next season when the environment becomes favorable. The oospores sticking on to the seed surface or with the crop debris reach the soil and initiate the disease.

10.7.5   MANAGEMENT

- Seed should always be obtained from diseased free crop and crop rotation should be followed.
- Since the disease is seed-borne, seed treatment with Ridomil followed by foliar spray can effectively manage the disease.

10.8   BLAST

10.8.1   INTRODUCTION

Blast on proso millet has been recorded from Dholi (Bihar) by Singh and Prasad (1981). Simmonds (1948) has recorded 20–60% reduction in yield because of this disease.

10.8.2   SYMPTOMS

First symptoms of the disease begin as small spindle shaped brown spots on leaves. Later, these spots coalesce to cover larger area of the leaf and in advanced form leaves dry up. Spots are 3–8 mm long and 0.5–1.5 mm wide (mean 5 × 1 mm). Center of the spots becomes ash grey with dark brown margins. Dark brown to black spots also develop on neck region which enlarge and girdle the neck. Often mycelial growth can occur on infected neck region in moist weather. The growth contains conidiophores and conidia of the fungus. Infected spikelets also become black. If neck infection occurs before grain formations it results in sterility and if the infection is delayed it results in half or shriveled seeds. At times neck infection results in breaking the infected area. At this stage there is great reduction in yield (Singh and Prasad, 1981). Humid weather is congenial for infection.

10.8.3 CAUSAL ORGANISM

Pyricularia grisea (Cooke) Sacc.
[Teleomorph: *Magnaporthe grisea* (T. T. Hebert) M. E. Barr]

Hyphae are hyaline and septate. However, as the fungus gets older, the hyphae become brown. Under high humidity large number of conidia and conidiophores are produced which give a dirty brown color to the lesion. Conidiophores are simple, septate, basal portion being relatively darker. Conidia are three celled, pyriform, the upper cell pointed and middle cell being wider and darker. Conidia measure 29.9–39.9 × 9.1–11.6 µm (mean 33.6 × 9.5 µm).

10.8.4 DISEASE CYCLE

The pathogen overwinters on collateral hosts and air borne conidia cause the infection on proso millet. It is seed-borne seeds (Mishra, 1983). A temperature of 25–30°C, humidity of 90% and above, cloudy days with intermittent rainfall, are favorable for the rapid spread of the disease. The disease prevails from the month of May to September, but it may extend even up to November, if favorable conditions are prolonged (Anonymous, 1979). The maximum disease incidence has been recorded in the crop sown in the month of August. The fungus spreads from field to field mainly by air borne conidia. The initial inoculum probably comes from weeds or some cereal plants acting as collateral hosts. The fungus may also perpetuates in crop debris and to some extent in the shriveled grains i.e., from the diseased ears which give rise to the initial infection in the nursery from where it may spread to the main field.

10.8.5 MANAGEMENT

• Not much work has been done on the management of the disease; however, the practices recommended for blast of finger millet can help in reducing the disease.

10.9   UDABATTA DISEASE

10.9.1   INTRODUCTION

Udbatta disease is very common in kodo millet and foxtail millet. However, sometimes it has been observed in proso as well as little millets. Diseased panicles are convereted into a compact agarbatti like structure, hence the name "Udbatta."

As the crop is of low returns and grown mostly on marginal poor fertility soils, no efforts are made on the management aspects of any disease.

10.9.2   SYMPTOMS

The most characteristic symptoms of the disease are that the diseased panicles are transformed into a compact 'Agarbatti' like structure, hence named "Udabatta."

10.9.3   CAUSAL ORGANISM

Ephelisoryzae Syd.
[Teleomorph: *Balansiaoryzae* (Syd.) Naras. and Thirum.].

10.9.4   MANAGEMENT

- Hot water seed treatment at 45°C for 10 min. effectively manages the disease.
- Removal of collateral hosts like *Isachne elegans, Eragrostis tenuifolia* and *Cynadon dactylon.*

10.10   BACTERIAL STRIPE

10.10.1   SYMPTOMS

Symptoms appear and narrow, brown to water soaked streaks on leaves extending down to the sheaths and also on culms, where many streaks coalesce on a leaf, the tissue turns brown and translucent. Abundant exudates become evident in the form of thin, white scales along the streaks (Ramakrishnan,

1971). Lesions also occur on the peduncles and pedicels of the panicle. Though, the disease severity may not kill the plants, but individual leaves are partly or entirely discolored. In some cases the entire top of the plant may be killed, tissue becomes soft and brown, especially where partly enclosed and protected by lower leaves and sheaths. In such case, new shoots or tillers may come out from the base (Charlotte Elliott, 1923).

10.10.2 CAUSAL ORGANISM

Pseudomonas syringae Van Hall pv. *panici*
[Syn. *Bacterium panici*]

The pathogen (bacterium) is a short rod with rounded ends, arranged singly or in pairs or occasionally in chains. Measure 3.1–1.5 µm in length and 0.3–0.45 µm wide. No spores, endospores, capsule are present, aerobic, have one polar flagellum but occasionally 2–3. All strains of the bacterium stain readily with carbol fuchsin and gentian violet, Gram negative and not acid fast. On beef peptone agar medium the colonies are round, white, smooth, shining, raised, margin at first entire and later undulate, gelatine is slowly liquefied, without coagulation milk is cleared in 5–6 weeks, on litmus milk turns blue in 3 days and reduction takes place in 7 days. Ammonia and hydrogen sulfide are produced, nitrates are reduced, nonindole producing. Optimum temperature 33–34°C, maximum >45°C and minimum 5.5°C. Thermal death point (TDP) is about 51°C. Sensitive to drying, 99% killed by freezing, 90% by sunlight. Vitality on culture media is for 14 months. The bacterium is transmitted through seed.

10.10.3 MANAGEMENT

- Treating seed with 5% Magnesium arsenate (1 g/liter of water) help in reducing the disease.

10.11 NEMATODE

It has been reported that proso millet seeds received for quarantine clearance, prior to export were found to be infested with *Aphelenchoides besseyi*. Nematodes were found localized underneath the glumes in anhydrobiotic state. Up to 16 nematodes with an average of 1.8 per seed were recorded.

Eradication by pre-soaking the seeds in 1% H_2O_2 for 3 h followed by hot water treatment at 48°C for 15 minutes else complete eradication was recorded by exposing at 50°C for 15 minutes in case of seeds without pre-soaking (Gokte and Mathur, 1993).

10.12   MINOR DISEASES

In addition to the above-listed diseases, wheat streak virus (Sill and Agusiobo, 1955), rice dwarf or stunt virus, leaf streak virus of maize (Markov, 1972; Seth et al., 1972) have also been reported to infect this crop.

KEYWORDS

- **bacterial stripe**
- **endospores**
- **gelatin**
- **nematode**
- **thermal death point**
- **udabatta disease**

REFERENCES

Baltensperger, D. D., (1996). Foxtail and proso millet. In: Janick, J., (ed.), *Progress in New Crops* (pp. 182–190). ASHS Press, Alexandria, VA.

Charlotte, E., (1923). A bacterial stripe disease of proso millet. *J Agric. Res., 26*(4), 151–159.

Gokte, N., & Mathur, V. K., (1993). Treatment schedule for denematization of seeds of *Setari-aitalica* and *Panicum miliaceum* infested with *Aphelenchoides besseyi. Nematologica., 39,* 274–276.

Kovacs, J., Koppnyi, M., Nemeth, N., & Prtroczi, I., (1997). Occurrence and prevention of head smut caused by *Sporisoriumdestruens* (Schlechtend) K. Vankey in millet. *Noventyermeles, 46*(4), 373–382.

Kumar, B., & Prasad, D., (2010a). A new record on banded sheath blight disease of proso millet from mid hills of Uttarakhand, India. *J. Mycol. Pl. Pathol., 40*(3), 331–333.

Lee-Du, H., (1997). Morphological characters and seed transmission of *Bipolarispanici-miliacei* causing leaf spot of common millet. *Korean J. Pl. Pathol., 13*(1), 18–21.

Lovik, V. I., & Dahlstrem, A. F., (1936). Improvement of methods for germination of wheat bunt spores in the laboratory. *Summ. Sci. Res. WK Inst. PL. Prot. Leningard*, 172–178. (cf. *Rev. Appl. Mycol., 15,* 786).

Markov, M., (1972). Studies on maize mosaic in Bulgaria-I. Identification of the virus. *Rastenievudni Nauki, 9*(8), 171–179.

Mishra, B., (1983). Diseases of cheena (*Panicum miliaceum*) losses caused by them and control. In: *State Level Training Programme on Minor Millets* (pp. 36–39). Rajendra Agril. Univ. Pusa (Samastipur).

Misra, A. P., (1973). *Helminthosporium Species Occurring on Cereals and other Gramineae* (pp. 240–242). U.S.P.L. 480. Project Report, Tirhut College of Agriculture, Dholi, Muzaffarpur.

Nagaraja, A., Kumar, B., Jain, A. K., & Sabalpara, A. N., (2016). Emerging diseases: Need for focused research in small millets. *J. Mycopathol. Res., 54*(1), 1–9.

Rajagopalan, B., Balakrishnan, B., & Das, L., (1992). Sheath rot of *Panicum miliaceaum* L. *Indian Phytopath., 45,* 279.

Ramakrishnan, T. S., (1971). *Diseases of Millets* (pp. 83–100). Indian Council of Agricultural Research, New Delhi.

Seth, M. L., Raychanduri, S. P., & Singh, D. V., (1972). Bajra (*Pearl millet*) streak: A leaf hopper-borne cereal virus in India. *Plant Dis. Reptr., 56,* 424–428.

Sharma, P. N., & Sugha, S. K., (1991). Effectiveness of seed-dressing fungicides for control of smut *Sphacelotheca destruens* of proso millet (*Panicum miliaceum*). *Indian J. Agric. Sci., 61*(9), 692–693.

Shirokov, A. I., & Maslenkova, L. I., (1983). Comparative characteristics of geographic populations of the pathogen of millet smut *Sphacelotheca panicimillacei* (Pers) Bub. *Mikologiya I. Fitopatologiya, 17,* 60–63.

Sill, W. H., & Agusiobo, P. C., (1955). Host range studies of the wheat streak-mosaic virus. *Plant Dis. Reptr., 39,* 633–642.

Simmonds, J. H., (1948). Report of the plant pathology section. *Rep. Dep. Agric. Ad.,* 33–35.

Singh, R. S., & Prasad, Y., (1981). Blast of proso millet in India. *Plant Disease, 65,* 442–443.

Sinha, A. P., & Upadhyay, J. P., (1997). *Millet Ke. Rog.* (p. 180) Directorate of Publication. G.B. Pant University of Agriculture and Technology, Pantnagar, Uttarakhand.

CHAPTER 11

Finger Millet or Ragi (*Eleusine coracana* Gaertn.) Diseases and Their Management Strategies

BIJENDRA KUMAR[1] and J. N. SRIVASTAVA[2]

[1]*College of Agriculture, Department of Plant Pathology,*
G.B. Pant University of Agriculture and Technology, Pantnagar–263145,
Udham Singh Nagar, Uttarakhand, India

[2]*Department of Plant Pathology, Bihar Agricultural University,*
Sabour–813210, Bhagalpur, Bihar, India

Finger millet or Ragi (*Eleusine coracana* Gaertn.)

(*Eleusine coracana* Gaertn.)

Vernacular names: *Ragi, Keppai, Mandia, Mandua, mandika, Marwa, and Nagli*

Finger millet or Ragi (*Eleusine coracana* Gaertn.) is a short and profusely tillering plant with characteristic finger-like inflorescences, producing small reddish-brown seeds. Depending on the variety and growing conditions, the crop matures in about 3–6 months. The crop adjusts well fairly reliable rainfall conditions. It is an annual plant with an extensive but shallow root system, extensively grown as a cereal in the drylands of India, especially south India. Grains are highly rich in calcium, protein, essential amino acids, Vitamin A, Vitamin B and phosphorous. In Karnataka, Ragi flour is mostly prepared into Ragi balls known as mudde, leavened dosa and thinner, chapatis, etc. High fiber content of Ragi prevents high blood cholesterol, intestinal cancer, and constipation.

Finger millet is also known as African millet, is believed to be originated in Ethiopian highlands later introduced in India about 4000 years ago. It is highly adaptive to high altitudes and grown in high hills of the Himalayas up to an altitude of 2300 m. One of its greatest qualities is that it can be stored for long periods without spoilage and thus is so important in times of famine.

Finger millet is one of the most important small millets in the tropics and is cultivated in more than 25 countries in Africa (eastern and southern) and Asia (from Near East to Far East), occupies 12% of global millet area. The countries like; Uganda, Srilanka, India, Nepal, China, parts of Africa, Madagascar, Malaysia, Uganda, and Japan are the major producers of finger millet (Sakamma et al., 2018). It has high yield potential (more than 10 t/ha under optimum irrigated conditions).

In India, Ragi is cultivated over an area of 1.19 million hectares with a total production of 1.98 million tonnes and productivity of 16.61 q/ha. Karnataka accounts for 56.21 and 59.52% of area and production, respectively. Tamil Nadu (9.94% and 18.27%), Uttarakhand (9.40% and 7.76%) and Maharashtra (10.56% and 7.16%), are some of the other leading states (Sakamma et al., 2018).

11.1 BLAST DISEASE

11.1.1 INTRODUCTION

Blast of Ragi is one of the major production constraints causing serious yield losses. It was recorded first in India by McRae (1920) from Tanjore delta of Tamil Nadu. Now, it is known to be prevalent during rainy season almost every year in all the Ragi cultivated areas and is considered as one of the major diseases causing heavy yield losses in almost every state of India. Presently, the disease is known to occur wherever the crop is cultivated with varying intensities depending on the climate, variety, and production practices.

The damage depends on time of onset and severity of the disease. The average loss around 28% and as high as 80–90% in endemic areas has been reported due to finger millet blast has been reported.

According to McRae (1922) the disease could cause grain loss over 56%, while, Venkatarayan (1947) recorded more than 80% loss in yield at Mysore. Sunil and Anilkumar (2004a) recorded 3 to 35% loss in 1000 grain weight in blast affected panicles in Bangalore.

11.1.2 SYMPTOMS

The symptoms of the disease appear at almost every stage of plant growth. Seedling death is very common if infected seed are used for sowing. When the young healthy seedlings catch the disease, patches of seedlings showing blight appearance because of severe blast.

On grown up seedlings the disease appears on leaves as typical spindle or diamond shaped spots (Figure 11.1a). Under favorable conditions, these spots further enlarge and coalesce to cover large area and leaves especially from the tip downwards give a blasty appearance. In the neck region 2–4 inches just below the panicle initially brown and later black (Figure 11.1b) lesions are formed. Under humid conditions, olive ceousgrey fungal growth of the pathogen may be seen on this area. The pathogen also infects fingers (Figure 11.1c) causing finger blast.

FIGURE 11.1a   Leaf blast of finger millet or Ragi.

FIGURE 11.1b   Neck blast of finger millet or Ragi.

FIGURE 11.1c   Finger blast of finger millet or Ragi.

Seedlings are more susceptible to leaf blast than are matured plants (Rachie and Peters, 1977) however, no relationship is known between the intensity of seedlings infection and that of head infection. Rath and Mishra (1975) reported that the most damaging stage is when the pathogen attacks the neck region, which significantly reduces grain number and grain weight and significantly, increases spikelet sterility. An olive grey growth of the fungus may be seen on this area. Symptoms also appear on fingers. The infection is usually from the apical portions running towards the base. The pathogen attacks seeds resulting in shriveled and blackened seeds, otherwise in an apparently healthy ear.

Pyricularia grisea also attacks barnyard and little millets causing leaf blast, while, *P. setariae* is responsible for blast disease of foxtail millet with only foliar symptoms.

11.1.3   CAUSAL ORGANISM

Pyricularia grisea (Cooke.) Sacc.
[Perfect stage: *Magnaporthe grisea* (Herbert) Barr]

P. grisea commonly survive in the glumes and on the straw. The pathogen is also observed in the seed outside the embryo, but embryal infection has never been observed (Viswanath and Seetharam, 1989). According to Anilkumar et al. (2004), the finger millet blast pathogenic population was highly diverse as revealed by Shahann diversity index of 0.98, suggesting that each and every individual in pathogen population tended to be a different pathotype.

11.1.4   DISEASE CYCLE

The primary inoculum of the pathogen generally comes from infected seeds. The pathogen may also perpetuate in crop debris and in seeds from infected fingers. One infected seed could result a disease epidemic (Pall, 1988).

Development of the disease is favored by minimum temperature of 15–25°C and relative humidity more than 85% with intermittent rainfall (Pall, 1987). Various factors such as susceptibility of the genotypes, sowing time and the corresponding climatic conditions influence severity of the disease.

From the studies conducted at Coimbatore and Paiyur, on the influence of climatic factors and blast disease development, Anil Kumar et al. (2004) reported that among the weather parameters, relative humidity exerted major influence on disease occurrence. High relative humidity during early hours of a day was favorable for the pathogen to infect the host and quick development. However, higher mean maximum and lower mean minimum temperatures were detrimental to disease/pathogen development as indicated by negative correlation values. However, maximum disease incidence has been recorded in the crop sown during second fortnight of July to first fortnight of August (Nagaraja et al., 2007a).

11.1.5 MANAGEMENT

- At present, cultivation of varieties with inbuilt resistance and need-based sprays of fungicides and bioagents are the most economic means of managing the disease.
- Since, infected seeds are the primary source of inoculum, seed treatment with Tricyclazole (8 g/kg seed) followed by two sprays of Ediphenphos or Kitazin or propiconazole (0.1%) or Carbendazim or tricyclazole (0.05%), first at the time of ear emergence and second after 10 days or an initial spray of 0.05% Carbendazim or tricyclazole followed by a spray of Mancozeb (0.2%) 10 days later are reported to be effective. Two sprays of Saaf (0.2%) or carbendazim 0.05% or tricyclazole 0.05% with first spray at 50% flowering followed by the second 10 days after were also advised by Madhukeshwara et al. (2004) to minimize the disease.
- Seed treatment with bioagents like *T. harzianum* or *P. fluorescens* @6 g/kg coupled with two sprays of *P. fluorescens*@ 0.3% first at flowering and second its 10 days later can manage leaf, neck, and finger blasts effectively (Patro et al., 2008). Nagaraja et al. (2012) reported that seed treatment to resistant variety with Carbendazim @ 2 gkg^{-1} or *P. flourescens* 6 gkg^{-1} reduced blast disease incidence by two and half times over control.
- Cultivating resistant varieties/genotypes like; GPU 26, GPU 28, GPU 45, VL 149, CO 13, etc. coupled with seed treatment with Carbendazim @ 2 g/kg has also been reported effective (Madhukeshwara et al., 2005 and Kumar, 2011b).
- Management of the disease through varietal mixture @ 1:1 ratio of PRM 1 (susceptible) and VL 149 (resistant) was also found economical in Uttarakhand (Kumar and Kumar, 2011).
- Nagaraja et al. (2012) suggested that the blast of finger millet can be managed by choosing a resistant cultivar coupled with seed dressing by Carbendazim (2 g Kg^{-1}) or bioformulation of *P. fluorescens* (6 g Kg^{-1}).
- Neck and finger blasts have been reported to be increased with increasing doses of nitrogen. Therefore, excess use of nitrogenous fertilizers should be avoided (Kumar and Yadav, 2012).

11.2 BROWN SPOT OR SEEDLING BLIGHT OR LEAF BLIGHT

11.2.1 INTRODUCTION

The disease was first reported by Butler (1918) causing seedling blight or leaf blight, foot rot of Ragi. The disease is known to prevalent in most parts of the finger millet growing areas in India (Govindu, 1982; Misra, 1979; Pall and Sharma, 1976; Mitra, 1931; Thomas, 1940; Narain et al., 1975). If crop suffers from nutrient deficiency or prolonged drought, the disease appears very severely. In terms of severity and distribution, it is next only to blast. Several hosts are infected by the fungus including *Setariaitalica, Eleusineindica, Dactyloctenium aegyptium, Echinochloa frumentacea, Panicum miliaceum, Pennisetum typhoides, Sorghum vulgare* and *Zea mays* (Mitra, 1931; Pall et al., 1980).

11.2.2 SYMPTOMS

The disease may occur throughout the field or in well defined patches sporadically. The pathogen affects almost every plant part such as base, culms, leaf sheath, leaves, peduncle, and fingers. Pre or post emergence rotting of the seedlings is seen if infected seeds are used for sowing.

The characteristic symptoms of the disease on the leaves appear as brown to dark brown spots. Such symptoms can also be observed on leaf sheath, particularly in older plants. Under high humidity conditions, woolly growth of the fungus can be seen in the center of these lesions. If conditions are favorable, foot rot symptom can also be noticed. Dark tan lesions seen initially in the neck region enlarge and extend both up and down the neck. In severe cases of disease, the peduncle breaks and hangs down the plant. The pathogen may infect earhead, fingers, and grains. The grains affected by the disease may not develop completely and shrivel, resulting in heavy crop losses. Severe infection causes chaffiness and seed discoloration (Figure 11.2).

11.2.3 CAUSAL ORGANISM

Drechslerano dulosa (Berk. and M. A. Curtis ex Sacc.) Subram. and B. L. Jain [Syn. *Helminthosporium nodulosum* Berk. and M. A. Curtis ex Sacc.] [Perfect stage: *Cochliobolus nodulosus*].

FIGURE 11.2   Brown spot or seedling blight or leaf blight of finger millet or Ragi.

11.2.4   DISEASE CYCLE

The disease is seed-borne in nature (Grewal and Pal, 1965). The conida carried on the seeds remain viable for one year (Narasimhan, 1933). The first infection is caused by the fungus on the seed. The fungus becomes systemic from the early stages. The fungus also remains viable on crop stubbles and dead host remains. Pathogen is able to survive in soil for 18 months (Vidhyasekaran, 1971). Secondary spread of the pathogen occur air borne conidia.

Infection can occur between 10 and 37°C temperature however, optimum is 30–32°C. The first sign of infection becomes visible in 24 h. Lesions are formed within 4 days. Young seedlings may be killed in 3 days and older in 15 days.

11.2.5   MANAGEMENT

- The disease is primarily seed borne; seed treatment with Agrosan G. N. had been reported to give complete control of pre-emergence damping-off seedling blight (Grewal and Pal, 1965).
- Need-based spraying of Mancozeb @ 0.2% has also been reported effective in managing the disease (Nagaraja et al., 2007).

11.3 CERCOSPORA LEAF SPOT

11.3.1 INTRODUCTION

Cercospora leaf spot is one of the major foliar diseases of finger millet although restricted to certain geographical regions. It occurs in the Himalayan foothills and mid-hills of Nepal. Muyanga (1995) also reported this disease as one of the most destructive diseases of finger millet in Zambia. The disease occurs at all the stages of plant growth, starting from seedling to grain setting. If the disease occurs just after earhead formation, up to 40% yield can be reduced. Reduction in 1000 seed weight by 21% has also been reported (Pradganang, 1994). However, when the disease incidence is around 25%, there will be no yield loss (Pradganang and Abington, 1993).

11.3.2 SYMPTOMS

The plants are susceptible to infection at all the stages of its growth, from seedling to grain filling. Under natural conditions, infection usually starts during mid-June in the early sown crop.

Symptoms are usually recorded on the matured leaves spreading to the younger leaves. Initially, the disease appears as reddish-brown specks surrounded by yellow halo. At this stage, the disease is easily confused with the *Helminthosporium* leaf spot. The lesions enlarge in their size and several such lesions coalesce to form large lesions usually surrounded with yellowish halo.

During rains or under high relative humidity conditions, fungus sporulates and produces grayish white growth in the center of these lesions; the lesions enlarge to become eye or diamond shaped spot measuring 15 × 3 mm and usually confused with those of blast. Such leaves give burnt appearance. At the end of August, severely infected leaves turn necrotic and begin to dry and shrivel and plants look completely blighted. Similar symptoms are also noticed on the leaf sheath, stem, peduncle (neck), and fingers (Figure 11.3).

11.3.3 CAUSAL ORGANISM

Cercospora eleusinis, Munjal, Lall, and Chona

Munjal et al. (1961) studied the specimen collected by J. N. Kapoor form Kathgodam, Nainital, Uttarakhand, and gave the descriptions and nomenclature of the fungus. They reported it a new species *Cercosporaeleusinis.*

However, Wallace and Wallace (1947) reported *Cercospora fusimaculans* Atk. causes leaf spot in Tanganyika.

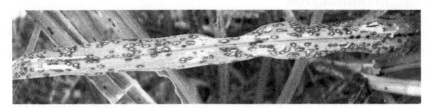

FIGURE 11.3   Cercospora leaf spot of finger millet or ragi.

Because of the characteristic slow growth of *Cercospora*, pathogen is difficult to isolate from the diseased leaves. Of the eight media tested for growth and sporulation of the *Cercospora*, Ragi flour lactose yeast agar (RFLYA) [Ragi flour 20 g, Lactose 5 g, Yeast, Lactose 5 g, Agar, 20 g] was found to be the best recorded maximum growth and sporulation of the fungus under *in vitro* conditions (Kumar, 2005).

When infected leaves were incubated in a moist chamber for 36–48 h at 25°C, fungal growth was noticed in the lesions. Fungus produces stroma from which conidiophores bearing and conidia arise. The condia germinate from both the ends or from several cells along the length of the conidia. Conidiophores are white, thread-like, straight to slightly curve and 3–12 septate. Mean dimension of 25 conidia is reported to be 1.5 µm (range 1–1.7 µm) × 69.6 µm (range 50–95 µm) (Kumar, 2005).

11.3.4   DISEASE CYCLE

The disease is generally confined to mid-hills of Nepal, where rainfall is generally high and the average daily temperature does not exceed 20°C. However, in India, the disease is found to be a serious problem in mid and high hills of Uttarakhand and is known to occur from 850 m to >1900 m altitude while the intensity of the disease was reported to be low in lower hills (Kumar et al., 2007a). The disease appears most severely during June in early sown crop.

11.3.5   MANAGEMENT

- Collection and destruction of crop debris and foliar sprays of Carbendazim @ 0.05% at 15 days intervals have been reported to provide satisfactory results (Kumar et al., 2007a).

11.4 LEAF SPOT

11.4.1 INTRODUCTION

Shaw (1921) recorded the occurrence of *Curvularialunata* (*Acrothecium lunatum*) on many small millets including finger millet from Pusa. He considered it, at that time, as a weak parasite that produced small spots. Subramanian (1953) reported it from Madras. Rao and Manivarghese (1981) reported it from Kerala, while Roy (1989) reported it from Assam. Rao et al. (1992) reported the occurrence of *Curvularia pallescens* on finger millet plants growing as weed in sugarcane fields. According to them, finger millet serves as a reservoir of the pathogen.

11.4.2 SYMPTOMS

According to Shaw (1921), it produces small spots on finger millet, *Setaria italica* and *Panicum frumentaceum*.

11.4.3 CAUSAL ORGANISM

Curvularialunata (Wakker) Boedijn
[Syn. *Acrothecium lunatum* Wakker]

The conidia are curved, Knee-shaped, 2–4 septate with large middle curved cell, measuring 18.22×5.7 µm. In non-synthetic media, greater development of aerial hyphae takes place, but the spores are usually smaller, whereas in synthetic media, lesser development of aerial hyphae takes place, but the spores are bigger.

11.4.4 MANAGEMENT

Details of management practices are not available

11.5 GREEN EAR OR DOWNY MILDEW OR CRAZY TOP

11.5.1 INTRODUCTION

Green ear or downy mildew disease of Ragi was reported by Venkatarayan (1947) for the first time in India from the old Mysore state. The disease occurred in a severe form in the erstwhile Mysore state in 1948. The disease has also been recorded in Tamil Nadu (Venkatarayan, 1947) and Uttarakhand in India (Kumar et al., 2007b).

But for its sporadic occurrence, the disease could be destructive leading to total failure of the crop due to malformation of the affected ears.

11.5.2 SYMPTOMS

The diseased plants are generally stunted with shortened internodes and excessive tillering become bunchy and bushy in appearance. Often, pale yellow translucent spots are seen on leaves of affected plants. The characteristic white cottony growth of the pathogen as seen in many other downy mildews is generally not visile in the downy mildew of finger millet. As a result, the asexual stage quite often remains unnoticed.

However, the green ear symptoms appear at the reproductive stage. The earheads in the affected plants completely or partially convert into narrow green leafy structures causing complete sterility. Floral structures including lemma, palea, and glumes partially or whole ear convert into narrow leafy structures. The proliferation occurs first in the basal spikelets and later others also get affected. Finally, the entire earhead gives a bushy appearance showing typical 'green ear' symptom.

Partially infected ears also fail to produce grains. However, leaves of affected plants does not show any change in their external morphology. Growth of auxiliary leafy structures from internodes has also been observed (Kumar et al., 2007b) (Figure 11.4).

FIGURE 11.4   Green ear of finger millet or ragi.

11.5.3   CAUSAL ORGANISM

Sclerophthora macrospora (Sacc.) Thirum.
[Syn. *Sclerosporamacrospora* Sacc.]

11.5.4   DISEASE CYCLE

The life cycle of *Sclerophthora macrospora* was described by Safeeulla (1955). The disease is both internally as well as externally seed borne (Raghavendra and Safeeulla, 1973), has very wide host range and thus the disease development becomes easy. The oospore germination is indirect, oospore geminate to form a big lemon shaped sporangium that liberates 24–48 zoospores (Safeeulla, 1976). Mycelium is located in the endosperm and embryo region of infected seeds. On seed germination, mycelium spreads to meristematic tissue.

The first sporangium develops at the apex which is followed by development of sporangia in lateral position. These are lemon-shaped with a distinct papilla and pedicel and measure 60–100 × 43–64 µ. They germinate and liberate many unequally biflagellate zoospores. The optimum temperature for their germination is 22–25°C (Safeeulla, 1976).

Sporangial germination readily occurs in water. Almost 100% sporangia germinate within 3 h. The sporangial viability is maximum at saturation, below which they lose viability very fast. During zoospores formation, delimitation of sporangial protoplasm occurs in such a way that each fragment contains a nucleus. The zoospores are pyriform, spherical, or irregular. They germinate by germ tube which form appressorium at the tip.

Kumar et al. (2007b) reported that the disease is favored by low temperature and heavy dew during the crop development period in *kharif*. Temperature around 20–25°C during night and early hours of morning favors rapid spore germination (Semeniuk and Maukin, 1964) and disease development. However, light appears to have very little effect on sporangial germination (Raghavendra and Safeeulla, 1971). During night, when the temperature is between 22 and 25C, large numbers of sporangia are produced that release zoospores.

11.5.5   MANAGEMENT

• As the disease is seed borne, seed treatment with Apron 35 SD @ 2.5–3.0 g/kg (Nagaraja et al., 2007b) or Ridomil MZ (Anilkumar et al., 2003) would control systemic infection and protect plants up to 30 days from aerial infections also.

- Selection of seed from healthy crop.
- Collect diseased plants, especially before oospores are formed, and burn them.
- Summer deep plowing.
- Rogueing out infected plants.
- Prolonged crop rotation.
- Spraying of Mancozeb@0.25% or Metalaxyl (RidomilMZ)@0.25% starting from 30 days after sowing in the field.

11.6   GRAIN SMUT

11.6.1   INTRODUCTION

The disease was reported for the first time from Malkapur in 1918 and reported by Kulkarni (1922) in the then princely state of Kolhapur. Later it was reported in Mysore state by Coleman (1920), McRae (1924), Narasimhan (1934), Mundkur (1939), and Venkatarayan (1947). The damage due to this disease is negligible but according to Mantur (1994) when disease appears in epidemic form as many as 200 grains may be infected in an ear. Although it was known to be a disease of summer crop, in the recent past has been recorded sparingly during *kharif* also.

11.6.2   SYMPTOMS

The smut sori can be seen generally a few days after heading scattered randomly in the ear. The infected ovaries are converted into velvety green galls (smut sori) which are many times bigger than the size of normal grains. The sorus gradually turns pinkish-green and finally dirty black on drying. The affected grains are located singly, or in some cases grouped into patches of varying size and are usually confined to one side or towards the apex or base of the head showing signs of rupturing at several places (Thirumalachar and Mundkur, 1947).

Ganapathi (1971) described certain other characteristic symptoms of the disease. At times, the inflorescence remains extremely reduced and shriveled, with a complete absence of the development of the spike or spikelets including glumes that were replaced by greenish globose to elongated sac-like structure containing sooty teliospores. Yet in other type, the sori were formed around the stalk of the main rachis of the inflorescence in the form of galls scattered or coalesced with each other containing mass of teliospores on the same ear and were abundant, small, rounded, and elongated sac-like smut sori, resembling phyllody instead of developing into normal grains (Figure 11.5).

FIGURE 11.5   Grain smut of finger millet or ragi.

11.6.3   CAUSAL ORGANISM

Melanopsichium eleusinis (Kulk.) Mundk. and Thirum.
[Syn. *Ustilago eleusines* Kulk.]

First, it was named *Ustilago eleusines* by Kulkarni. Later, Mundkur, and Thirumalachar, 1946) changed it to the genus *Melanopsichium* and named it as *M. eleusinis*.

The fungal spores are globose or subglobose, echinulate, and measure 7–11 μm with a mean of 9.5 μm. The spores germinate to form septate promycelium producing both lateral and terminal sporidia. Promycelium first protrudes out as a small papilla which gradually elongates into promycelium. Two to three transverse septa are formed, dividing the promycelium into 3–4 cells from which terminal and lateral sporidia are produced. The primary sporidia by budding produce secondary sporidia immediately.

11.6.4   DISEASE CYCLE

Infection of flowers takes place through air borne spores. Thirumalachar and Mundkur (1947) could obtain infection by painting the sporidial suspension on the ears at during ear emergence. Not much information is available on the disease cycle.

11.6.5   MANAGEMENT

Owing to air borne nature of the pathogen, chemical control is neither economical nor warranted. However, Mantur et al. (1995) tested several fungicides both singly and in combination as spray to the inflorescence. Least incidence was noticed where two sprays of fungicides, first with Difolatan at panicle initiation followed by second with mancozeb at flowering were given.

11.7   FOOT ROT

11.7.1   INTRODUCTION

In India, the occurrence of footrot caused by *Sclerotium rolfsii* was reported for the first time by Coleman (1920) from the then princely state of Mysore. Up to 50% loss was recorded at Rampur, Nepal (Basta and Tamang, 1983). It was reported from Coimbatore (Anon., 1954), Odisha (Narain, 1972), Uttarakhand (Kumar and Prasad, 2010b) and Gujarat (Waghunde et al., 2011). It is mostly a problem in heavy rainfall areas and irrigated ragi (Nagaraja and Reddy, 2009).

11.7.2   SYMPTOMS

The disease appears in patches in the field. The infected plant becomes pale and stunted. The infection occurs around the collar region, the infected area being restricted to two to three inches above ground level. Normally, the plants are attacked at a stage when plants are flowering or setting seeds, due to the debility of the stem as the movement of photosynthates is towards sink.

The collar region of affected plant initially appears water soaked due to infection by the pathogen. Later affected region turns brown and subsequently dark brown followed by the concomitant shrinking of the stem. Profuse white cottony mycelial growth occurs in this area. Mycelium soon disappears leaving behind small roundish white velvety grain-like structures in the fungal matrix. They grow in their size become mustard seed-like, color turn brown. These are the sclerotial bodies of the causal fungus. At this stage, the leaves lose their luster, droop, and dry. Ultimately, the entire plant dries up prematurely (Figure 11.6).

FIGURE 11.6   White cottony mycelial growth due to foot rot and sclerotial bodies on the infected tissue.

11.7.3   CAUSAL ORGANISM

*Sclerotium rolfsii*Sacc.
[Perfect Stage: *Atheliarolfsii* (Curzi) C. C. Tu and Kimbr.].

11.7.4   DISEASE CYCLE

The pathogen survives mainly in soil-borne and becomes active in the rainy season. Sandy loam soils favor disease incidence and the pathogen survive better at low soil moisture levels. *S. rolfsii* has very wide host range and known to parasitize more than 500 hosts (Aycock, 1966). At the end of crop season, enormous sclerotial bodies are produced from the growth that had occurred on the host plant. These find their way to soil and move through rain water from field to field. Generally, *S. rolfsii* is considered as a weak or an opportunist organism, capable of attacking the host plant when it is debilitate. The infection occurs around the collar region, the infected area being restricted to two to three inches above ground level. Normally, the plants are attacked at flowering or seed setting satge. Though, only imperfect stage is seen in the field, fertile sporophores are produced under ideal situation.

11.7.5   MANAGEMENT

- Cultural methods viz., deep plowing, early sowing, rotating crop with non-graminaceous crops and maintaining optimum soil condition, etc., are some ideal ways of managing the disease.
- Use of resistant varieties/Genotypes GPU 28, RAU 8, L-49-1, MR 6, OEB 82, PR 202 and OEB 10. While, PPR 1735 showed stable resistance to foot rot over 3 consecutive years whereas GPU 16 and PPR 2350 were reported moderately resistant (Jain et al., 1994).
- Field trials conducted in sick soils at Waghai (Gujarat) and Mandya (Karanataka) have shown that all the plots receiving bio-agent seed treatment or soil application or value added product had lesser disease incidence. However, soil application of 1 kg *T. viride* talc formulation or *P. fluorescens* + *T. viridae* 500 g each mixed with 25 kg compost incubated for 15 days and spread over an acre at the time of weeding resulted in least foot rot and higher Ragi yields (AICSMIP, 2013).
- Chemical control may not be economical as the pathogen is a soil inhabitant. However, Channamma et al. (1980) reported that soil incorporation of Vitavax@ 10 kg/ha being effective in managing the foot rot.

11.8   RUST

11.8.1   INTRODUCTION

Although, rust of finger millet as of now is negligible, however, it can cause extensive reduction in grain yield under favorable environmental conditions for pathogen perpetuation. However, in African countries like Ethiopia, out of the more than 30 crop diseases recorded as the major staple food crops like finger millet in the region only rusts (locally called '*humodia*') are considered by farmer to be highly important (Ayimut and Mathew, 2008). The first report of rust incidence on finger millet was from Meerut in Uttar Pradesh (Dublish and Singh, 1976). After a gap of twenty years, it was reported from Bangalore in Karnataka (Channamma et al., 1996) on samples from *Eleusine coracana*. Recently, severe incidence of rust was reported on various varieties at Agricultural Research Station, Vizianagaram, and Andhra Pradesh.

11.8.2 SYMPTOMS

Symptoms of the rust appear as minute to small, dark brown, broken pustules arranged in rows on the upper surface of the leaves. The symptoms are more severe towards the top 1/3 portion of the upper leaf as compared to the lower and middle leaves (Figure 11.7).

FIGURE 11.7   Rust of finger millet or ragi.

11.8.3 CAUSAL ORGANISM

Uromyceseragrostidis Tracy.

The pathogen was confirmed as *Uromyces eragrostidis* Tracy. It was earlier reported on *Eragrostis cynosuroides* by Sydow and Butler (1906). The spores measure 24 µm × 26.25 µm, with 3–4 germ pores approximately equatorial (Channama et al., 1996).

11.8.4 DISEASE CYCLE

The uredospores infect the host and produce uredia in about ten days, thus assuring several cycles of the uredial stage during the crop season. If the crop is grown throughout the year as in many parts of tribal hills, the fungus can perpetuate in its uredial stage.

11.8.5 MANAGEMENT

- Four varieties *viz.*, SEC 915, 314, 712 and ICMV-221 developed in Uganda which are reported to be resistant to finger millet rust. The millet can be ground into flour for millet bread, porridge, and can also be used in the fermentation process for alcohol. These four varieties also intended to boost diet for diabetics and people with HIV/AIDS as they are with high (14%) protein concentration (Wanyera, 2008).
- Foliar spray of Mancozeb @ 0.2% effectively controls the rust.
- Spraying of propiconazole @ 2.2 ml/Liter of water is also effective in the management of rust.

11.9 BANDED BLIGHT

11.9.1 INTRODUCTION

It is one of the emerging problems on all the small millets. The disease was first reported in severe form at Vellayani, Kerala, India (Lulu Das and Girija, 1989). Subsequently, the disease was recorded in a severe form in the experimental plots of Birsa Agricultural University, Ranchi, mostly on exotic genotypes (Dubey, 1995). Rice sheath blight pathogen *Rhizoctonia solani* can infect finger millet crop but the finger millet banded blight pathogen *Rhizoctonia solani* AG1 cannot infect rice crop (Kannaiayan and Prasad, 1978). Recently during 2006 *kharif*, in Karnataka also the disease was noticed on popular Ragi variety GPU 28 in Narasipur village, Kolar dist (Nagaraja and Reddy, 2010a; Nagaraja et al., 2016).

11.9.2 SYMPTOMS

The disease is characterized by the formation of oval to irregular light grey to dark brown lesions on the lower leaf and leaf sheath. As the lesions enlarge in their size, the central portions of the lesions become white with narrow reddish brown border. At later stages, these spots may be distributed irregularly on leaf lamina. The white to brown sclerotia along with mycelial growth can be observed on and around the lesions.

Infection may occur in the temperature range from 23–30°C and 80% or above relative humidity favor the rapid development of the disease. These lesions enlarge rapidly and coalesce to cover large area of the sheath and leaf lamina. At this stage, the symptoms are characterized by a series of brown or copper colored bands across the leaves giving a characteristic banded appearance. In severe cases, the leaves dry up and the plants are completely blighted.

In severe cases, symptoms also appear on peduncles, fingers, and glumes as irregular to oval, dark brown to purplish brown necrotic lesions. Early infection on neck region is similar to neck rot resulting in poor grain setting. Affetecglumes produce smaller and shriveled grains. As the symptoms produced on every plant part, give characteristic banded appearance, hence the disease has been named banded blight (Dubey, 1995).

It has been observed that, early varieties suffered more than late ones, as late maturing varieties have more chance of escaping the disease than early varieties because of the low temperature in autumn, when the upper leaves are not damaged (Figure 11.8).

11.9.3 CAUSAL ORGANISM

Rhizoctonia solani Kuhn.
[Basidial Stage: *Thanatephorus cucumeris* (Fr.) Donk.].

11.9.4 DISEASE CYCLE

Pathogen is a soil inhabitant, high humidity (above 80%) and moderate temperature (26 ± 2°C) favors the disease (Dubey, 1995).

FIGURE 11.8   Bunded blight symptoms of finger millet or ragi.

11.9.5 MANAGEMENT

- Spraying of propiconazole @ 1 ml l⁻¹ water is reported effective for the management of banded blight pathogen in finger millet (Patro, 2013).
- Clean cultivation, proper drainage, and removal of weed grassy weeds from the bunds can prevent the disease (Patro, 2013).

11.10   BACTERIAL LEAF SPOT

11.10.1   INTRODUCTION

Leaf spots were observed during the rainy season of 1960 in Chidambaram taluk, South Arcot district of Tamil Nadu that were found to be caused by a bacterium. Rangaswami et al. (1961) who studied the disease and the causal agent reported it to be due to hitherto unknown species of *Xanthomonas*, which they named as *X. Eleusinae*.

11.10.2 SYMPTOMS

Linear spots are seen both on upper and lower surfaces of the leaves spreading along the veins. The spots measure 2–4 mm long, but often extends up to one inch or more. In the beginning, spots are light yellowish brown, but soon become dark brown. In advanced stage, the leaf splits along the streak giving a shredded appearance. All the leaves, including the tender shoots, in a plant are affected. The bacterium, mainly affects the leaves, but at times characteristic streaks may be found on the peduncle. These streaks are narrow, 5–10 mm in length and appear sub-cuticular.

11.10.3 CAUSAL ORGANISM

Xanthomonas eleusinae Rangaswami, Prasad, Eswaran

The bacterium is short rod, $1.8–2.7 \times 0.8–1.0$ μm with a single polar flagellum, aerobic, Gram negative, non-capsulated, non-spore forming, and non-acid fast. It forms dull yellow slimy and shiny colonies on nutrient agar and growth in nutrient broth is turbid with pellicle formation. Gelatine is rapidly liquefied but starch is not utilized. Litmus milk turns neutral and is coagulated. Nitrate is reduced, H_2S produced but ammonia and indole not produced. It gives positive lipolytic activity and negative MR and VP tests. It utilizes lactose as carbon source, with acid production and little or no gas formation.

11.10.4 MANAGEMENT

Details of management practices are not available

11.11 BACTERIAL BLIGHT

11.11.1 INTRODUCTION

Desai et al. (1965) opined that the bacterial pathogen reported by Rangaswami et al. (1961) on Ragi was not a species of *Xanthomonas* and reported a bacterial blight disease on Ragi caused by *Xanthomonas coracanae* that was widely prevalent in Gujarat.

11.11.2   SYMPTOMS

The all stages of plants are susceptible to infection. If infection takes place during early stages of growth, the plants become yellow and show premature wilting. Infection first appears as water soaked, translucent, linear, pale yellow to dark greenish-brown streaks, 5–10 mm long along the midrib of the lamina. The hyaline streak later develops into a broad yellowish lesion measuring 3–4 cm and turns brown. When the infection is heavy, especially in the early stages, the entire leaf turns brown and withers away.

11.11.3   CAUSAL ORGANISM

Xanthomonas axonopodis Starr and Garces pv. *coracanae* Desai, Thirumalachar and Patel
[Syn. *Xanthomonas coracanae* Desai, Thirumalachar, and Patel]

Bacterium appear as short rods with rounded ends, usually single, occasionally in pair, measuring 1.1–1.8 × 0.5–0.7 μm, motile by a polar flagellum, Gram negative, encapsulated, nonendospore and non-acid fast. Colonies on PDA plates are circular with entire margin, smooth, pulvinate, butyrous, and glistering yellow. Growth on nutrient agar and potato dextrose agar slants is moderate to abundant, filiform, convex, glistering, smooth opaque, butyrous, and lemon yellow, medium unchanged.

11.11.4   MANAGEMENT

Details of management practices are not available

11.12   LEAF STRIPE

11.12.1   INTRODUCTION

Billimoria and Hegde (1971) observed a disease on Ragi during 1969–70, which was found to be due to bacterium. The disease appeared in serious proportion in and around Bangalore, Karnataka, India.

11.12.2   SYMPTOMS

The characteristic symptom of the disease is brown coloration of the leaf sheath particularly from the base upwards. The affected portion of the leaf invariably involves the midrib and appears straw colored. This symptom spreads to about three fourths the lamina and then abruptly stops or occasionally reaches the leaf tip. Occasionally the strips of the infected areas are seen to proceed along the margin of the leaf, leaving the central portion, including the midrib healthy. The bacteria are readily detected in the phloem vessels. Diseased plants can be recognized from a distance by characteristic drooping of the leaves.

Infected culms show a light brown discoloration along one side. In some cases, discoloration begins from the base, but in most cases, it begins 5–7 cm above the base and extends to leaf sheath proper. There is, however, no apparent reduction in girth or turgidity of the affected culms as compared to the healthy ones. Young plants especially less than a month old are usually free from the disease. The bacterium is systemic and soil borne.

11.12.3 CAUSAL ORGANISM

Pseudomonas eleusinae

The bacterium is a short rod, found singly or in pairs, measuring 0.83–2 × 0.31–0.42 μm, capsulated with one or two monotrichous flagella. Gram negative, non-spore forming, and non-acid fast. It strongly hydrolyses starch, produces acid but no gas from glucose, turns plain milk alkaline, and shows mild lipolytic activity. The bacterium does not liquefy gelatine, reduces nitrate, V. P. and Methyl red test positive. Does not produce indole and does not hydrolyze casein.

Other information *viz.*, disease cycle, management etc. on bacterial diseases is not available as they have not reported during other years to warrant further investigations.

11.12.4 MANAGEMENT

Details of management practices are not available

11.13 VIRUSES DISEASES

11.13.1 RAGI SEVERE MOSAIC

11.13.1.1 INTRODUCTION

The Ragi crop in Southern Karnataka and the border districts of Andhra Pradesh was affected by a severe mosaic in *kharif* 1966. In certain pockets of Hiriyur in Chitradurga district and Devanahally of Bangalore district, the disease was so severe that the farmers abandoned their crop as heavily diseased plants failed to set seed (Joshi et al., 1966). The epidemic that occurred in 1960's was mainly due to continuous cropping of Ragi, coupled with abnormal weather factors, which had favorable impact on vector population (Keshavamurthy and Yaraguntaiah, 1977).

11.13.1.2 SYMPTOMS

The virus induces mosaic symptoms, which are more clear and pronounced on young leaves. Infected plants remain stunted and the ears of severely affected plants malformed. Such plants produce few seeds of smaller size, which reduces the yield considerably. In addition, the affected plants appear pale yellow due to severe chlorosis and in severe cases become brownish-white. Thus, the entire field appears yellow and can be readily distinguished from non-infected stands from a distance. Stunted plants do not recover, develop roots at nodes, generally do not produce ears and if produced are mostly sterile.

Subbayya and Raychaudhuri (1970) observed severe mottling, chlorotic streak, general chlorosis, and yellowing of leaves. They also observed profuse lateral shoots and aerial adventitious roots, stunting of plants, fewer flower, and poor seed formation.

11.13.1.3 CAUSAL ORGANISM

Sugarcane mosaic virus (SCMV)

Particles were flexuous rods with an average length of 667± 8 nm and an approximate diameter of 12–14 nm. The virus, thus, was identified as a strain of sugarcane mosaic virus (SCMV) (Subbayya and Raychaudhuri, 1970).

11.13.1.4 DISEASE CYCLE

The virus is neither seed borne nor soil borne.

11.13.1.4.1 *Transmission by Insects*

Paul Khurana et al. (1973) studied the virus vector relationship employing *Longiunguis sacchari* as vector. The optimum acquisition feeding period was five minutes and optimum transmission feeding period was reported to be one hour.

The aphid acquires the virus in one minute. Pre-acquisition fasting increases the efficiency of the vector and maximum transmission is obtained within 1½ to 2 h fasting. Even a single aphid can transmit the virus with an optimum of 10 aphids per plant. Post-acquisition fasting decreases the vector efficiency and the virus is found to be non-persistent in *R. maidis* since it was retained only for an hour after acquisition. The incubation period is found to be influenced by temperature but not by the age of the host. Ragi plants of all ages are susceptible, but the severity of infection decreases significantly with increase in the age of the host (Subbayya and Raychaudhuri, 1970).

11.13.2   RAGI MOTTLE STREAK

11.13.2.1   INTRODUCTION

During mid-1960's there was a severe disease problem in epidemic proportions all through Southern Karnataka. Govindu et al. (1966) studied this disease and thought it was due to combined effect of virus and *Helminthsporium* sp. Later, Mariappan et al. (1973) reported a Ragi streak disease from Tamil Nadu, which was also transmissible by *Sogatella* sp. and they regarded this virus to be a strain of the virus reported from Karnataka.

Maramorosch et al. (1977) observed 50–100% losses in certain areas due to mottle streak.

11.13.2.2   SYMPTOMS

The infected plants exhibit regular dark-green areas all along the leaf veins when the plants are 4–6 weeks old. Other symptoms on leaf include chlorosis and streak. In some cases, occasional yellowing to almost albino symptoms are also observed. However, in the lower leaves, the symptoms are of mottle type in the form of white specks and the affected plants are generally stunted bearing small ears.

The characteristic symptoms on leaves are chlorotic mottling, streaking, striping, and yellowing. The diseased plants have more nodal branches and unproductive tillers (Saveetha et al., 2006). Blackening of collar region, yellow collar rot, formation of adventitious roots, blackened, and stunted lateral roots at upper node were also noticed in diseased plants (Govindu and Shivanandappa, 1967). The virus is reported to cause leaf chlorosis, streaking, albino type discoloration, striping, and mottling leading to severe yellowing also (Maramorosch et al., 1977). In Tamil Nadu, a profound seasonal influence on the incidences of mottle streak and streak diseases on finger millet was observed (Narasimhan, 1980) and the prevalence of higher vector population was responsible for the maximum disease incidence.

11.13.2.3   CAUSAL ORGANISM

Ragi mottle streak virus.

Short rod-like bacilliform particles were seen in the perinuclear portion in cells from all parts examined, including the epidermis, mesophyll, and conducting element. The particles were very abundant and therefore easily detected. The particles measure 80 nm in cross-section and 285 nm length-wise. The particles were enveloped, bacilliform, and spiked corresponding to the morphology of rhabdoviruses.

11.13.2.4   DISEASE CYCLE

Ragi mottle streak virus is transmitted by two species of jassids *viz.*, *Cicadulina bipunctella*, and *C. chinai*. *C. bipunctella* was able to transmit up to 82%. The minimum acquisition feeding period was 48 h and minimum inoculation feeding period was 24 h. The virus can persist in the insect for 8 days. Only a section of the population of the vector *C. bipunctella* transmitted the virus in high percentage and the virus is carried in the leafhopper in a persistent manner.

The adverse climatic conditions *viz.*, high maximum-minimum temperature, relatively low rainfall and less average relative humidity in all the affected districts from August to November during 1945 to 1965 were presumed to have led to enormous increase in the vector population.

Among the weather parameters, rainfall and relative humidity were positively correlated while wind velocity was negatively correlated with disease incidence and vector population. A significant positive correlation has been reported between the vector population and disease incidence (Saveetha et al., 2006).

According to Saveetha et al. (2007), the leaf hopper *Cicadulinabipunctella* was found to transmit the virus up to 80%. The virus vector relationship revealed that 72 hours of acquisition access period, 5 days of inoculation access period and about 30 days of incubation in host are required for attaining efficient transmission.

11.13.2.5 MANAGEMENT

Seed extract of *Harpullia cupanioides* @ 10% was reported to reduce the incidence of mottle streak by 61.42% with maximum yield of 2.28 t/ha. Spraying of Acephate@ 0.01% and neem oil @ 1% were also found effective, reducing the disease incidence up to 39.6 and 36.4% with 2.23 and 2.18 t/ha of grain yield, respectively. AVP sprayed plants showed enhanced activation of defense-related enzymes such as peroxidase (PO), polyphenol oxidase (PPO) and phenylalanine ammonia lyase (PAL) as compared to control (Saveetha et al., 2006).

11.13.3 RAGI STREAK

11.13.3.1 INTRODUCTION

During the year 1974–75, a virus disease producing streaking and yellowing of leaves and stunting of ragi plants in the fields around Bangalore was observed and the virus was found not to be transmitted either through mechanical sap inoculation or aphid species tested (Anon., 1975).

The loss in grain yield depends on the age at which the virus infects the crop. Similarly, depending upon the plant stage, the virus infection resulted in a drastic reduction in 1000-grain weight. The loss in 1000-grain weight was 84, 63, 27, and 24% when the infection occurred at 30, 40, 50 and 60 days old seedlings. However, there was no significant change in number of tillers except where infection occurred on 10 days old seedlings when there was a significant increase in seedling number (Nagaraju et al., 1982). The virus has several host plants; it can infect crop plants like bajra, sorghum, maize, wheat, barley, and oats (Nagaraju and Viswanath, 1981). 100% yield loss was observed when 30 days old seedlings were inoculated. However, the yield loss is as high as 60–70% when inoculated even during ear emergence (Nagaraju et al., 1981). Ragi plants affected with streak, yellowing, and stunting symptoms were noticed in a wider geographical area in the districts of Chitradurga, Mandya, Bangalore, Tumkur, and Hassan in the subsequent surveys during 1977–78 and 1978–79. The incidence ranged from 5 to 45%. Subsequent studies revealed this virus to be different from all those reported earlier on finger millet and was found to be a strain of maize streak virus.

11.13.3.2   SYMPTOMS

Symptoms appear on unfolding young leaves as pale specks or stripes of different size. The specks coalesce involving larger areas resulting in chlorotic bands running almost the entire length of the leaf parallel to the midrib. These bands are occasionally interrupted by dark green areas. The new emerging leaves of both the main shoot and the tillers show number of well defined chlorotic streaks having almost uniform width along the midrib throughout the length of the leaf lamina.

The infected plants in the field produce comparatively more number of tillers and bear yellowish sickly ears, often bearing few shriveled seeds. The plants infected very early in the crop growth stage die before they bloom.

11.13.3.3   CAUSAL ORGANISM

Eleusinestrain of maize streak virus.

11.13.3.4   DISEASE CYCLE

Leaf hopper *Cicadulina chinai* is able to transmit the disease.

11.13.3.5   MANAGEMENT

Details of management practices are not available

11.14   NEMATODE DISEASES

Details of management practices are not available

11.14.1   INTRODUCTION

Several nematodes viz; *Helicotylenchus, Trichodorus, Pratylenchus, Roty-lenchulus, Heterodera, Criconemoides, Macroposthonia, Meloidogyne* have been reported to parasitize finger millet (Krishnappa et al., 1992).

11.14.1.1   HETERODERA SPP.

During 1972, occurrence of a cyst nematode was recorded at Hebbal, Bangalore, India by Setty (1975) for the first time on Ragi. However, earlier, *Heterodera-marioni* was reported on this crop (Ayyar, 1933, 1934). Krishnaprasad et al. (1980) tested the reaction of many genotypes of Ragi as well as certain other crops. Maize, bajra, foxtail millet were found to be good hosts but, Sorghum, and *Echinocloa* were better.

11.14.1.1.1   Symptoms

The main symptoms are yellowing of leaves in patches and stunting of plants. The affected plants show unthrifty growth even under optimum conditions of moisture and nutrition, and can easily be pulled out. The cysts embedded or attached to the roots of the infected plants can be seen by naked eye.

11.14.1.1.2   Causal Organism

Heteroderagambiensis

The cysts are lemon shaped, measure 800–960 × 450–600 μm while the second stage larvae are 400–540 μm. Vulva distinctly protruded, cuticle thick exhibiting zigzag pattern with punctuations. The vulval cone is characterized by scattered bullae, enface view of the vulva ambitenestrate vulval slit 30–35 μ long and underbridge well developed.

11.14.1.2   ROTYLENCHULUS RENIFORMIS (LINFORD AND OLIVEIRA)

11.14.1.2.1   Introduction

Chandrasekaran (1964) in his survey observed finger millet to be susceptible to *R. reniformis*. Increased population of this nematode had positive correlations with the reduction in height of the plants, top weight, root weight, and yield. Rajagopal (1965) found the association of high population of *R. reniformis* with stunted grassy patches of finger millet.

According to Krishnappa et al. (2002), 4.8% of cropped area to Ragi is affected by *R. reniformis* in Karnataka and green manuring was highly effective in reducing the nematode population.

In addition to *Heterodera* and *Rotylenchulus*, Narayanaswamy, and Govindu (1966) reported the natural occurrence of *Helicotylenchus, Trichodorus, and Pratylenchus* species from several places in Karnataka.

Mohanthy and Das (1976) studied the physiology of parasitism of *Criconemoides oranatus*, the ring nematode in Ragi. In the nematode infested roots 9 amino acids were identified as against eight in healthy, the additional one being L-proline, which they considered as defense mechanism against the invading nematode.

11.14.1.2.2   Management

The nematodes are best controlled by soil amendments like poultry manure or neem cake or by applying granular insecticides *viz.*, Phorate 10G or Carbofuron 3G.

KEYWORDS

- **maize streak virus**
- **nematode diseases**
- **phenylalanine ammonia lyase**
- **polyphenol oxidase**
- **Ragi flour lactose yeast agar**
- **Ragi streak**

REFERENCES

AICSMIP, (2013). *Annual Report of the All India Coordinated Small Millets Improvement Project, Project Coordinating Unit.* ICAR, GKVK, Bangalore.

Anilkumar, T. B., Kumar, J., Shashidhar, V. R., Sudarshan, L., Ramanathan, A., Madhukeshwara, S. S., & Mantur, S. G., (2004). Finger millet blast (*Pyriculariagrisea* (Cke.) Sacc.) and its management, NATP (RNPS-4). *Project Coordination Cell* (p. 127). ICAR, GKVK, Bangalore.

Anilkumar, T. B., Mantur, S. G., & Madhukeshwara, S. S., (2003). *Diseases of Finger Millet* (p. 126). Project coordinating cell, All India coordinated small millets improvement project ICAR, GKVK, Bangalore.

Anonymous, (1954). *Mem. Dep. Agric. Madras. Bot. Ser., 30,* 117–194.

Anonymous, (1975). *Annual Report of the Virologist.* AICRP on Small Millets, UAS, Bangalore.

Aycock, R., (1966). *Stem Rot and Other Diseases Caused by Sclerotium rolfsii.* North Carolina Agril. Expt. Sta. Bull. No. 174.

Ayimut, K. M., & Mathew, M. A., (2008). Farmers' knowledge of crop diseases and control strategies in the regional state of Tigrai, northern Ethiopia: Implications for farmer–researcher collaboration in disease management. *Agriculture and Human Values, 25*(3), 433–452.

Ayyar, P. N. K., (1933). Some experiments on the root gall nematode *Heteroderama rioni* in South India. *Madras Agric. J., 21,* 97–107.

Ayyar, P. N. K., (1934). Further experiments on the root gall nematode *Heteroderama rioni* in South India. *Indian J. Agric. Sci., 3,* 1064–1071.

Basta, B. K., & Tamang, D. B., (1983). Preliminary report on the study of millet diseases in Nepal. In: *Maize and Finger Millet.* 10th Summer Workshop, held at Rampur, Chitwan.

Billimoria, K. N., & Hegde, R. K., (1971). A new bacterial disease of Ragi, *Eleusinecoracana* (Linn) Gaertn in Mysore state. *Curr. Sci., 40,* 611–612.

Butler, E. J., (1918). *Fungi and Diseases in Plants* (p. 547). Thaker Spinck and Co. Calcutta.

Chandrasekaran, N. M., (1964). Studies on the reniform nematode *Rotylenchulus reniformis* with special reference to its pathogenicity on castor (*Ricinus communis*) and Ragi (*Eleusinecoracana*). *M.Sc. (Agri.). Thesis.* University of Madras.

Channamma, K. A. L., Hiremath, P. C., & Viswanath, S., (1980). Efficacy of some fungicides in controlling foot rot of Ragi caused by *Sclerotium rolfsii. Curr. Res., 9,* 142, 143.

Channamma, K. A. L., Viswanath, S., & Mathur, S. G., (1996). New record of *Uromyces* sp. on Ragi from Karnataka. *Curr. Res., 25,* 97.

Coleman, L. C., (1920). The cultivation of Ragi in Mysore. *Bull. Dep. Agric. Mysore. Gen. Ser.,* 11–12.

Desai, S. G., Thirumalachar, M. J., & Patel, M. K., (1965). Bacterial blight disease of *Eleusinecaracana* Gaertn. *Indian Phytopath., 28,* 384–386.

Dubey, S. C., (1995). Banded blight of finger millet caused by *Thanetephorus cucumeris. Indian J. Mycol. Pl. Pathol., 25,* 315–316.

Dublish, P. K., & Singh, P. N., (1976). Phytopathogenic fungi of Meerut, some new records of India. *Curr. Sci., 45,* 168.

Ganapathi, T. K., (1971). Studies on the biology of smut disease of *Eleusinecoracana* (L.) Gaertn. (Finger millet or Ragi) caused by *Melanopsichium eleusinis* (Kulk.) Mundkur and Thirum. *M.Sc. (Agri.) Thesis.* UAS, Bangalore.

Govindu, H. C., (1982). Green revolution, its impact on plant disease with special reference to cereals and millets. *Indian Phytopath., 35,* 363–375.

Govindu, H. C., & Shivanandappa, N., (1967). Studies on an epiphytotic Ragi disease in Mysore state. *Mysore J. Agric. Sci. 1,* 142–148.

Govindu, H. C., Shivanandappa, N., & Renfro, B. L., (1966). Observations on diseases of *Eleusinecoracana* with reference to *Helminthosporium* disease. *Abstr. Internat. Symp. Pl. Path* (pp. 48, 49). New Delhi.

Grewal, J. S., & Pal, M., (1965). Seed microflora I. Seed borne fungi of Ragi (*Eleusinecoracana* Gaertn) their distribution and control. *Indian Phytopath., 18,* 33–37.

Jain, A. K., Gupta, J. C., & Yadav, H. S., (1994). Stability of resistance to foot rot in finger millet. *Bhartiya Krishi Anusandhan Patrika, 9,* 109–112.

Joshi, L. M., Raychaudhuri, S. P., Batra, S. K., Renfro, B. L., & Ghosh, A., (1966). Preliminary investigations on a serious disease of *Eleusinecoracana* in the states of Mysore and Andhra Pradesh. *Indian Phytopath., 19,* 324–325.

Kannaiayan, S., & Prasad, N. N., (1978). Reaction of certain crop plants to sheath blight of rice. *Indian Phytopath., 31,* 541.

Keshavamurthy, K. V., & Yaraguntaiah, R. C., (1977). *Virus Disease of Ragi in Symposium on Ragi (Eleusinecoracana)* (pp. 100–105). Post graduate college, UAS, Bangalore.

Krishnappa, K., Ravichandra, N. G., & Reddy, B. M. R., (1992). Nematode pests of millets, In: Bhatti, D. S., & Walia, R. K., (eds.), *Nematode Pests of Crops* (pp. 70–76). C.B.S. Publishers and Distributors, Delhi-32.

Krishnappa, K., Ravichandra, N. G., & Reddy, B. M. R., (2002). Role of green manuring in the management of nematodes. In: Shankar, M. A., Manjunath, A., Gajanana, G. N., Mariraju, H., & Lingappa, B. S., (eds.), *Potentials of Green Manuring in Rainfed Agriculture* (pp. 45–55).

Krishnaprasad, K. S., Krishnappa, K., Setty, K. G. H., Reddy, B. M. R., & Reddy, H. R., (1980). Susceptibility of some cereals to ragi cyst nematode, *Heteroderadelvi. Curr. Res., 9*(6), 114–115.

Kulkarni, G. S., (1922). The smut of nachani or Ragi (*Eleusinecoracana*). *Ann. Appl. Biol., 9,* 184–186.

Kumar, B., & Yadav, R., (2012). Influence of nitrogen fertilizer dose on blast disease of finger millet caused by *Pyriculariagrisea. Indian Phytopath., 65*(1), 52–55.

Kumar, B., (2011b). Management of blast disease of finger millet (*Eleusinecoracana*) in mid-western Himalayas. *Indian Phytopath., 64*(2), 154–158.

Kumar, B., & Kumar, J., (2011). Management of blast disease of finger millet (*Eleusinecoracana*) through fungicides, bioagents and varietal mixture. *Indian Phytopath., 64*(3), 272–274.

Kumar, B., & Prasad, D., (2010b). A new record on foot rot or wilt disease of finger millet (*Eleusinecoracana*) caused by *Sclerotium rolfsii* from mid hills of Uttarakhand. *J. Mycol. Pl. Pathol., 40*(3), 334–336.

Kumar, B., Kumar, J., & Srinivas, P., (2007b). Occurrence of downy mildew or green ear disease of finger millet in mid hills of Uttarakhand. *J. Mycol. Pl. Pathol., 37*(3), 532–533.

Kumar, J., (2005). *All India Coordinated Small Millets Improvement Project (QRT Report, 2000–2004, Ranichauri Centre)* (p. 151). G.B. Pant University of Agriculture and Technology, Ranichauri, Tehri Garhwal, Uttarakhand.

Kumar, J., Kumar, B., & Yadav, V. K., (2007a). *Small Millets Research at G.B. Pant University,* (p. 58). G.B. Pant University of Agriculture and Technology, Hill Campus, Ranichauri, Uttarakhand.

Lulu, D., & Girija, V. K., (1989). Sheath blight of Ragi. *Curr. Sci., 58,* 681–682.

Madhukeshwara, S. S., Mantur, S. G., Krishnamurthy, Y. L., & Babu, H. N. R., (2004). Control of blast (*Pyriculariagrisea* (Cke.) Sacc) of finger millet (*Eleusine coracana* (L.) Gaertn) by fungicides. *Environment and Ecology, 22*(4), 824–826.

Madhukeshwara, S. S., Mantur, S. G., Ramanathan, A., Kumar, J., Shashidhar, V. R., Jagadish, P. S., Seenappa, K., & Anilkumar, T. B., (2005). On-farm adaptive management of the blast of finger millet. *International Sorghum and Millets Newsletter, 46,* 111–114.

Mantur, S. G., (1994). Studies on the management of the smut disease of finger millet caused by *Melanopsichium eleusinis* (Kulk) Mundk. and Thirum. *M.Sc. (Agri.) Thesis* (p. 55). UAS, Bangalore.

Mantur, S. G., Viswanath, S., Channama, K. A. L., & Somasekar, Y. M., (1995). Evaluation of some fungicides against finger millet smut *Melanopsichium eleusinis. Pl. Prot. Bull., 47,* 12–14.

Maramorosch, K., Govindu, H. C., & Kondo, F., (1977). Rhabdovirus particles associated with mosaic disease of naturally infected *Eleusinecoracana* (finger millet) in Karnataka state (Mysore) South India. *Plant Dis. Reptr., 61,* 1029–1031.

Mariappan, V., Natarjan, C., & Kandaswamy, T. K., (1973). Ragi streak disease in Tamil Nadu. *Madras Agric. J., 60,* 451–453.

McRae, W., (1920). *Detailed Administration Report of the Government Mycologist for the Year* 1919–1920.

McRae, W., (1924). Economic botany, Part III - mycology. *Annul. Rept. Bot. Sci. Advice, India - 1922–1923,* pp. 31–35.

Misra, A. P., (1979). Variability, physiologic specialization and genetics of pathogenicity in graminicolous *Helminthosporia* affecting cereal crops. *Indian Phytopath., 32,* 1–22.

Mitra, M., (1931). *Report of the Imperial Mycologist* (pp. 58–71). Scient. Reports Agri. Res. Int. Pusa.

Mohanthy, K. C., & Das, S. N., (1976). Free amino acids in the roots of finger millet plants infected with ring nematodes. *Indian Phytopath., 29,* 434–436.

Mundkur, B. B., (1939). A contribution towards knowledge of Indian Ustilaginales. *Trans. Br. Mycol. Soc., 23,* 105.

Mundkur, B. B., & Thirumalachar, M. J., (1946). Revisions and addition to Indian fungi. *Mycol. Paper, 16,* CMI, Kew.

Munjal, R. L., Lall, G., & Chona, B., (1961). Some Cercospora species from India VI. *Indian Phytopath., 14,* 179–190.

Muyanga, S., (1995). Production and research review of small cereals in Zambia. In: Danial, D. L., (ed.), *Breeding for Disease Resistance with Emphasis on Durability.* Porc. Reg. Workshop for Eastern, Central and Southern Africa. Njaro, Kenya, Wageniya, Netherland.

Nagaraja, A., Kumar, B., Jain, A. K., & Sabalpara, A. N., (2016). Emerging diseases: Need for focused research in small millets. *J. Mycopathol. Res., 54*(1), 1–9.

Nagaraja, A., Kumar, B., Raguchander, T., Hota, A. K., Patro, T. S. S. K., Devaraje, G. P., Savita, E., & Gowda, M. V. C., (2012). Impact of disease management practices on finger millet blast and grain yield. *Indian Phytopath., 65*(4), 356–359.

Nagaraja, A., & Anjaneya, R. B., (2009). Foot rot of finger millet- an increasing disease problem in Karnataka. *Crop Res., 38*(1–3), 224, 225.

Nagaraja, A., & Anjaneya, R. B., (2010a). Banded blight: A new record on finger millet in Karnataka. *J. Mycopathol. Res., 48*(1), 169, 170.

Nagaraja, A., Hanumanthe, G. B., & Krishne, G. K. T., (2008). Effect of long-term use of organic and inorganic fertilizers on the incidence of finger millet blast and brown spot. *Environment and Ecology, 26*(1), 236–237.

Nagaraja, A., Jagadish, P. S., Ashok, E. G., & Krishne, G. K. T., (2007a). Avoidance of finger millet blast by ideal sowing time and assessment of varietal performance under rainfed production situations in Karnataka. *J. Mycopathol. Res., 45*(2), 237–240.

Nagaraja, A., Kumar, J., Jain, A. K., Narasimhudu, Y., Raghuchander, T., Kumar, B., & Hanumanthe, G. B., (2007b). *Compendium of Small Millets Diseases* (p. 80). Project Coordination Cell, All India Coordinated Small Millets Improvement Project, UAS, GKVK Campus, Bangalore.

Nagaraju, A., & Viswanath, S., (1981). Studies on the relationship of Ragi streak virus and its vector *Cicadulinachinai. Indian Phytopath., 34,* 458–460.

Nagaraju, A., Viswanath, S., Lucy, C. K. A., & Reddy, H. R., (1981). Effect of streak virus on growth and yield of Ragi. *Indian Phytopath., 34,* 256–258.

Nagaraju, A., Viswanath, S., Reddy, H. R., & Lucy, C. K. A., (1982). Ragi streak a leaf hopper transmitted virus disease in Karnataka. *Mysore J. Agric. Sci., 16,* 301–305.

Narain, A., (1972). Foot rot disease of Ragi in Orissa. *Curr. Sci., 41,* 823–824.

Narain, A., Barua, S. C., & Rajan, R., (1975). A quick laboratory method for testing the reaction of Ragi to *Drechsleranodulosa*. *Indian Phytopath.*, *29*, 330, 331.

Narasimhan, M. J., (1933). *Adm. Rep. Mycol. Mysore Dept. Agric.*, 1931, 1932.

Narasimhan, M. J., (1934). Report of the work done in mycology section. *Dept. Agri., Mysore*, 1932, 1933.

Narasimhan, V., (1980). Studies on two virus diseases of finger millet (*Eleusinecoracana* Gaertn.). *PhD Thesis*. Tamil Nadu Agricultural University, Coimbatore, India.

Narayanaswamy, B. C., & Govindu, H. C., (1966). A preliminary note on the plant parasitic nematodes of the Mysore state. *Indian Phytopath.*, *19*, 239–240.

Pall, B. S., (1987). Epidemiological studies on neck blast of finger millet (*Eleusinecoracana* (L.) Gaertn). *Narendradev J. Agric. Res.*, *2*, 187–189.

Pall, B. S., (1988). Effect of seed borne inoculum of *Pyriculariasetariae* Nisikado on the finger millet blast. *Agric. Sci. Digest*, *8*, 225, 226.

Pall, B. S., & Sharma, Y. K., (1976). Change in the nutritive value of Ragi due to Helminthosporiose infection. *Food Farming Agri.*, *10*, 86.

Pall, B. S., Jain, A. C., & Singh, S. P., (1980). *Diseases of Lesser Millet* (p. 69). JNKVV, Jabalpur (MP), India.

Patra, A. K., (1996). Ragi for Orissa. *Intensive Agri.*, *34*, 6, 7.

Patro, T. S. S. K., Anuradha, A., Madhri, J., Suma, Y., & Sowjanya, A., (2013). Management of banded blight of finger millet incited by *Rhizoctonia solani* (Kuhn.) *Plant Dis. Res.*, *28*(2), 235, 236.

Patro, T. S. S. K., Rani, C.H., & Kumar, G. V., (2008). *Pseudomonas fluorescens*, a potential bioagent for the management of blast in Eleusinecoracana. *J. Mycol. Pl. Pathol.*, *38*(2), 298–300.

Paul, K. S. M., Rayachaudhuri, S. P., & Sundaram, N. V., (1973). Further studies on Ragi mosaic in Delhi. *Indian Phytopath.*, *26*, 554–559.

Pradganang, P. M., (1994). Quantification of the relationship between cercospora leaf spot disease (*Cercosporaeleusine*) and yield loss of finger millet. *Tech. Paper-Lumle Agricultural Centre No. 94/3*, *12*, 7.

Pradganang, P. M., & Abington, J. B., (1993). Cercospora leaf spot disease (*Cercosporaeleusine*) FO finger millet in Nepal. In: Riley, K. W., Gupta, S. C., Seetharam, A., & Mushonga, J. N., (eds.), *Advances in Small Millets*. Oxford I.B.H. Pub. Co., New Delhi.

Rachie, K. O., & Peters, L. V., (1977). *The Eleusines: A Review of World Literature* (p. 179). ICRISAT, Hyderabad.

Raghavendra, S., & Safeeulla, K. M., (1971). Sporulation & infection of zoospores of *Scleropththora macrospora* on *Eleusine coracana*. In: *Abstr. Epidemiology conf. Netherlands* (Vol. 3, p. 244).

Raghavendra, S., & Safeeulla, K. M., (1973). Investigation on the Ragi downy mildew. *J. Mysore Univ.*, *26*, 138–155.

Rajagopal, B. K., (1965). Further studies on the damage caused by the reniform nematode Rotylenchulus Reniformis Linford, and Oliveira 1940 to Ragi and castor. *M.Sc. (Agri.) Thesis*. University of Madras.

Rangaswami, G., Prasad, N. N., & Eswaran, K. S. S., (1961). Two new bacterial diseases of sorghum. *Andhra Agric. J.*, *8*(6), 269–272.

Rao, A. N. S., (1990). Estimates of losses in finger millet (*Eleusinecoracana*) due to blast disease (*Pyriculariagrisea*). *Journal of Agricultural Sciences*, *24*, 57–60.

Rao, G. P., Singh, S. P., & Singh, M., (1992). Two new alternative hosts of *Curvularia pallescens*, the leaf spot causing fungus of sugarcane. *Trop. Pest Management*, *38*, 218.

Rao, V. G., & Manivarghese, K. I., (1981). Fungi imperfecti from Kerala, South India III, *Biovigyanam, 6,* 167–172.

Rath, G. C., & Mishra, D., (1975). Nature of losses due to neck blast infection in Ragi. *Sci. Cult., 41,* 322.

Roy, A. K., (1989). Further record of plant diseases from Karbi Anglong district, *Assam. J. Res., 10,* 88–91.

Safeeulla, K. M., (1955). Comparative morphological and cytological studies in some species of the genera Albugo, Sclerophthora and Sclerospora. *PhD Thesis* (p. 179). University of Mysore, India.

Safeeulla, K. M., (1976). *Biology and Control of the Downy Mildews of Pearl Millet Sorghum and Finger Millet* (p. 304). Mysore University, Mysore.

Sakamma, S., Umesh, K. B., Girish, M. R., Ravi, S. C., Satishkumar, M., & Veerabhadrappa, B., (2018). Finger millet (*Eleusinecoracana* L. Gaertn.) production system: Status, potential, constraints, and implications for improving small farmer's welfare. *Journal of Agricultural Science, 10*(1), 162–179.

Saveetha, K., Sankaralingam, A., Pant, R., & Ramanathan, A., (2007). Etiology and transmission of mottle streak disease of finger millet (*Eleusinecoracana* Gaertn.). *Arch. Phytopathol. Plant Protect, 40*(1), 53–60.

Saveetha, K., Sankaralingam, A., Ramanathan, A., & Pant, R., (2006). Symptoms, epidemiology, and management of finger millet mottle streak disease. *Arch. Phytopathol. Plant Protect, 39*(6), 409–419.

Semeniuk, G., & Maukin, C. J., (1964). Occurrence and development of *Sclerophthora macrospora* on cereals and grasses in South Dakota. *Phytopathology, 53,* 887.

Setty, K. G. H., (1975). Studies on Ragi cyst nematode *Heterodera* sp. *Report of UNDP/ICAR Research Project.* UAS, Hebbal, Bangalore.

Shaw, F. J. F., (1921). *Report of the Imperial Mycologist Scient* (pp. 34–40). Reports Agric. Res. Inst. Pusa.

Subbayya, J., & Raychaudhuri, S. P., (1970). A note on a mosaic disease of Ragi (*Eleusinecoracana*) in Mysore, India. *Indian Phytopath., 23,* 144–148.

Subramanian, C. V., (1953). Fungi imperfecti from Madras-V. *Proc. Indian Acad. Sci., 38*(B), 27–29.

Sydow, H., & Butler, E. J., (1906). Fungi *Indiae Orientalis*, Pars I. *Ann. Mycol., 4,* 424–445.

Thirumalachar, M. J., & Mundkur, B. B., (1947). Morphology and mode of transmission of Ragi smut. *Phytopathology, 37,* 481–486.

Thomas, K. M., (1940). *Detailed Administrative Report of the Government Mycologist* (p. 18). Madras for the year 1939–1940.

Venkatarayan, S. V., (1947). Diseases of Ragi (*Eleusinecoracana* L.). *Mysore Agric. J., 24,* 50–57.

Vidhyasekaran, P., (1971). Saprophytic survival of *Helminthosporium nodolosum* and *H. tetramera* in soil. *Indian Phytopath., 24,* 347–353.

Viswanath, S., & Seetharam, A., (1989). Diseases of small millets and their management in India. In: Seetharam, A., Riley, K. W., & Harinarayana, G., (eds.), *Small Millets in Global Agriculture* (pp. 237–253). Oxford and I.B.H. Publishing Co. Pvt. Ltd., New Delhi, India.

Waghunde, R. R., Sabalpara, A. N., Chaudhary, P. P., & Solanky, K. U., (2011). Foot rot of finger millet in Gujarat-a new record. *J. Pl. Dis. Sci., 6*(1), 70.

Wallace, G. B., & Wallace, M., (1947). Second supplement to the revised list of plant diseases in Tanganyika territory. *E. Afric. J. B., 61,* 4.

Wanyera, N., (2008). *New Millet Varieties Developed in Uganda.* www.unffe.org.

CHAPTER 12

Foxtail Millet or Italian Millet or Halian Millet or Moha Millet or Kakun (*Staria italika* L.) Diseases and Their Management Strategies

BIJENDRA KUMAR[1] and J. N. SRIVASTAVA[2]

[1]*College of Agriculture, Department of Plant Pathology,*
G.B. Pant University of Agriculture and Technology, Pantnagar–263145,
Udham Singh Nagar, Uttarakhand, India

[2]*Department of Plant Pathology, Bihar Agricultural University,*
Sabour–813210, Bhagalpur, Bihar, India

Foxtail millet or Italian Millet or Halian millet or Moha Millet

(*Staria italika* L.)

Vernacular names: *Kakun, Kangni, Kang, Rala, Korra, Navane, Tenai, Kaon, Kanghu*

Foxtail millet (*setariaitalika*) is one of the oldest cultivated small millest. The crop is generally cultivated in semi-arid regions; require less water, however, it does not recover well from drought conditions due to its shallow root system. It is an important staple food for the millions of population in Asia and southern Europe (Reddy et al., 2006). Foxtail millet is considered to be the native of China; it is one of the oldest grown crops cultivated mainly for hay, pasture, and grains. In terms of total world production of millets, foxtail millet ranks second and has an important place in the world agriculture providing about 6 million tons of food to large population, mainly on marginal or poor soils in southern Europe and in tropical, subtropical, and temperate Asia. It grows up to an altitude 2000 m above mean sea level.

It is fairly tolerant to drought and can escape drought because of it maturs early. However, it cannot tolerate water logging. Because of its quick growth, it is cultivated as a short-term catch crop. It adjusts well in a wide range of soils, temperatures, and elevations. Its grain is used for human consumption and as feed for poultry and cage birds. It is rich in phytates and poly phenols that have important functional properties.

In India, foxtail millet's major area is mostly in Karnataka, Andhra Pradesh, and Tamil Nadu and in North Eastern states.

12.1   BLAST DISEASE

12.1.1   INTRODUCTION

The disease was reported in 1919 from Tamil Nadu, though Nishikado reported blast disease in 1917 from Japan. In India, it has been reported from Maharashtra, Andhra Pradesh (Sinha and Upadhyay, 1997), and Uttarakhand (Kumar, 2013).

In epidemic conditions, the disease can cause up to 30–40% loss in the grain yield (Nagaraja et al., 2007b). *Setaria* isolate easily infects other plants like; finger millet, pearl millet, wheat, and *Dactyloctaenium aegyptium* (Viswanath and Seethram, 1989).

12.1.2   SYMPTOMS

Plants up to the age of 40 days are highly susceptible to blast. Initial symptoms on leaves appear as small pinhead sized water-soaked yellowish dots that within 2–3 days turns circular or oval shaped with grayish at the center usually surrounded by dark brown margins. The spots measure about 2–5 mm in diameter. Several nearby spots coalesce to cover large area and make the leaves to dry up. If node is affected, it becomes black and breaks at the joints. As the disease starts from lower to upper leaves, the lower leaves should be examined for the appearance of disease symptoms. However, like blast of finger millet, neck blast symptoms of the disease do not appear (Palaniswami et al., 1970) (Figure 12.1).

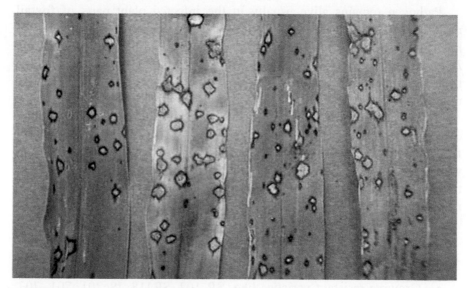

FIGURE 12.1   Blast symptoms on leaf of foxtail millets.

12.1.3   CAUSAL ORGANISM

Pyriculariasetariae Y. Nisik.
[*Pyriculariaoryzae* Cavara]

Four different physiological races of fungus have been reported on the basis of cultural, morphological, and physiological characters and pathogenicity of the fungus (Kulkarni and Patel, 1956). However, Yan et al. (1985) in China described 7 races of the fungus. Gaikwad and D' Souza (1987) on the basis of conidia size in culture medium and on host concluded that *Pyricularia setariae* that infects foxtail millet is different from the isolates that infect rice, pearl millet, and Ragi.

12.1.4   DISEASE CYCLE

According to Palaniswami et al. (1970) it is externally seed borne, although, it's seed-borne nature was first observed by Goel et al. (1967). Seeds inoculated with the solution of 2–4 million spores per ml produced 86–97% diseased plants under favorable environmental conditions. Fungus lost its germinability and infectivity when *Pyricularia setariae* infested seeds were stored for 75 days (Palaniswami et al., 1970). Therefore, the chances of seed transmission of the disease in nature from one year to next are questionable.

Pathogen remains alive usually in the infected plant parts of foxtail millet and to some extent in soil. Survival of pathogen through collateral hosts is possible. *Pyricularia setariae* isolate infects pearl millet, Ragi, and *Dactylotaenium aegyptium* but not rice. Primary infection occurs through these sources under favorable environment. Secondary infection takes place rapidly through air.

The factors affecting blast epidemics are susceptible variety, amount of inoculum available to initiate the disease, application of high nitrogen fertilizer, cloudy, and drizzling weather or dew resulting in continuous leaf wetness for more than 10 h, night temperature between 15–24°C and relative humidity above 90%. Neutral pH (7.0) and 28–30°C temperature are reported best for the growth of fungus (Ramakrishnan, 1971).

12.1.5   MANAGEMENT

- Growing resistant varieties like, SR 102, SR118, ISc201, 701, 703, 709, 710, JNSc 33, 56, RS 179 and ST 5307 helps in minimizing the yield loss.
- Spraying Carbendazim 50 WP @ 1 g/l or Ediphenphos 50 EC @ 1 m¹/l or Carbendazim + Mancozeb @ 1 g/l of water at the first appearance of symptoms should be given to check the further disease development.
- If top dressing of nitrogen is to be done it should be taken up after the fungicidal spray.

12.2   DOWNY MILDEW OR GREEN EAR

12.2.1   INTRODUCTION

Downy mildew or green ear of *Setaria* has been reported from Japan, China, Russia, Manchuria, the south-eastern countries of Europe, America, and India (Ramakrishnan, 1971). In certain years, the disease causes loss up to 50%. Takasugi and Akaisahi (1933) reported that the losses go up to 20% in Manchuria. In India, it is known to occur in Tamil Nadu, Bihar, Maharashtra, Andhra Pradesh, Karnataka, and Kashmir (Sinha and Upadhyay, 1997). According to Monondo (1957), foxtail millet is infected by this downy mildew or green ear disease every year in Italy and sometimes the yield losses can go up to 30%. It is a major pathogen of pearl millet in India that led to devastation of many bajra hybrids.

12.2.2 SYMPTOMS

The primary infection occurs at seedling as systemic and secondary infection on older plants as local infection. Primary infection results in chlorosis of the leaves which turn whitish. The terminal spindle fails to unfold, becomes chlorotic and later converts brown, and is shredded. Sporangiophores and sporangia develop as whitish bloom, on the surface of the leaves infected with the pathogen under humid conditions. The affected plant rarely comes to flower. If the infection is mild, the plants may produce ears, but the inflorescence is proliferated into green leafy structures, hence the name "green ear." In a spikelet all, the parts are transformed into green leafy structures of variable size. Sometimes ears may be only partially affected; the remaining may be normal and produce grains. Secondary infection causes chlorotic lesions on the younger leaves on which downy growth of the fungus may be observed under humid conditions (Figure 12.2a–c).

FIGURE 12.2   Various symptoms of downy mildew/green ear disease of foxtail millet: (a) Green ear of foxtail millets, (b) Downy growth on the leaves, and (c) Leaf shredding at later stage.

12.2.3   CAUSAL ORGANISM

Sclerospora graminicola (Sacc.) J. Schröt.

The fungus produces both the conidial and oospore stages on this host. The conidial stage is common on the infected portions and the oospores develop in the shredded tissues and other parts with advanced stages of infection.

12.2.4   DISEASE CYCLE

S. graminicola is an obligate parasite. The pathogen survives either on seeds or in soil through oospores. The initial infection is mainly through soil borne oospores that germinate at the time of seed germination and infect the seedlings systemically and transform the heads into green leafy structures. Secondary infection occurs through conidia disseminated either by air or water. Since, the conidia are short-lived the favorable environment is essential for their infection. Oospores survive in soil for long time and serve as primary source of inoculums. Oospores remain viable up to 8 years (Takasugi, 1935).

The fungus infects *Setaria viridis, S. Verticillata, Zea mays, Panicum miliaceum* and teosinite. The same fungus on *Pennisetum typhoides* fails to infect *S. Italica* and *vice versa*, indicating high physiologic specialization. In Japan, four physiologic races of this species have been identified (Rangaswami and Mahadevan, 1999).

The infection intensity is affected by the moisture content and temperature of the soil. Soil temperature around 20–21°C is optimum for infection, however, 12–13°C is the minimum temperature and 30°C is the maximum temperature. Time of sowing also influence the intensity of infection. In the early sown crop, the intensity will be high than in late sown crop. High soil moisture content and high relative humidity favor rapid disease development.

12.2.5   MANAGEMENT

- Growing resistant varieties, collection, and destruction of infected plant debris are ideal.
- Treating seeds with Ridomil MZ 72WP @ 3 g Kg^{-1} protects the seed from seed borne infection and from soil borne inoculum.

12.3   RUST

12.3.1   INTRODUCTION

The disease was first reported from Japan by Yoshino and is very common on foxtail millet. It has been reported from Asia, Africa, and Europe. In India, it is known to occur in Madhya Pradesh, Maharashtra, Tamil Nadu, Andhra Pradesh, Karnataka, Uttar Pradesh, and Bihar.

In some years, it appeared in epiphytotic forms and caused severe losses in grain yield. The disease occurred in a very severely in Andhra Pradesh and Karnataka states, in 1944.

12.3.2   SYMPTOMS

The infection occurs at all the stages of crop growth, but it is more serious, if infection occurs before flowering. The disease characterized by the formation of numerous minute, brown uredopustules on both the surfaces of the leaf. Pustules are usually arranged in linear rows; however, in severe cases they cover almost the entire leaf. Similar pustules can also occur on other plant parts such as leaf sheath, culms, and stem. In advanced cases of disease premature drying of leaves and poor grain filling has been observed. The teliotopustules, are larger than the uredia, and formed on the leaf, leaf sheath as well as on stem (Figure 12.3).

FIGURE 12.3   Foxtail millet leaves affected with rust pustules.

12.3.3   CAUSAL ORGANISM

Uromycessetariae-italicae Yoshino

The fungus has both uredial and telial stages. Uredospores are pedicillate, oval or globose, echinulate, and yellowish-brown having 3–4 germ pores. The teliospores are pedicellate, single celled, globose, oblong, yellowish brown with smooth and thick walls. Walls are comparatively thicker at the apex than at the base.

12.3.4   DISEASE CYCLE

The urediospores infect the host and produce uredia in 7–10 days. If the crop is grown throughout the year, as in many parts of India, the pathogen can easily perpetuate on the collateral hosts, in its uredial stage. The symptoms usually appear within 20–25 days after sowing subsequently the severity increases as the plants are aged. The telial stage occurs at the time of crop maturity.

Studies conducted in indicate that there are several races of this rust on *S. Italica* and other species. *Eriochloa procera* is said to be a common collateral host (Rangaswami and Mahadevan, 1999).

High relative humidity and low temperature favors the rapid disease development. The intensity of the disease will be high during December and January months, and cause extensive reduction in grain yield.

12.3.5   MANAGEMENT

- Growing resistant varieties and destruction of collateral hosts.
- Spraying of Mancozeb @ 2.5 g^{-1} of water, when first symptoms of the disease noticed in the field are some adopted measures.
- Since, foxtail millet is of minor importance, not much attention is paid to manage the disease.

12.4   SMUT

12.4.1   INTRODUCTION

The disease is common in China, Europe, Manchuria, and India, causes heavy losses to the crop. In Romania, the damage due to the disease is very high. The loss in grain yield due to smut varies from 8–50% in China, and up to 50% in Manchuria. Sundararaman in 1921 recorded 75% infection in grains of foxtail millet. Inhalation of spores during threshing may cause Asthma in laborers (Fischer, 1953). In India, it has been observed in Andhra Pradesh, Tamil Nadu, Karnataka, Maharashtra (Ramakrishnan, 1971) and Uttarakhand (Kumar, 2011a; Nagaraja et al., 2016).

12.4.2 SYMPTOMS

Usually most of the grains in an ear are affected; sometimes terminal portion of the spike may remain healthy. The pale grayish smut sori, measuring 2–4 mm in diameter are seen in the flowers and basal parts of palea. The sori rupture at maturity producing dark powdery mass of teliospores of the fungus.

12.4.3 CAUSAL ORGANISM

Ustilagocrameri Koern.

The teliospores are dark brown may be angular or round in shape have smooth wall and measure 7–10 μm in diameter. The spores readily germinate in water, producing promycelia, but rarely sporidia. The fungus grows well on agar media, producing terminal and intercalalry chlamydospores. At least six physiologic races have been differentiated in other countries (Figure 12.4).

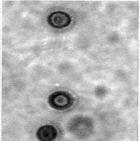

FIGURE 12.4   Grain smut of foxtail millet: the affected panicle, microscopic view of smut balls and individual sclerotium (from left to right).

12.4.4   DISEASE CYCLE

Although it is mainly seed born in nature, however, to some extent infection through soil borne inoculum has also been reported in dry areas (Ramakrishnan, 1971). The spores germinate during seed germination and infect the mesocotyl region and enter into the host. The invading hyphae are systemic mainly concentrated towards the apical portions and during flowering replace the ovaries producing septate hyphae, which are converted into chlamydospores. Spores stick on to the surface of the seed at the time of harvesting or threshing and reach the soil at the time of sowing. The pathogen can also infect a related grass, *Setariaviridis* (Rangaswami and Mahadevan, 1999).

High relative humidity and low temperature favors the rapid disease development. Although, it is externally seed borne however, soil borne infection has also been reported.

12.4.5   MANAGEMENT

- Growing resistant varieties, removal of infected ears and treating seed with Carbendazim @ 2 g/kg^{-1} of seed can help in managing the disease.

12.5   LEAF SPOT OR BLOTCH

12.5.1   INTRODUCTION

The disease was first recorded in Japan and Farmosa in the year 1906 (Ramakrishnan, 1971). Later, the disease appeared in severe form in New Jersey (Haenseler, 1941). During favorable weather, the disease causes considerable loss in grain yield. Cool weather is favorable for the incidence of disease.

12.5.2   SYMPTOMS

Leaf spots are brown in color, which enlarge in their size. Several nearby spots coalesce and cover the entire leaf blade. Finally, the leaves dry up. Lesions also appear as blotches. Rotting of the secondary roots has also been observed (Sprauge, 1950).

12.5.3 CAUSAL ORGANISM

Cochliobolus setariae (S. Ito and Kurib.) Drechsler ex Dastur,
[Syn. *Bipolarissetariae* (Sawada) Shoemaker; *Drechslera setariae* (Sawada) Subram. and B. L. Jain; *Helminthosporium setariae* Sawada; *Helminthosporium setariae* Lind; *Ophiobolus setariae* S. Ito and Kurib.]

The conidiophores are simple, erect, cylindrical brown, slightly swollen at the base and geniculate at the apex. They measure 72–199 μm × 5.6–9 μm. The conidia are acrogenous ellipsoid to obclavate, fusoid, straight or slightly curved, pale to moderately dark brown and thin walled. The spores measure 39–120 μm × 10–18 μm. Four to ten septa are present and there is no connection at the septum. Other aspects of the disease and pathogen including management have not been studied thoroughly.

12.5.4 MANAGEMENT

Details of management practices are not available

12.6 LEAF AND SHEATH BROWN SPOT

12.6.1 INTRODUCTION

The disease is caused by *Bipolaris australiensis* and was reported for the first time by Mirzaee et al. (2010) from southern Khorasan, Iran in 2008.

12.6.2 SYMPTOMS

The disease manifest as oblong to irregular spots of varying size, (3–20 mm in diameter). The spots are dark-brown with a lighter border and an indistinct margin. In some cases causes rupturing of the leaf veins. Severe lodging has also been observed on severely infected plants (Mirzaee et al., 2010).

12.6.3 CAUSAL ORGANISM

Bipolarisaustraliensis (M. B. Ellis) Tsuda and Ueyama

[Syn. *Helminthosporium australiense* Bugnic.; *Drechslera australiensis* (Bugnic.) Subram. and B. L. Jain; *Drechslera australiensis* M. B. Ellis]

According to Mirzaee et al. (2010), the morphological characteristics of fungal colonies emerging from the host tissue are typical of *Bipolaris* spp. On PDA, the colonies are initially white, but after 3 days turn grey to blackish brown, effuse, velvety, and black from the bottom side of the Petri plates. Conidiophores are simple or branched, septate, solitary, geniculate with sympodial growth, smooth, reddish brown, 95–205 mm long (average 142 mm), 3–7 mm thick (average 5 mm). Conidia are pale brown to reddish brown, straight, ellipsoidal, or oblong, rounded at both the ends, mostly three pseudoseptate and measure 7.5–10.0 × 15.0–27.5 mm. A flattened hilum is seen on the basal cell of the conidia. Other information are not available.

12.6.4 MANAGEMENT

Details of management practices are not available

12.7   UDABATTA DISEASE

12.7.1 INTRODUCTION

Udbatta disease has been reported on foxtail and kodo millets. Sometimes, it has been observed on little as well as prosomillets. Infected panicles are converted into a compact agarbatti like structure; hence, it is named "Udbatta."

The disease has been reported to cause significant yield losses in areas where it is endemic, but generally its occurrence is sporadic and it is of minor importance.

12.7.2 SYMPTOMS

The earheads affected by *Ephelis* sp. become somewhat mummified with partially formed buds which later become darker in color and more stromatic as conidial acervuli develop on the surface. These conidial acervuli appear gelatinous when wet and produce a saucer-shaped fructing body bearing conidiophores in palisade layer. The diseased plants are usually stunted and sometime prior to panicle emergence, the white mycelium and conidia form narrow stripes on the flag leaves along the veins. The sheath and flag leaf of diseased tillers are sometimes slightly distorted and the upper leaves (including the flag leaf) may appear silvery. First symptoms of the disease appear at the time of panicle emergence. While still within the sheaths, earheads become matted together by the mycelium of the fungus. The earheads emerge as small, single, cylindrical rods enveloped with white mycelium. Later, they become hard and sclerotium-like, bearing several black dots. In place of a normal inflorescence, an erect, grayish-white axis emerges from the leaf sheath.

12.7.3 CAUSAL ORGANISM

Ephelisoryzae Syd.
[Teleomorph: *Balansiaoryzae* (Syd.) Naras. and Thirum.].

12.7.4 MANAGEMENT

No work has been done as it is sporadic and minor disease.

12.8 BACTERIAL BLIGHT OR SPOT

12.8.1 INTRODUCTION

The bacterial blight or spot disease has been reported from the USA.

12.8.2 SYMPTOMS

The disease manifests as small, grayish-green spots with brown margin or light to dark brown spots on the leaf blade and leaf sheath. Sometimes, the symptoms also appear in the form of streaks of irregular sizes (Ramakrishnan, 1971).

12.8.3 CAUSAL ORGANISM

Pseudomonas alboprecepitans Rosen.

The bacterium is rod measuring 0.6–1.8 μm, occur singly or in pairs, no spores are formed, capsulated, motile with one polar flagellum, Gram negative and aerobic, optimum temperature 30–35°C with maximum and minimum at 0 and 40°C, respectively. Thermal death point (TDP) is 41–43°C.

12.8.4   MANAGEMENT

Details of management practices are not available

12.9   BACTERIAL BROWN STRIPE

12.9.1   INTRODUCTION

The disease is known to occur in Japan and Formosa.

12.9.2   SYMPTOMS

The disease is seen as long, narrow, dark brown streaks on the leaf blade. Central shoot is usually involved in top rot accompanied by foetid smell.

12.9.3   CAUSAL ORGANISM

Acidovorax citrulli
[Syn. *A. avenae subsp. citrulli; Pseudomonas setariae* (Okabe) Savulescu]

The bacterium is rod shaped with rounded ends, occurs singly or in pairs or as short chains, measures 1.8–4.4 μm × 0.4–0.8 μm, motile with one or two flagella, Gram negative and aerobic, optimum temperature 31–34°C with maximum and minimum at 5 and 42°C, respectively. TDP is 55–56°C. Other details are not available.

Other details on bacterial diseases like; disease cycle, epidemiology, management etc. are not worked out as these are only of minor importance.

12.9.4   MANAGEMENT

Details of management practices are not available

12.10   MINOR DISEASES

In addition to the above-listed diseases, leaf spot (*Cochliobolus setariae*), bacterial blight, bacterial brown stripe, viral streak and many fungi associated with seeds in storage, especially *Aspergillus flavus, A. terreus, Penicillium citrinum, Alternaria alternata, Curvularialunata, C. maculans, Fusarium* sp. *Rhizopus* sp. *and Mucor* sp. (Kavitha and Vijayalakshmi, 2007) have been observed on this crop, but are of little importance. Recently sheath blight caused by *R. solani* has also been reported.

KEYWORDS

- **bacterial blight**
- **bacterial brown stripe**
- **blotch**
- **sheath brown spot**
- **smut**
- **udabatta disease**

REFERENCES

Fischer, G. W., (1953). *Manual of the North American smut Fungi*. Ronald Press Co., New York.

Gaikwad, A. P., & D' Souza, T. F., (1987). A comparative study on *Pyricularia* spp. *J. Maharashtra Agril. Univ.*, *12*, 134, 135.

Goel, L. B., Mathur, S. B., & Joshi, L. M., (1967). Seed borne infection of *Pyriculariasetariae* in *Setariaitalica*. *Plant Dis. Reptr.*, *1*, 138.

Haenseler, C. H., (1941). Helminthosporium leaf spot on millet in New Jersey. *Plant Dis. Reptr.*, *25*, 486.

Kavitha, A., & Vijayalakshmi, M., (2007). Phytotoxic effect of seed mycoflora associated with the genotypes of foxtail millet. *Seed Research*, *35*(2), 268–271.

Kulkarni, S., & Patel, M. K., (1956). Study of the effect of nutrition and temperature on the size of spores in *Pyriculriasetariae* Nishikado. *Indian Phytopath.*, *9*, 31–38.

Kumar, B., (2011a). First record of smut disease of foxtail millet caused by *Ustilagocrameri* Korn. *J. Mycol. Pl. Pathol.*, *41*(3), 459–461.

Kumar, B., (2013). Diseases of small millets in Uttarakhand and their management. In: Singh, K. P., Prajapati, C. R., & Gupta, A. K., (eds.), *Innovative Approaches in Plant*

Disease Management-Crop Diseases and Their Management (pp. 257–287). LAP Lambert Academic Publishing. Germany.

Mirzaee, M. R., Zare, R., & Nasrabad, A. A., (2010). A new leaf and sheath brown spot of foxtail millet caused by *Bipolarisaus traliensis. Australasian Plant Disease Notes, 5*, 19, 20.

Monondo, F., (1957). Downy mildew of millet. *Agriculture Italy, 57,* 99–111.

Nagaraja, A., Kumar, B., Jain, A. K., & Sabalpara, A. N., (2016). Emerging diseases: Need for focused research in small millets. *J. Mycopathol. Res., 54*(1), 1–9.

Nagaraja, A., Kumar, J., Jain, A. K., Narasimhudu, Y., Raghuchander, T., Kumar, B., & Hanumanthe, G. B., (2007b). *Compendium of Small Millets Diseases* (p. 80). Project coordination cell, All India Coordinated Small Millets Improvement Project, UAS, GKVK Campus, Bangalore.

Palaniswami, A., Govindaswami, C. V., & Vidyasekaran, P., (1970). Studies on blast disease of tenai. *Madras Agric. J., 57,* 686–693.

Ramakrishnan, T. S., (1971). *Diseases of Millets* (pp. 83–100). Indian Council of Agricultural Research, New Delhi.

Rangaswami, G., & Mahadevan, A., (1999). Diseases of cereals. In: *Diseases of Crop Plants in India* (4th edn., pp. 160–264) Prentice-Hall of India, Pvt. Ltd. New Delhi.

Reddy, V. G., Upadhyaya, H. D., & Gowda, C. L. L., (2006). *Characterization of World's Foxtail Millet Germplasm Collections for Morphological Traits.* Internet resource: http:// www.icrisat.org/journal/cropimprovement/v2i1/v2i1characterization.pdf (accessed on 25 January 2020).

Sinha, A. P., & Upadhyay, J. P., (1997). *Millet Ke. Rog.* (p. 180). Directorate of publication. G.B. Pant University of Agriculture and Technology, Pantnagar, Uttarakhand.

Sprauge, R., (1950). *Diseases of Cereals and Grasses in North America* (p. 538). Ronald Press Co. New York.

Sundararaman, S., (1921). *Ustilagocrameri* Koern on *Seatriaitalica.* Beauv. *Bull. Agric. Res. Inst., Pusa. 97,* 11.

Takasugi, H., (1935). *The Relation of Environmental Factors and the Treatment of Oospores of Sclerospora graminicola* (studies in Nipponense Peronosporales IV) ibid, 2, 459–480.

Takasugi, H., & Akaisahi, Y., (1933). Studies on downy mildew of Italian millet in Manchuria. *Res. Bull. S. Manchuria Rly. Co., 11,* 1–20 (cf. *Exp. Sta. Rec., 70,* 489–490, 1934).

Viswanath, S., & Seetharam, A., (1989). Diseases of small millets and their management in India. In: Seetharam, A., Riley, K. W., & Harinarayana, G., (eds.), *Small Millets in Global Agriculture* (pp. 237–253). Oxford and I.B.H. Publishing Co. Pvt. Ltd., New Delhi, India.

Yan, W. Y., Xie, S. Y., Jin, L. X., Lin, H. J., & Hu, J. C., (1985). A preliminary study on the physiological races of *Pyriculariasetariae. Scientia Agricultura Sinica, No., 3,* 57–62.

CHAPTER 13

Kodo Millet or Kodo (*Paspalum scorbiculatum* L.) Diseases and Their Management Strategies

BIJENDRA KUMAR[1] and J. N. SRIVASTAVA[2]

[1]*College of Agriculture, Department of Plant Pathology, G.B. Pant University of Agriculture and Technology, Pantnagar–263145, Udham Singh Nagar, Uttarakhand, India*

[2]*Department of Plant Pathology, Bihar Agricultural University, Sabour–813210, Bhagalpur, Bihar, India*

Kodo millet or Kodo

(*Paspalum scorbiculatum* L.)

Vernacular names: *Kodon, Kodra, Arikelu, Haarka, Varagu, Kodo, Kodua*

Kodo millet, one of the oldest crops of the world, is a draught resistant plant, believed to be originated from Africa and later domesticated in India few thousand years ago. It is cultivated in arid and semi-arid regions of Asian and African countries. In India, it mostly cultivated in the Deccan region and cultivation spread to the foothills of Himalayas (Deshpande et al., 2015).

In India, it is cultivated in Madhya Pradesh, Chhattisgarh, Tamil Nadu, Karnataka, Maharashtra, and Jharkhand. Its grains are rich in minerals like iron, antioxidant, and dietary fiber. Though, the phosphorus content is lower compared to any other millet however, its antioxidant potential is much higher than any other millet and major cereals. Processing such as parboiling and debranning affects the minerals and fiber; however, it reduces anti-nutritional factors like phytate. Many Indian food items have been prepared traditionally solely from kodo millet or blended with other cereal and legume flours to increase the nutritional value, palatability, and functionality (Deshpande et al., 2015).

13.1 HEAD SMUT

13.1.1 INTRODUCTION

Head smut of kodo millet was first reported by McAlpine (1910) from Queensland, Australia. Subsequently, it has also been reported from eastern and southern Asia (Butler, 1918; Teng, 1947). In India, the disease was first reported by Butler in 1918 from Chamtaghat and Monghyer. Later, reported from Karnataka, Andhra Pradesh, Tamil Nadu, Madhya Pradesh, and Bihar. Butler (1918) recorded heavy losses in grain yield due to head smut. Viswanath (1992) reported 30–40% loss in grain yield due to this head smut. With the increase in disease incidence, the losses in grain yield increased linearly and estimated loss of 13.15 to 32.98% at 13.15 to 40.15% smut incidence was recorded by Jain and Yadava (1997). The disease has also been reported from Mysore (Mundkur and Thirumalachar, 1952), Bihar (Thirumalachar and Mishra, 1953), Madhya Pradesh (Mishra et al., 1976), Karnataka, Andhra Pradesh, and Tamil Nadu (Viswanath and Seetharam, 1989). Presently, the disease is known to occur in all the kodo millet growing areas.

13.1.2 SYMPTOMS

The diagnostic symptoms of the disease occur at flowering to dough stage of the crop. Diseased plants are stunted in their growth. Almost all the panicles in infected plants are converted into a long smut sorus that varies from 2.1 to 14.6 cm in length and 0.1 to 0.6 cm in breadth. Initially, the developing smut sorus remains covered by a creamy membrane. At maturity, the membrane is ruptured and exposing the black mass of teliospores. In some cases, the smut sori remain enclosed in the boot leaf and fail to emerge fully. Entire inflorescence except the fibro-vascular bundles is destroyed by smut sori. In few varieties of kodo millet, the boot leaf covering infected panicles show necrotic streaks Ahmed (1991) (Figure 13.1).

FIGURE 13.1   Head smut disease symptoms of kodo millet.

13.1.3   CAUSAL ORGANISM

Sorosporium paspali-thunbergii (Henn.) S. Ito
[Syn. *Sorosporium paspali* Mc Alp.]

Spores of the fungus known as teliospopres are borne on loose spore ball like masses of 60×30 μ in diameter. Spore ball disintegrate into individual spores after applying little pressure. The individual spores are globose, angular to roughly peared shaped, dark to yellowish brown with thick smooth wall of $11–18 \times 8–12$ μ in diameter. The spore germinates by producing septate, single, or branched hyphae constricted at septum, which bears lateral and terminal sporidia.

13.1.4   DISEASE CYCLE

The disease is both seed as well as soil borne in nature. For getting uniform infection, seedling inoculation with viable sporidial suspension resulted into maximum disease (Ahmed, 1991). Significantly higher smut incidence was recorded by mixing the seed with viable teliospores of the fungus @ 3 g kg⁻¹

before sowing (Jain, 2000). Seedling infection occurs by germ tube through the cell wall. After entry, the mycelium spreads inter and intra cellularly in the host tissues and becomes systemic. It reaches the meristematic tissues and ultimately infects the flower primordia. No sign of hypertrophy or hyperplasia of infected cells was observed. Hyphae are established in all the infected tissues except xylem vessels (Ahmed, 1991).

13.1.5 MANAGEMENT

- The cheapest way of managing the head smut is use of resistant varieties. KMV 8, KMV 20, JK 41, JK 62, JK 65, JK 106, and JK 13 were reported resistant to head smut by several workers.
- Seed treatment with fungicides such as Carbendazim, Mancozeb, Carboxin, Chlorothalonil, and Thiram @ 2 g kg^{-1} and Raxil @ 1.5 g kg^{-1} has been reported to be effective in managing the disease (Parambaramani et al., 1973; Pall, 1985; Chalam et al., 1989; Jain and Gupta, 1993; Jain, 1995, 1999; Mantur et al., 1997).
- The disease was observed more in black soil as compared to red gravel and red soils. Shallow sowing also results in early emergence of seedlings and less incidence of head smut (Jain, 1995).

13.2 LEAF BLIGHT

13.2.1 INTRODUCTION

The disease was first reported in India from Kanpur (U.P.) during 1980 by Gupta et al. (1982). Further, the disease was observed in 20 districts of Uttar Pradesh with higher severity in Allahabad, Faizabad, Gorakhpur, and Varanasi regions (Gupta et al., 1994). Shriveled and light weight grains are formed in abnormal and twisted earheads of infected plants that results in economical loss of grain yield.

13.2.2 SYMPTOMS

Pale and straw colored small scattered lesions are formed on the leaf blade. Severely affected plant showed a blighted appearance causing premature drying of leaves from tip to downwards.

13.2.3   CAUSAL ORGANISM

Alternaria alternata (Fr.) Keissl.

13.2.4   MANAGEMENT

- Use of resistant cultivars and foliar spray of Brestan 60 (0.1%) and Mancozeb (0.2%) controlled the disease effectively (Gupta and Singh, 1994).

13.3   ERGOT OR SUGARY DISEASE

13.3.1   INTRODUCTION

Butler and Bisby (1931) were first to record the ergot infection in kodo millet from Burma. In India, it has been was first observed at Kodaikanal in South, Assam in the North and Gwalior in Madhya Pradesh (Ramakrishnan and Sundaram, 1950).

The disease directly reduces the grain yield by replacing grain with sclerotia of the fungus and lowering quality of produce by contaminating grains with sclerotia, but no authentic report on crop loss is available. Sclerotia contain alkaloids which are not harmful to humans, but it may cause paralysis or sometimes death in cattles.

13.3.2   SYMPTOMS

The disease manifests only during emergence of panicle. Few or all flowers in an inflorescence may be infected. The most obvious external sign of the disease is the exudation of thin to viscous, sweet, sticky fluid from the infected flowers that give the name 'sugary' or 'honeydew' disease and the fluid overflows on lemma and palea. The honey dew soon becomes dry and hard reddish brown crusts. The ovary is replaced by a plectenchymatous mass of fungus. At panicle maturity, kernels are replaced by dark grey sclerotia or ergot which is the most characteristic symptom of the disease.

13.3.3   CAUSAL ORGANISM

Clavicepspaspali Stevens and Hall.

The disease is air borne. The conidia or ascospores infect the spike at early stage. A large number of conidia are present in the honey dew. They are produced at the tip of closely arranged conidiophores. Conidia are hyaline, oblong, and granular or guttulate. The size of condia ranged from 9–18 × 3–6 μ with a mean of 15 × 5 μ. Conidia germinate within 3 hr by producing germ tube which bears secondary condia on tips. Secondary condia are smaller in size. The sclerotia are dark, oblong, and measure 1.5–2.3 × 1.2 mm². Sclerotia projects out between the lemma and palea. The sclerotial germination on *P. scrobiculatum* has not been observed but developmental stages of the strains are reported on other species of *Paspalum*. A number of pin head like stromatoid structures bearing numerous sunken perithecia in the terminal swollen portion are produced on germination from each sclerotium. The asci produced in perithecia, which are hyaline, cylindric clavate with average size of 130 × 3.3 μ. The asci contain eight ascospores, which are hyaline and slender with average size of 101 × 0.75 μ.

13.3.4   MANAGEMENT

Use of clean seeds, roguing of infected panicles, eradication of the collateral hosts, long crop rotation with noncereal crop and repeated summer plowing can minimize the primary inoculums. Seed lots containing sclerotia can be separated by immersing them in 20% solution of commercial salt (Yadava and Jain, 2006) and no other control measure has been tried so far.

13.4   RUST

13.4.1   INTRODUCTION

Sydow and Butler (1906) recorded this disease for the first time on kodo millet from Himalayan Hills at Kanaighat (Sylhet) and Kumaon Hills in India. Later, it has also been observed at Coimbatore in India and Ceylon.

13.4.2 SYMPTOMS

The disease is characterized by the formation of oval, brown erumpent uredia on the upper surface of the leaves and leaf sheath. The teliopustules which are brown colored formed on the lower surface of the leaves and leaf sheath during December - February in Coimbatore condition. Flora long time the teliaremain covered by the epidermis. Uredia are present throughout the year on the grass hosts under sheltered conditions, from where it can pass readily on the cultivated kodo millet.

13.4.3 CAUSAL ORGANISM

Puccinia substriata Ellis and Barthol.
[Syn. *Uredo paspali-scrobiculati* Syd.]

The uredospores measuring an average size of 31 × 25 μ are subglobose to elliptical, light brown and finely echinulate. They germinate readily in water drops within 3 hours and produce one or more stouted germ tube. The incubation period is 8–12 days. Four equatorial germ pores are present in uredospores. Teliospores are two celled, oblong, rounded at both ends with smooth uniform thick wall and size of 40 × 22 μ ranging from 28–47 × 19–28 μ. The pedicels are colored and short. Paraphyses are present in both uredia and telia.

13.4.4 MANAGEMENT

• Collection and destruction of the grass hosts is partly useful in the management of the disease.
• Foliar spray of Zineb (0.25%) is also reported effective in reducing the disease (Yadava and Jain, 2006).

13.5 UDBATTA DISEASE

13.5.1 INTRODUCTION

Udbatta disease is very common in kodo millet and foxtail millet. Sometimes it has been observed on little as well as prosomillets. Diseased earheads are converted into a compact agarbatti like structure, hence it is named "Udbatta."

In India, Butler, and Bisby (1931) were first to report this disease. Later, in 1966 it was observed at Koraput and Kalahandi in Odisha and in 1974 at Jabalpur in Madhya Pradesh. Recently it was also noticed in Bangalore, Karnataka (Nagaraja et al., 2010). There is no authentic study regarding crop loss assessment owing to sporadic occurrence of the disease.

13.5.2 SYMPTOMS

The disease is characterized by the transformation of infected panicles into a compact stick like structure agarbatti; hence, it is named "Udbatta" (Figure 13.2).

FIGURE 13.2   Udbatta disease symptoms on kodo millet.

13.5.3 CAUSAL ORGANISM

Butler and Bisby (1931) recorded *Ephelis japonica* P. Henn. However, Mohanty, and Mohanty (1957) reported *Ephelisoryzae* Syd. [Teleomorph: *Balansiaoryzae* (Syd.) Naras. and Thirum.] on the same host from India.

13.5.4 MANAGEMENT

- Collection and destruction of diseased panicles and keeping bunds weed free especially from graminaceous weeds believed to serve as the collateral hosts.
- The information about the management of Udbatta disease in kodo millet is meager. Pall and Nema (1976) at Jabalpur, Madhya Pradesh screened fifty kodo millet genotypes under natural conditions and found IPS 45, 196, 342, 365, 368, 381, 387, 140 and Niwas 1 to be highly resistant to Udbatta.
- Pre-sowing seed treatment with carbendazim @ 4 gkg⁻¹ seed may be followed.

13.6   SHEATH ROT

13.6.1   INTRODUCTION

Sheath rot is very common in Vriddachalam area of Cuddalore dist. in Tamil Nadu where kodo millet is grown on a large area after the paddy fallows.

13.6.2   SYMPTOMS

The disease manifest by the formation of oblong or somewhat irregular spots, measuring about 0.5–1.5 cm in length. The spots gray to light brown at centers surrounded by distinct dark reddish brown margins. Later, the spots enlarge in their size and coalesce to form lesions and several nearby lesions further coalesce and may cover most of the leaf sheath. Lesions may also comprise of diffuse reddish-brown discoloration in the sheath. Profuse whitish powdery growth of the fungus can be observed inside the sheaths; though it may look normal from outside. If severe infection occurs at early stage, the earheads fail to emerge completely. The young panicles do not come out of the sheath or emerge only partially. The panicles that have not emerged begin to rot, and florets become reddish-brown to dark brown. Most grains remain sterile, become shriveled, grains are partially filled or unfilled and discolored. The leaf sheaths enclosing the young panicles rot.

13.6.3   CAUSAL ORGANISM

Sarocladium oryzae (Sawada) W. Gams and D. Hawksworth
[Syn. *Acrocylindrium oryzae* Sawada; *Sarocladium attenuatum* W. Gams
and D. Hawksw.]

Sarocladium oryzae produces white, sparsely branched, septatemyce-lium, measuring 1.5–2 mm in diameter. Conidiophores arising from the mycelium are slightly thicker than the vegetative hyphae, branched once or twice, each time with 3–4 branches in a whorl. Conidia are produced simply at the tip, produced consecutively, hyaline, smooth, single-celled, cylindrical, and are 4–9 × 1–2.5 mm in size.

13.6.4   MANAGEMENT

- Seed treatment with either Validamycin 2 ml or *P. fluorescens* 10 g or Hexaconazole 2 ml/kg seed that effectively minimized sheath rot incidence and producing highest yields (AICSMIP, 2012).

13.7   KODO MILLET POISONING

13.7.1   INTRODUCTION

Kodo millet poisoning a toxic syndrome, causing health hazards to humans as well as cattle and other animals after consumption in the cooked as well as raw state has been reported from different parts of the country (Swaroop, 1922; Ayyar and Narayanaswamy, 1949; Nayak and Mishra, 1962; Dwivedi and Tyagi, 1990). The husk and leaves acquire poisonous character which may be due to heavy rainfall. Locally kodo millet poisoning is also known as *Matona*, *Mina*, and *Kiruku Varagu*. Instances of stock poisoning have also been reported. This phenomenon is found only in ears produced during damp cloudy weather during harvest time or if premature grains are harvested.

Ayyar and Narayanaswamy (1949) developed a method to distinguish between poisonous and non-poisonous grains of kodo millet. The petrol extract from the poisonous grains gives red color with conc. HSO, whereas the extract from innocuous grains does not give this color.

Mishra et al. (2000) reviewed the chemistry and biological activity of kodo millet and reported that crude extracts and the pure isolates were reported to have nutritive, antifungal, transquilizing properties and sometimes showed toxic effects when consumed as food.

13.7.2 SYMPTOMS

A whitish fungal growth appeared in the affected plant parts like leaf, stem, ears, and grains during reproductive stage of the crop under cloudy damp weather. Initially, small patches of fungus are seen, which spreads rapidly in humid environment or just after rains in the crop standing in the field and in the threshing floor (Jain and Sharma, 2010). The toxins produced by the fungi are reported to be responsible for the poisoning. However, systematic work on actual cause and their remedy needs immediate attention.

The symptoms of kodo poisoning observed within twenty minutes of taking food in humans were tremors, giddiness, and perspiration, inability to speak or swallow. So far, no fatalities have been recorded and the symptoms disappeared after twenty four hours. Poisoning symptoms were also observed in dogs, monkeys, and crow (Ayyar and Narayanaswamy, 1949). Bhide and Aimen (1959) reported that the active fraction of the husk of poisonous grain had a tranquillizing effect.

Nayak and Mishra (1962) categorized the symptoms of cattle poisoning in three groups. Severely affected one could not remain standing and lay down with flexed limbs and erected neck and head with nodding movement. The moderately affected ones remained standing with to and fro nodding of the head and neck along with in-coordination movement and respiratory difficulties. In light affected ones the nodding of the head was present but respiratory difficulty was not observed.

13.7.3 CAUSAL ORGANISM

A number of fungal species are known to cause poisoning in human and cattle's. *Phomopsis paspali* was reported to produce two toxins named Paspalin P I and Paspalin P II (Pendse, 1974; Patwardhan et al., 1974 and Deshmukh et al., 1975). *Aspergillus flavus* produces cyclopiazonic acid (Lalitha Rao and Husain, 1985) and *A. tamari* was reported to produce cyclopiazonic acid and aflatoxin B 1 (Lalitha Rao and Husain, 1985; Ansari and Shrivastava, 1991; Antony et al., 2003). Janardhanan et al. (1984) reported that *A. tamarii* isolated from seeds of *Paspalum scrobiculatum* is found to produce ergot alkaloids (fumigaclavine A) in submerged culture. The head smut causing fungus

Sorosporium paspali that are almost invariably present in the outer husks of the grains are responsible for scorbic toxicity (Baki and Ipor, 1992).

13.7.4   MANAGEMENT

Harvested heaps should be protected from rains. Traditional practice of threshing by pre moistening the plants should be avoided and only dried harvest should be threshed. Unripe or premature grains should not be harvested. Storage of premature crop harvest and its storage as cylindrical heap before threshing also make the grain/ straw moldy.

Intravenous injection of 1% methylene blue (single dose of 50–60 cc) and subcutaneous injection of 1% atropine sulfate (3 cc) was reported to control cattle poisoning by kodo millet (Nayak and Mishra, 1962).

Use of some antidotes like juice of banana stem, the astringent juice of the guava or the leaves of *Nyctanthus arbortristis*, 'tamarind' water, butter milk and pickles are recommended to combat with kodo millet poisoning in humans and animals (Jain and Sharma, 2010).

13.8   OTHER FUNGAL DISEASES

A large number of leaf spot diseases incited by graminicolous *Drechslera* Ito (with sympodial and indeterminate conidiophore) have been reported under the generic name *Helminthosporium* Link ex Fries (with determinate conidiophore and obclavate conidia developing laterally often in verticils). Various species involved are given below with their present nomenclature status:

Drechslera state of *Trichometa sphaeriaholmii* Luttrell = *D. holmii* (Luttrell) Subram. and Jain (*Helminthosporium holmii*) causing leaf spots. *Drechslera*state of *Trichometa sphaeria turcica* Luttrell (*Helminthosporium turcicum* Pass.) is the cause of leaf blight disease. *Drechslera victoriae* (Meehan and Murphy) Subram. and Jain causing leaf spot on kodo millet from Bagha, Champaran, Bihar, India. *Alternaria alternate* (Fr.) Keissler causing leaf blight. *Dimmerosporium* (= *Asterina*) *erysiphoides* Ell. and Ev. causing leaf spot from Bassein, Maharashtra, India (Uppal et al., 1935). *Macrophomina phaseolina* (Tassi) Goid reported to cause root rot. *Rhizoctonia solani* Kuhn, the incitant of sheath blight disease of rice also infects the *Paspalum scrobiculatum* and produces symptoms (Nagaraja et al., 2016). *Uredo paspali-scrobiculati* Syd. Leaf rust from Kumaon, U.P., India. *Ustilagora benhorstiana* Kuehn, infects inflorescence from Shillong, Meghalaya, India.

13.9 BACTERIAL DISEASES

13.9.1 *BACTERIAL LEAF STREAK*

13.9.1.1 *INTRODUCTION*

In India, the disease was reported for the first time by Nema et al. (1978) from Jabalpur, Madhya Pradesh. Till date, it has not been observed at other kodo millet growing areas of the country. Quantitative yield losses have been reported (Pall, 1979).

13.9.1.2 *SYMPTOMS*

Bacterial leaf streak disease is characterized by the formation of pale yellow streaks, measuring 0.5 to 1.0 mm running parallel to the leaf veins. These streaks further enlarge to 1.0–1.5 mm × 3–4 cm lesions, and later turn brown. In severe cases of infection, the entire leaf turns brown and withers away. There may be shredding of leaves along the length. Streaks may also appear on stems and peduncles (Nema et al., 1978).

13.9.1.3 *CAUSAL ORGANISM*

Xanthomonas species.

13.9.1.4 *MANAGEMENT*

- Several varieties such as CO 2, CO 3, T 1, JK 41, JK 62, and IPS 14 posses fair resistance against bacterial leaf streak.
- Pall (1983) reported foliar spraying of Streptomycin sulfate @ 300 ppm twice, first when symptoms of the disease are noticed and second 15 days later to be effective against the disease, increasing the yield by 2.5 folds.

13.9.2 *BACTERIAL LEAF BLIGHT (BLB)*

Xanthomonas campestris (Pam.) Dowson pv. *oryzae* Uyeda and Ishiyama has been reported as epiphytic on leaf of kodo millet from Cuttack causing leaf blight (Trimurthy and Devadath, 1981).

13.10   PHANEROGAMIC PARTIAL ROOT PARASITE

13.10.1   INTRODUCTION

Witch weed (*Striga* spp.) has a very wide host range. It has been noticed associated with the kodo millet plants. Losses in grain yield due to *Striga* species depend primarily on level of host resistance and the number of *Striga* plants attacking the crop. Jain and Tripathi (2005) recorded 42.4–65.8% loss in kodo millet grain yield/plant due to *Striga densiflora*. In India, it has been reported from Andhra Pradesh, Tamil Nadu, Karnataka, and Madhya Pradesh. In a field survey programme, average incidence of *Striga* species varied from 1.6 to 2.0% with 66.7 to 100.0% frequency of incidence recorded in Rewa, Satna, and Sidhi districts of Madhya Pradesh during the years 2001 and 2003.

13.10.2   SYMPTOMS

The *Striga* infestation appears in the field when they emerge from the soil. The underground parts of *Striga* plant remain attached to the host roots by haustoria, from which it absorbs water and nutrients from host plant. The affected plants remain stunted with reduced aerial growth bearing lanky panicles. If the plants are attacked in early stages, the plants are dried before flowering (Yadava and Jain, 2006) (Figure 13.3).

13.10.3   CAUSAL ORGANISM

Striga asiatica (L.) Kuntze and *S. densiflora* Benth.

13.10.4   MANAGEMENT

- Hand weeding or pulling *Striga* plants before the flower is the most effective and cheapest way of its eradication (Yadava and Jain, 2006).
- Some of the improved varieties of kodo millet such as GPUK 1, GPUK 3, GPUK 5 and JK 41 were reported to be least affected by *Striga* (Reddy and Dastagiraiah, 1987; Jain and Tripathi, 2002).
- Infestation of *Striga* species is reduced by applying nitrogenous fertilizers (Pesch et al., 1983).

FIGURE 13.3   Witch weed (*Striga* spp.) associated with kodo millet.

13.11   NEMATODE DISEASES

Meloidogyne incognita (Kofoid and White) Chitwood produces galls of different sizes in *Paspalum scrobiculatum* and has been reported from Aligarh (U.P.) as a new host (Alam et al., 1973).

Kodo millet was reported a good host for *Tylenchorhynchus vulgaris* (Vaishnav and Sethi, 1977) and *Rotylenchulus reniformis* also (Sahoo and Padhi, 1986).

KEYWORDS

- **bacterial leaf blight**
- **bacterial leaf streak**
- **Kodo millet poisoning**
- **phanerogamic partial root parasite**
- **rust**
- **sheath rot**

REFERENCES

Ahmed, N. N., (1991). Biology of smut fungi of the minor millets *Paspalum scrobiculatum* L. and *Echinochloa frumentacea* (Robx.) Link in Karnataka. *M.Sc. (Agri.) Thesis* (p. 91). University of Agricultural Sciences, Bangalore, India.

AICSMIP, (2012). *Annual Report of the All India Coordinated Small Millets Improvement Project, Project Coordinating Unit.* ICAR, GKVK, Bangalore.

Alam, M. M., Khan, M. A., & Saxena, S. K., (1973). Some new host records of root knot nematode, *Meloidogyne incognita* (Kofoid & White, 1919) Chitwood, 1949. *Indian J. Nematology, 3,* 159–160.

Ansari, A. A., & Shrivastava, A. K., (1991). Susceptibility of minor millets to *Aspergillus flavus* for aflatoxin production. *Indian Phytopath., 44*(4), 533, 534.

Antony, M., Shukla, Y., & Janardhanan, K. K., (2003). Potential risk of acute hepatotoxicity of kodo poisoning due to exposure to cyclopiazonic acid. *J. Ethnopharmacol., 87*(2/3), 211–214.

Ayyar, K. V. S., & Narayanaswamy, K., (1949). Varagu poisoning. *Nature, 163,* 912–913.

Baki, B. B., & Ipor, I. B., (1992). *Paspalum scrobiculatum* L. record from proseabase. In: Mannetje, L. T., & Jones, R. M., (eds.), PROSEA (Plant Resources of South-East Asia) Foundation. Bogor, Indonesia.

Bhide, N. V., & Aimen, R. R., (1959). Pharmacology of a tranquilizing principle in *Paspalums crobiculatum* grains. *Nature, 153,* 1735, 1736.

Butler, E. J., (1918). *Fungi and Diseases in Plants* (p. 547). Thaker Spinck and Co. Calcutta.

Butler, E. J., & Bisby, G. R., (1931). *Fungi of India. Sci. Monogra., XVIII,* Calcutta.

Chalam, T. V., Reddy, K. S., & Rao, G. N., (1989). Seed treatment for the control of head smut of *Paspalums crobiculatum* (kodo millet). *Millets Newsl., 8,* 56.

Deshmukh, P. G., Kanitkar, U. K., & Pendse, G. S., (1975). A new fungal isolate from *Paspalum scrobiculatum* Linn. with new biologically active metabolites. *Acta Microbiologica. Academiae Scientiarum Hungaricae, 12*(3), 242–253.

Deshpande, S. S., Mohapatra, D., Tripathi, M. K., & Sadvatha, R. H., (2015). Kodo millet-nutritional value and utilization in Indian foods. *Journal of Grain Processing and Storage, 2*(2), 16–23.

Dwivedi, P. K., & Tyagi, R. P. S., (1990). Studies on fungal and mycotoxin contamination of stored millets kodon (*Paspalums crobiculatum*) and kutki (*Panicum miliare*). *J. Food Sci. Tech., 27*(2), 111–112.

Gupta, S. P., & Singh, B. R., (1994). Control of leaf blight disease of kodo. *Madras Agric. J., 81*(8), 440–442.

Gupta, S. P., Narain, U., Singh, M., & Shukla, T. N., (1982). A new leaf blight of kodon (*Paspalums crobiculatum* L) from India. *National Academy Science Letters, 5*(2), 41, 42.

Gupta, S. P., Singh, B. R., Tripathi, D. P., & Yadav, M. D., (1994). Prevalence of leaf blight disease on kodo in U.P. *Madras Agric. J., 81*(3), 166, 167.

Jain, A. K., (1995). Management of head smut in kodo millet (*Paspalums crobiculatum* L.). *Ann. Agric. Res., 16*(2), 172–178.

Jain, A. K., (1999). Evaluation of fungicides for the control of head smut in kodo millet. *Indian Phytopath., 52*(4), 423, 424.

Jain, A. K., (2000). Effect of inoculum load on incidence of head smut in kodo millet. *J. Mycol. Pl. Pathol., 30*(1), 121, 122.

Jain, A. K., & Gupta, J. C., (1993). Chemical control of *Sorosporium paspali thunbergii* McAlp causing head smut of kodo millet. *Adv. Plant Sci., 6*(1), 164–167.

Jain, A. K., & Sharma, N. D., (2010). Fungal disease problems of kodo millet: Diagnosis and management. In: Trivedi, P. C., (ed.), *Plant Diseases and its Management* (pp. 15–28). Pointer Publishers, Jaipur (India).

Jain, A. K., & Tripathi, S. K., (2002). Occurrence of Striga and its influence on yield of kodo millet. *Crop Research, 23*(3), 532–535.

Jain, A. K., & Tripathi, S. K., (2005). Management of grain smut (*Macalpinomyces sharmae*) in little millet. *Indian Phytopath., 60*(4), 467–471.

Jain, A. K., & Yadava, H. S., (1997). Recent approaches in disease management of small millets. *Proc. Nat. Symp. on Small millets-Current Research Trends and Future Priorities as Food, Feed and in Processing for Value Addition* (pp. 31–33). TNAU, Coimbatore.

Janardhanan, K. K., Sattar, A., & Husain, A., (1984). Production of fumigaclavine A by *Aspergillus tamarii* Kita. *Can. J. Microbiol., 30*(2), 247–250.

Lalitha, R. B., & Husain, A., (1985). Presence of cyclopiazynic acid in kodo millet causing kodua poisoning in man and production by associate fungi. *Mycopathologia., 89*(3), 177–180.

Mantur, S. G., Viswanath, S., & Chanamma, K. A. L., (1997). Control of head smut of kodo millet through seed treatment. *Curr. Res., 26*(8/9), 146.

McAlpine, D., (1910). *The Smuts of Australia* (p. 285). Melbourne Government Printer.

Mishra, M., Shukla, Y. N., & Kumar, S., (2000). Chemistry and biological activity of *Paspalum scrobiculatum*: A review. *Journal of Medicinal and Aromatic Plant Sciences, 22*(2–3), 288–290.

Mishra, R. P., Pall, B. S., & Nema, K. G., (1976). Ustilaginales of Jabalpur, Madhya Pradesh II. *JNKVV Res. J., 10*(2), 189.

Mohanty, U. N., & Mohanty, N. N., (1957). Ephelis on some new hosts. *Sci. and Cult., 22,* 434–436.

Mundkur, B. B., & Thirumalachar, M. J., (1952). *Ustilaginales of India* (p. 34). CMI, Kew Surrey.

Nagaraja, A., Kumar, B., Jain, A. K., & Sabalpara, A. N., (2016). Emerging diseases: Need for focused research in small millets. *J. Mycopathol. Res., 54*(1), 1–9.

Nagaraja, A., Anjaneya, R. B., & Govindappa, M. R., (2010). Occurrence of Udabatta disease on Kodo millet (*Paspalum scrobiculatum* L.): A new report from South India. *J. Mycopathol. Res., 48*(1), 163–164.

Nayak, N. C., & Mishra, D. B., (1962). Cattle poisoning by *Paspalum scrobiculatum* (kodua poisoning). *The Indian Veterinary Journal, 39,* 501–504.

Nema, A. G., Kulkarni, S. N., & Pall, B. S., (1978). Bacterial leaf streak of kodo (*Paspalum scrobiculatum* L.). *Sci. and Cult., 45*(9), 365–366.

Pall, B. S., (1979). Quantitative losses due to bacterial leaf streak of kodo (*Paspalum scrobiculatum* L.). *Food Farming and Agriculture, 11*(1), 1.

Pall, B. S., (1983). Bacterial streak of kodo millet: Varietal resistance and chemical control. *MILWAI Newsl., 2,* 12.

Pall, B. S., (1985). Chemical control of kodo smut. *MILWAI Newsl.,* (pp. 61–64). JNKVV, Jabalpur, M.P., India.

Pall, B. S., & Nema, A. G., (1976). Screening of kodo (*Paspalum scorbiculatum* L.) varieties against udbatta disease. *JNKVV Res. J., 10*(2), 194.

Parambaramani, C., Guruswamy, R. V. D., & Subramanian, R., (1973). Smut disease of varagu (*Paspalum scrobiculatum*) and method of control. *Farm and Factory, 7*(8), 26–27.

Patwardhan, S. A., Pandet, R. C., Dev, S., & Pendse, G. S., (1974). Toxic cytochalasins of *Phomopsis paspali*, a pathogen of kodo millet. *Phytochemistry.*, *13*, 1985–1988.

Pendse, G. S., (1974). A note of new chemical compound isolated from a fungus hitherto unknown. *Experientia*, *30*, 107–108.

Pesch, C., Pieterse, A. H., & Stoop, W. A., (1983). Inhibition of germination in *Striga* by means of urea. *Proceedings of the 2ⁿᵈ International Workshop of Striga, ICRISAT* (pp. 37–38).

Ramakrishnan, T. S., & Sundaram, N. V., (1950). Ergot on two grasses from South India. *Sci. and Cult. 16*, 214.

Reddy, P. L., & Dastagiraiah, P., (1987). Preliminary screening of kodo millet varieties against witch weed. *Millets Newsl.*, *6*, 57.

Sahoo, H., & Padhi, N. N., (1986). Susceptibility of plant to *Rotylenchulus reniformis*. *Indian J. Nematol.*, *16*(1), 97, 98.

Swaroop, A., (1922). Acute kodo poisoning. *Indian Med. Gazette.*, *57*, 257–258.

Sydow, H., & Butler, E. J., (1906). Fungi *Indiae Orientalis*, Pars I. *Ann. Mycol.*, *4*, 424–445.

Teng, S. C., (1947). Addition to the myxomycetes and the carpomycetes of china. *Bot. Bull. Acad. Sinica* (pp. 25–44) (RAM 26, 421).

Thirumalachar, M. J., & Mishra, J. N., (1953). Contribution to the study of fungi of Bihar, India-I. *Sydowia*, *7*, 29, 30

Trimurthy, V. S., & Devadath, (1981). Studies on epiphytotic survival of *Xanthomonas campestris pvoryzae* on graminaceous weeds. *Indian Phytopath.*, *34*, 279–281.

Uppal, B. N., Patel, M. K., & Kamat, M. N., (1935). *The fungi of Bombay* (Vol. VII, pp. 1–56). Private Publication.

Vaishnav, M. U., & Sethi, C. L., (1977). Reaction on some graminaceous plants to *Meloidogyne incognita* and *Tylenchorhynchus vulgaris*. *Indian J. Nematol.*, *7*, 176–177.

Viswanath, S., (1992). Management of biotic factors (diseases). In: *6ᵗʰ Annual Small Millets Workshop at BAU.* Ranchi-Kanke (Bihar).

Viswanath, S., & Seetharam, A., (1989). Diseases of small millets and their management in India. In: Seetharam, A., Riley, K. W., & Harinarayana, G., (eds.), *Small Millets in Global Agriculture* (pp. 237–253). Oxford & IBH Publishing Co. Pvt. Ltd., New Delhi, India.

Yadava, H. S., & Jain, A. K., (2006). *Recent Advances in Kodo Millet* (p. 84). ICAR, New Delhi.

CHAPTER 14

Little Millet or Gundali or Goudli or Gondola (*Panicum miliare* L.) Diseases and Their Management Strategies

BIJENDRA KUMAR[1] and J. N. SRIVASTAVA[2]

[1]*College of Agriculture, Department of Plant Pathology,*
G.B. Pant University of Agriculture and Technology, Pantnagar–263145,
Udham Singh Nagar, Uttarakhand, India

[2]*Department of Plant Pathology, Bihar Agricultural University,*
Sabour–813210, Bhagalpur, Bihar, India

Little Millet or Gundali or Goudli or Gondola

(*Panicum miliare* L. / *Panicum sumatrense* subsp. *psilopodium*)

Vernacular names: *Kutki, Samalu, Same, Samai, Sava, Gajro, Sama, Suan*

Little millet (*Panicum sumatrense*), belonging to the family Poaceae (Gramineae) is native to India (de Wet et al., 1983) and is also called Indian millet. It is an annual herbaceous plant grows straight or with folded blades and 30 cm to 1 m tall. It is widely cultivated as a cereal across India, Nepal, and western Myanmar. The crop is grown by poor and tribal farmers in soils with low fertility or low or no cash input. It rejuvenates exceptionally well compared to other cereals. In India, the crop is cultivated in Andhra Pradesh, Chhattisgarh,

Madhya Pradesh, Odisa, Tamil Nadu, Karnataka, Jharkhand, and Gujarat are major little millet growing states in the country. Little millet is tolerant to both drought and water logging conditions. The crop is well adapted to varied soil and agro-ecosystems. It can be cultivated up to 2000 m above sea level. It is superior or at par with other cultivated cereals (Anonymous, 2009; Gopalan et al., 2010; Balasubramanian et al., 2007) in terms of nutritional and medicinal properties, recommended for the patients suffering from cardio-vascular and diabetes diseases. This millet is a good source of fat (4.7 g), crude fiber (7.7 g), iron (9.3 mg) and phosphorus (220 mg) per 100 g and the amino acid is well balanced (Gopalan et al., 2010). It contains about 65% carbohydrate, a good proportion of which is in the form of non-starchy polysaccharides and dietary fiber (Menon, 2004). Little millet has low glycaemic index and exhibits hypoglycaemic effect due to its higher proportion of dietary fiber (Itagi et al., 2013). It plays an important role in providing significant amounts of antioxidants and phytochemicals in the diet (Ushakumari and Malleshi, 2007; Pradeep and Guha, 2011). The total phenol content and ferrous ion chelating activity of soluble extract from little millet is found to be higher than other millets. Besides, grains have excellent storage properties and can be stored easily under ordinary storage conditions for several years without any fear of storage pests (Kumar, 2015; Vetriventhan and Upadhyaya, 2016).

Two subspecies of little millet have been described:

- *Panicum sumatrense* Roth ex Roem. and Schult. subsp. *psilopodium* (Trin.) Wet.
- *Panicum sumatrense* Roth ex Roem. and Schult. subsp. *sumatrense*

14.1   GRAIN SMUT

14.1.1   INTRODUCTION

Grain smut disease is quite prevalent in early maturing genotypes of little millet. In India, the grain smut was recorded for the first time from Dindori district of Madhya Pradesh by Sharma and Khare (1987). Subsequently, it was reported in sporadic form from other parts of the country. Yield loss studies are very limited. Up to 50% of plants or grains are reported to be infected by the pathogen (Sharma and Khare, 1987). However, Jain et al. (2006) recorded 4.2–16.6% reduction in plant height, 6.4–38.9% in panicle length and 9.8–53.5% in grain yield per plant.

14.1.2 SYMPTOMS

Symptoms are evident at grain filling stage. The infected ovaries are transformed into smut sori. The size of ovaries does not increase in size as compared to the normal grain. The glumes get spread apart by the transformed sori. The sorus is initially covered by a thin dull delicate membrane, later get ruptured exposing the smut spores. The spores are easily disseminated by wind leaving behind only the glumes. Some of the grains which develop late remain green with slightly increased size as compared to the normal grains. Such greenish healthy looking grains if pressed release spores (Sharma and Khare, 1987; Nagaraja et al., 2016) (Figure 14.1).

FIGURE 14.1   Grain smut of little millet or gundali.

14.1.3 CAUSAL ORGANISM

Macalpinomyces sharmae Vanky
[Syn. *Tolyposporium* sp.].

14.1.4 MANAGEMENT

- The disease can be managed by selecting resistant cultivars. Some of the genotypes viz; DPI 2386, DPI 2394, PLM 202, OLM 203 and CO 2 were reported to be resistant to grain smut.

- Cultural practices like; early sowing (First fortnight of July) although recorded the maximum smut incidence and susceptibility index but, resulted in highest grain yield.
- Early maturing cultivars were reported to be susceptible to grain smut as compared to late maturing.
- Seed treatment with Carbendazim or Carboxin @ 2 g kg^{-1} seed has been reported effective and economical in managing the disease (Nagaraja et al., 2007b).

14.2   RUST

14.2.1   INTRODUCTION

The disease has been reported from India, Philippines, and Ceylon (Cummins, 1971; Pall et al., 1980; Haider, 1997).

14.2.2   SYMPTOMS

Numerous, minute, narrow brown pustules arranged in linear rows appear on the upper surface of the leaves.

14.2.3   CAUSAL ORGANISM

Uromyceslinearis Berk. and Broome
[Syn. *Ustilagolinearis* (Berk. and Broome) Petch].

14.2.4   MANAGEMENT

No work has been carried out to control the rust disease in little millet.

14.3   UDBATTA DISEASE

14.3.1   INTRODUCTION

In India, the disease was first reported from Bhubaneswar (Odisha) in 1965 and about 40–50%, plants were found infected. Later, during 1966 and 1967 the disease was found at a greater intensity.

14.3.2 SYMPTOMS

The diseased plants are conspicuous by their malformed inflorescence bearing grayish white fructifications of the fungus. In the infected panicles, the spikelets are found to be glued to one another and to the main rachis by the viscid spore masses, which harden into a crust. Black sclerotial masses are formed on mature panicles. The inflorescence of the healthy plant is a lose panicle measuring 30 to 40 cm long whereas in the diseased plant, the spikelets become glued into a cylindrical structure and the length of the panicle gets reduced to 15–23 cm long.

14.3.3 CAUSAL ORGANISM

Ephelisoryzae Syd.
[Teleomorph: *Balansiaoryzae* (Syd.) Naras. and Thirum.].

14.3.4 MANAGEMENT

No work has been carried out as it is sporadic and minor disease.

14.4 OTHER FUNGAL DISEASES

Drechslera state of *Cochliobolus nodulosus* Luttrell = *D. nodulosa* (Berk. and Curt.) Subram. and Jain (*Helminthosporium nodulosum* Berk and Curt. apud Sacc.) isolate of finger millet infects producing blight symptoms. *Drechslera* state of *Cochliobolus miyabeanus* (Ito and Kuribayashi) Drechsler ex Dastur (*Helminthosporium oryzae* Breda de Hann.) infects little millet. *Rhizoctonia solani* produces sheath blight symptoms.

14.5 NEMATODE DISEASES

Little millet was reported very good host for spiral nematode, *Helicotylenchus dihystera* (Rao and Swaroop, 1974) and *H. abunaamai* (Padhi and Das, 1982). The life cycle of *H. abunaamai* from egg to egg was completed in 32–40 days (Padhi and Das, 1986).

Little millet was also reported as a good host for *Meloidogyne incognita* and *Tylenchorhynchus vulgaris* (Vaishnav and Sethi, 1977). Ragi cyst nematode, *Heteroderadelvi* can infect the little millet crop (Krishnaprasad et al., 1980).

KEYWORDS

- **carbendazim**
- **grain smut**
- **nematode diseases**
- ***Panicum sumatrense***
- **rust**
- **udbatta disease**

REFERENCES

Anonymous, (2009). *Millet Network of India*. FIAN, India: Deccan Development Society.

Balasubramanian, S., Vishwanathan, R., & Sharma, R., (2007). Post harvest processing of millets: An appraisal. *Agriculture Engineering Today*, *31*(2), 18–23.

Cummins, G. B., (1971). *The Rust of Cereals, Grasses, and Bamboos*. Springer Verlag, Berlin.

De Wet, Prasada, J. M. J., Rao, K. E., & Brink, D. E., (1983). Systematics and domestication of *Panicum sumatrense* (Gramine). *J. D'Agriculture Tradit. Bot. Appliquee*, *30*, 159–168.

Gopalan, C., Ramasastri, B. V., & Balaubramanian, S. C., (2010). *Nutritive Value of Indian Foods*. ICMR, Hyderabad: National Institute of Nutrition.

Haider, Z. A., (1997). Little millet in Indian Agriculture: Progress and Perspectives. In: *Nat. Semi. on Small Millets: Current Trends and Future Priorities as Food, Feed and in Processing for Value Addition*. TNAU, Coimbatore.

Itagi, S., Naik, R., & Yenagi, N., (2013). Versatile little millet therapeutic mix for diabetic and non diabetics. *Asian J. Food Sci. Tech.*, *4*(10), 33–35.

Jain, A. K., Tripathi, S. K., & Singh, R. P., (2006). *Macalpinomyces sharmae*: A new threat for the cultivation of little millet in Madhya Pradesh. *Proc. Nat. Symp. on Emerging Plant Diseases, Their Diagnosis and Management* (pp. 31–32). N.B.U., Siliguri (W.B.).

Krishnaprasad, K. S., Krishnappa, K., Setty, K. G. H., Reddy, B. M. R., & Reddy, H. R., (1980). Susceptibility of some cereals to Ragi cyst nematode. *Heteroderadelvi. Curr. Res.*, *9*(6), 114–115.

Kumar, A., (2015). Studies on grain smut of little millet (*Panicum sumatrense* Roth ex Roemer and Schultes) caused by *Macalpinomyces sharmae* K. Vanky. *M.Sc. Thesis* (p. 80). Jawaharlal Nehru Krishi Vishwa Vidyalaya, Jabalpur.

Menon, M. V., (2004). Small millets call for attention. *Kisaan World.*, *4*, 63, 64.

Nagaraja, A., Kumar, J., Jain, A. K., Narasimhudu, Y., Raghuchander, T., Kumar, B., & Hanumanthe, G. B., (2007b). *Compendium of Small Millets Diseases* (p. 80). Project coordination cell, All India Coordinated Small Millets Improvement Project, UAS, GKVK Campus, Bangalore.

Nagaraja, A., Kumar, B., Jain, A. K., & Sabalpara, A. N., (2016). Emerging diseases: Need for focused research in small millets. *J. Mycopathol. Res., 54*(1), 1–9.

Padhi, N. N., & Das, S. N., (1982). Host range of the spiral nematode, *Helicotylenchus abunaamai*. *Indian J. Nematol., 12*(1), 53–59.

Padhi, N. N., & Das, S. N., (1986). Biology of *Helicotylenchus abunaamai*. *Indian J. Nematol., 16*(2), 141–148.

Pall, B. S., Jain, A. C., & Singh, S. P., (1980). *Diseases of Lesser Millet* (p. 69). JNKVV, Jabalpur (MP), India.

Pradeep, S. R., & Guha, M., (2011). Effect of processing methods on the nutraceutical and antioxidant properties of little millet (*Panicum sumatrense*) extracts. *Food Chem., 126,* 1643–1647. doi: 10.1016/j.foodchem.2010.12.047. [PubMed] [Cross Ref].

Rao, V. R., & Swaroop, G., (1974). Susceptibility of plants to the spiral nematode *Helicotylenchus dihystera*. *Indian J. Nematol., 4*(2), 228–230.

Sharma, N. D., & Khare, M. N., (1987). Two new smut diseases in little millet (*Panicum sumatrense*) from India. *Acta Botanica Indica, 15,* 143–144.

Ushakumari, S. R., & Malleshi, N. G., (2007). Small millets: Nutritional and technological advantages. In: Krishnegowda, K., & Seetharam, (eds.), *Food Uses of Small Millets and Avenues for Further Processing and Value Addition*. All India Coordinated Small Millets Improvement Project, ICAR, UAS, Bangalore.

Vaishnav, M. U., & Sethi, C. L., (1977). Reaction on some graminaceous plants to *Meloidogyne incognita* and *Tylenchorhynchus vulgaris*. *Indian J. Nematol., 7,* 176, 177.

Vetriventhan, M., & Upadhyaya, H. D., (2016). *Little millet, Panicum Sumatrense, an under-utilized multipurpose crop*. In: *1ˢᵗ International Agrobiodiversity Congress*. New Delhi, India.

CHAPTER 15

Present Scenario of Diseases in Cotton and Their Management

NIRANJAN CHINARA and KAILASH BEHARI MOHAPATRA

Department of Plant Pathology, College of Agriculture, Odisha University of Agriculture and Technology, Bhubaneswar – 751 003, Odisha, India, Tel.: (0) 9439212557, E-mail: niranjanchinara@gmail.com (N. Chinara)

15.1 INTRODUCTION

Cotton (*Gossypium* spp.) known as king of fibers or white gold, belongs to the order Malvales and family Malvaceae. The four cultivated species namely, *Gossypium arboreum, G. herbaceum, G. hirsutum,* and *G. barbadense* contain almost all the varieties of cotton cultivated in India. The most important part of the cotton plant is the fiber or lint, which is used in making cotton cloth. Linters-the short fuzz on the seed-provide cellulose for making plastics, explosives, padding mattresses, furniture, and automobile cushions. Cotton oil is used for edible purpose and the cake used as cattle feed. Major cotton growing states are Gujarat, Maharashtra, Andhra Pradesh, Haryana, Punjab, Madhya Pradesh, and Rajasthan producing more than 80% cotton in India. However, the national productivity is still far below than the global. Among the different constraints of low productivity, disease is one of the major factors.

15.2 ROOT ROT

15.2.1 INTRODUCTION/ECONOMIC IMPORTANCE

Root rot disease of cotton occurs in several countries including Egypt, Greece, India, Israel, Pakistan, Sudan, Trinidad, Uganda, USA, Venezuela, and Republic of Congo. This is one of the most devastating diseases of cotton. The disease affects both the *hirsutum* and *arboreum* species of cotton as it is serious in desi cottons. In India, as loss of 3% has been reported. However, it can go as high as 90% under suitable condition.

15.2.2   SYMPTOMS

The symptoms of the disease are of three types: seedling disease, sore shin (seedling stem canker) and root rot. One or two weeks old seedlings are generally attacked at the hypocotyls leading to black lesions, girdling of stem and seedling death. In sore shin stage, the collar region (stems near the soil surface) of four to six week old plants girdle with dark reddish brown canker, which later turn into dark black. In later stage, the leaves turn bronzing to brown, dry prematurely and the entire plants break at the collar region. The root rot symptom appears generally at the time of maturity of plants. As above-ground symptoms, all the leaves droop suddenly and plants get die within a day or two without showing any symptoms. The most prominent symptom is sudden and complete death of plants in more or less circular patches. The disease starts much early and its aboveground manifestation in the form of wilting is a very late symptom. The entire root systems decay leaving the tap and few lateral roots, help the plant to stand on soil and can be easily pulled out. The bark of roots is broken into shreds and gives yellowish appearance as compared to healthy plants. In severely affected plants, the woody portion of the plant may change to black and brittle. The bark of root as well as the collar region of the affected plants shred and sclerotia is found underneath the bark (Figure 15.1).

15.2.3   CAUSAL ORGANISM

This disease is caused by soil borne fungi, *Rhizoctonia solani* Kuhn and *R. bataticola* (Perfect stage: *Thanatephorus cucumeris* Frank).

15.2.4   DISEASE CYCLE

The pathogens overwinter usually as mycelium or sclerotia in the soil and in or on infected plant parts. The fungi spread with rain, irrigation, or flood water; with tools and anything else that carries contaminated soil; and with infected or contaminated propagative materials.

FIGURE 15.1   Root rot of cotton.

15.2.5   *EPIDEMIOLOGY*

Disease is more severe in soils that are moderately wet than soils under water-logged or dry condition. Infection of young plants is more severe as plant growth is slow in adverse environmental conditions. Dry weather following heavy rains, high soil temperature (35–39°C), low soil moisture (15–20%), cultivation of favorable hosts like vegetables, oil seeds and legumes preceding cotton and wounds caused by ash weevil grubs, nematodes, and agricultural implements induce the aggressiveness of the pathogen.

15.2.6 MANAGEMENT

- Adjustment of sowing time to early (first week of April) or late sowing (last week of June) escapes the crop from high soil temperature conditions.
- Sorghum or moth bean can be adopted as intercrop in cotton to reduce the soil temperature.
- Seed treatment with carbendazim or captan @ 2 g /kg or *Trichoderma viride@* 4 g/kg of seed protects the plants for an initial period of about one month.
- Soil amendment with FYM 10 t/ha or neem cake or mustard cake 150 kg/ha effectively manage the disease.
- Spot drench with 0.1% carbendazim or thiophanate methyl minimizes the further spread of disease.

15.3 SEEDLING BLIGHT

15.3.1 INTRODUCTION/ECONOMIC IMPORTANCE

Seedling disease (damping-off, seedling blight, seed rot hypocotyl rot) of cotton is important in almost all cotton growing areas. It is estimated that the average yield loss is 5% due to seedling blight. The disease may be initiated either by a single fungus species or by a complex of soil borne fungi including *Fusarium* spp., *Pythium* spp., *Rhizoctonia solani* Kuhn, *Thielaviopsis basicola* Berk. and Br. Ferr. These disease causing organisms can attack the seed before or at the time of germination. They can also attack the young seedlings before or after emergence. Disease incidence and severity in a given field are determined by environmental factors such as soil temperature and moisture conditions and by other factors such as seed quality and vigor.

15.3.2 SYMPTOMS

The symptoms include pre germination decay of seed (pre-emergence damping off), decay of seedlings (post-emergence damping off), and partial or complete girdling of the emerged seedlings at or near the soil surface and seedling root rot. Pre-emergence damping-off occurs when seeds are killed before germination or germinating seeds are killed prior to emergence from the soil. But, post-emergence damping-off occurs when the germinating seed has emerged from soil, and then dies. *Rhizoctonia solani* is the most prevalent post-emergence pathogen that produces brown or black lesions on the stems. The lesions may be sunken, which is a symptom known as "sore shin" or the lesions may girdle and pinch the stem at the soil surface, which is otherwise known as "wire stem." *Thielaviopsis basicola* is also a post-emergence pathogen that does not usually cause mortality, but may stunt plants and delay to flower. The fungus blackens the tap root and cortex (exterior) of the hypocotyls and lateral roots of the older plants get killed. *Phoma exigua* causes post-emergence damping off, from the time the seedlings emerge until they are about six inches tall which is characterized by premature dying of cotyledons that turn brown and shrivel. *Pythium* spp. causes pre-emergence damping off in which the hypocotyl of the emerged seedlings gets rot below the soil surface. *Fusarium* also attacks the seed and below-ground parts of the young seedlings.

15.3.3   CAUSAL ORGANISM

The disease is caused by different fungal soil borne plant pathogens, *Rhizoctonia solani, R. bataticola (Macrophomina phaseolina), Phoma exigua (Ascochyta gossypii), Pythium* spp., *Thielaviopsis basicola, Fusarium* spp.

15.3.4   DISEASE CYCLE

Rhizoctonia overwinters usually as mycelium or sclerotia in the soil and in or on infected plant parts and spreads with rain, irrigation, or flood water; with tools and anything else that carries contaminated soil; and with infected or contaminated propagative materials. *Pythium* survives in the soil as oospores and spreads through zoospores. Similarly, *Fusarium* forms chlamydospores as surviving structure and spreads through conidia.

15.3.5   EPIDEMIOLOGY

Poor drainage facilities, susceptible host help in aggressiveness of the disease.

15.3.6   MANAGEMENT

- Use of healthy seeds sowing, shallow planting or planting on raised bed with irrigation facilities.

- Avoidance of waterlogged area and provision of drainage facilities minimize the epidemic nature of disease.
- Basal application of antagonistic fungi and bacteria like *Trichoderma viride, T. harzianum, T. hamatum, Gliocladium virens, Chaetomium globosum, Arthobacter globiformis, Enterobacter cloacae, Erwinia herbicola* and *Pseudomonas fluorescence* with compost or FYM eliminate the inoculum potential in the field.
- Application of fungicides like propiconazole @ 0.1 or ridomyl mz @ 0.1 also minimize the disease.

15.4   FUSARIUM WILT

15.4.1   INTRODUCTION/ECONOMIC IMPORTANCE

This is one of the major diseases of cotton. It is found to occur in all cotton growing tracts. It is a sporadic disease. The loss of crop is due to poor crop stand, small bolls, and poor quality of lint.

15.4.2   SYMPTOMS

The disease attacks all the stages during plant growth. The symptom appears on the cotyledons of the seedlings, which become yellow to brown. The base of the petiole turns to brown followed by wilting and drying of the seedlings. In young and adult plants, discoloration starts from the margin and spreads towards the midribs. The leaves lose their turgidity, gradually turn brown, droop, and fall down. Generally, the symptom starts from older leaves of the base and gradually progresses upward till complete wilting of plant. Yellowing begins at the edge of the leaf; spreading inwards and affected leaves eventually wilt and die. In some cases, partial wilting occurs, where only one branch of the infected plant get affected remaining others are free. The transverse section of vascular system shows brown to black discoloration. Black stripes may be seen upwards to the stem and downward to the lateral roots. The discoloration of internal tissue of stem is similar to that of *Verticillium* wilt but usually appears as continuous browning rather than flecking. Tap root is usually stunted with less number of lateral roots. Often the diseased plants are small with smaller leaves, fewer smaller bolls, prematurely opened. The affected plants are found to be in patches in the field. Symptoms can appear as individual plants or as a small patch, often but not always, near the tail drain or low-lying areas of the field.

15.4.3 CAUSAL ORGANISM

Fusarium wilt of cotton caused by a soil borne fungus *Fusarium oxysporum* Schlechtend. f. sp. *vasinfectum* (Atk.) Snyd. and Hans.

15.4.4 DISEASE CYCLE

The fungus can survive in soil as saprophyte and chlamydospores as resting spores at least 10 years even in absence of cotton crop (Smith et al., 2000). The pathogen is also both externally and internally seed-borne. The primary infection of the crop is due to dormant hyphae and chlamydospores in the soil where as the secondary spread is through conidia and chlamydospores which are disseminated by wind and irrigation water. The pathogen also attack many alternate and collateral hosts that include tobacco, coffee, capsicum, okra, pigeon pea, cowpea, sesame, and rubber.

15.4.5 EPIDEMIOLOGY

The soil temperature of 20–30°C, the optimum being 24–28°C; hot and dry periods followed by rains; heavy black soils with an alkaline reaction; high doses of nitrogen and phosphatic fertilizers; wounds caused by nematode (*Meloidogyne incognita*) and grubs of Ash weevil (*Myllocerus pustulatus*) induce the development of wilt by the fungal pathogen.

15.4.6 MANAGEMENT

- Removal and burning of the infected plant debris in the soil after deep summer plowing during June-July reduce inoculum potential in the field.
- Crop rotation has been recommended with non-host species to reduce presence of *Fusarium* wilt pathogen in the soil; however, long rotation periods (several years) are more effective than short rotations.

- Disease resistant varieties of *G. hirsutum* and *G. barbadense* like Varalakshmi, Vijay, Pratap, Jayadhar, and Verum may be grown in wilt affected areas.
- Application of increased doses of potash along with balanced dose of nitrogenous, phosphatic fertilizers as well as heavy doses of farm yard manure or other organic manures reduce the incidence of disease.
- Treatment of the acid delinted seeds with carboxin or carbendazim at 2 g/kg of seed reduce the initiation of disease at early stage of crop.
- Spot drenching with carbendazim 0.1% reduce further spread of disease.

15.5   VERTICILLIUM WILT

15.5.1   INTRODUCTION/ECONOMIC IMPORTANCE

This disease is widespread in the cotton growing area and is most severe during cool, wet growing seasons. It was prevalent in egg plants and not dominant upon cotton crop due to high temperature during cropping season. The disease spreads in circular patches that increases year by year unlike *Fusarium* wilt which is sporadic in nature. It is aggressive in alkaline and heavy soils. The disease causes a loss of about 90% of the crop and sometimes up to total loss. It not only reduces the yield but also weakens the fibers that fetch low market value.

15.5.2   SYMPTOMS

Plants of all stages of growth are affected but the symptoms appear mostly at blossoming time following rainy weather. It is a typical vascular wilt disease and plants get severely stunted if affected in early stages. The leaf vein of affected plant turns to brown with yellow interveinal areas. The lamella becomes mottled and cupping. The leaves dry up from the margin and tissues between veins undergo necrosis producing a typical "tiger stripe" symptom. Later, the affected leaves fall off leaving bare branches, the characteristic feature of the disease. Longitudinal section of the infected stem close to soil surface shows a slight brown discoloration which is usually more evenly distributed at the center of the stem and lighter brown than discoloration in case of *Fusarium* wilt. The infected plants shed all their leaves leaving young mature ones. The plants bear a few bolls that are smaller, open prematurely producing poor quality of fiber. *Verticillium* wilt reduces the rate of seed germination, boll weight, yield, ginning turnout, and weight of 1000 seeds, plant height, and number of leaves, number of bolls, number of nodes, oil percentage in wet seed, elongation, fiber fineness, length of fiber, fiber strength, fiber uniformity, and percentage of oil in dried seed. Sprouts or new shoots may develop near the base of infected plants (Figure 15.2).

FIGURE 15.2   Verticillium wilt of cotton.

15.5.3   CAUSAL ORGANISM

The disease is caused by fungus, *Verticillium dahliae* Kleb.

15.5.4 DISEASE CYCLE

The pathogen has a broad host range of more than 400 plant species including brinjal, chili, tobacco, and okra and can survive extremely long periods of time (up to 14 years) in the soil as microsclerotia. The primary spread is through the microsclerotia or conidia on seeds and/or in the soil. The secondary spread is through the contact of diseased roots to healthy ones and dissemination of infected plant parts through irrigation water and other implements.

15.5.5 EPIDEMIOLOGY

Several factors, including variety, plant density, pathogen aggressiveness, inoculum density, low temperature of 15–20°C, low lying and ill-drained soils, heavy soils with alkaline reaction and heavy doses of nitrogenous fertilizers influence *Verticillium* wilt development.

15.5.6 MANAGEMENT

- Removal and burning of the infected plant debris in the soil after deep summer plowing during June-July reduce inoculum potential in the field.
- Crop rotation by growing paddy or lucerne or chrysanthemum for 2–3 years reduces initiation of disease.
- Disease resistant varieties like Sujatha, Suvin, and CBS 156 and tolerant variety like MCU 5 WT may be grown in wilt affected areas.
- Application of increased doses of potash with a balanced dose of nitrogenous and phosphatic fertilizers as well as heavy doses of farm yard manure or other organic manures reduce the incidence of disease.
- Treatment of the acid delinted seeds with carboxin or carbendazim at 2 g/kg of seed reduce the initiation of disease at early stage of crop.
- Spot drenching with carbendazim 0.1% or benomyl 0.2% reduce further spread of disease.

15.6 ANTHRACNOSE

15.6.1 INTRODUCTION/ECONOMIC IMPORTANCE

Anthracnose of cotton is widespread in different cotton growing tracts of India. However, it is occasionally noticed in Eastern Maharashtra and summer cotton growing areas in South India. Anthracnose is known to farmers as boll spot or boll rot. The average yield loss due to disease is about 1 to 3%.

15.6.2 SYMPTOMS

All the above ground parts of the plants are attacked. The disease mainly attacks seedling, bracts, and bolls. The pathogen infects the seedlings and produces small reddish or dark brown circular spots on the cotyledons and primary leaves. Later, the lesions develop on the collar region, stem may be girdled, causing seedling to wilt and die. It often causes damping-off of the little plants or kills parts of the cotyledon during cool, wet weather, which is unfavorable for the growth of cotton. This damping-off is similar to that caused by the "sore shin" fungus, *Rhizoctonia,* but the wilting is more sudden and the water soaked appearance is more pronounced in the case of anthracnose attack. In mature plants, the fungus attacks the stem, leading to stem splitting and shredding of bark. The most common symptom is boll spotting. On the bract small water soaked, circular, reddish brown depressed spots appear which may spread to the bolls. On bolls, disease first appears as small round water soaked spots. Later the spots may enlarge and spread almost all the bolls which turn into black and finally have reddish border with pink center. As the disease area extent to the base of the boll, the bolls shed away. The lint becomes brittle and stains to yellow or brown color. The infected bolls do not grow further, burst, and dry up prematurely.

15.6.3 CAUSAL ORGANISM

The disease is caused by *Colletotrichum gossypii* (South-w) and *Colletotrichum capsici.*

15.6.4 DISEASE CYCLE

The pathogen has ability to survive as dormant mycelium in the seed or conidia on the surface of seeds for about a year. The pathogen also perpetuates on the rotten bolls and other plant debris in the soil. Air-borne conidia produced on the infected parts of plant help in further spread of the disease. The pathogen also survives in the weed hosts viz. *Aristolachia bractiata* and *Hibiscus diversifolius.*

15.6.5   EPIDEMIOLOGY

Prolonged rainfall at the time of boll formation, closer planting, dense canopy with warm (29–33°C) humid condition help rapid growth of the disease.

15.6.6   MANAGEMENT

- Removal and burning of the infected plant debris and bolls in the soil; roguing out the weed hosts reduces the chance of pathogen survival in the field.
- Seed treatment with carbendazim or carboxin or thiram or captan at 2 g/kg helps reduction of disease spread at seedling stage of plant.
- Foliar spraying of mancozeb @ 0.2% or copper oxychloride @ 0.3% or carbendazim @ 0.1% at boll formation stage suppresses the aggressiveness of the disease.

15.7   ALTERNARIA LEAF SPOT

15.7.1   INTRODUCTION/ECONOMIC IMPORTANCE

This is one of the most important foliar diseases in almost all cotton growing areas. It is generally found more severe in *hirsutum* and *arboreum* cottons. It is estimated that the crop loss due to the disease is more than 20–30%. Under poor crop management i.e., poor nutrition and ill drainage the disease becomes more severe.

15.7.2   SYMPTOMS

The disease occurs in almost all growing stages of crop. However, it is more severe in the 45 to 60 days old plants. Tiny, dull brown circular spots with reddish border develop on the leaves, which enlarge near about 1 cm in diameter. The peculiar symptom of the disease is development of spots with concentric rings. Mature spots have dry, grayish centers which may crack and drop. Later, the spots may coalesce to form larger lesions followed by extensive defoliation. Development of cankers on stems lead to cracking and breaking of stems. In India, circular lesions develop on the bolls resembling those on the leaves. The seeds may get infected and carry the infection for next generation of the crop (Figure 15.3).

FIGURE 15.3   Alternaria leaf spot of cotton.

15.7.3   CAUSAL ORGANISM

Several species of *Alternaria* infect cotton crop. Among those, *A. alternata* (Fr.) Keissler has been reported from India (Singh et al., 1984).

15.7.4   DISEASE CYCLE

The pathogen is both externally and internally seed borne in nature. The primary source of inoculum is infected seeds. However, secondary spread is through wind and rain splashes. The pathogen also attacks the bolls and grows on exposed lint if bolls open in wet weather, giving rise to contaminated seed. The disease cycle is completed when infected leaves fall to the ground.

15.7.5   EPIDEMIOLOGY

The disease is severe in the fields with poor drainage and inadequate nutrition. High humidity, intermittent rains rather than continuous one and moderate temperature of 25–28°C favor the development of disease. Sporulation of pathogen on the older leaves is more than the younger ones with intermittent wetting of leaves.

15.7.6 MANAGEMENT

- Deep summer plowing, crop rotation with nonhost, use of disease free seed, avoidance of rattoning helps in reducing disease inoculum potential in the field.
- Soil application of *Trichoderma viride,* balanced dose of fertilizers and foliar application of *Pseudomonas fluorescence* reduces the intensity of disease.
- The crop may be sprayed at 15 days interval with mancozeb @ 0.2% or copper oxy chloride @ 0.3% or propiconazole @ 0.1% to protect the crop from the disease.

15.8   CERCOSPORA LEAF SPOT

15.8.1   INTRODUCTION/ECONOMIC IMPORTANCE

The disease occurs in different cotton growing areas of the world. The disease has been reported from Argentina, Brazil, China, Fiji, India, Nigeria, Sri Lanka, USA, and West Indies. It has been found to cause severe damage of cotton in China but has minor importance in India. The disease is associated with physiological stress in the late season.

15.8.2   SYMPTOMS

The disease is mostly found on the older leaves of the mature plant. The spots are small, round to irregular with white center surrounded by purple border. As the lesion enlarges, the center may fall off leaving a ragged margin. The lesions are circular and may vary in size; concentric zones are often present with a red color at the margins. Severely affected leaves become paler, brown, and dry up resulting partial or complete defoliation. Conidial mass appear as dusky shading on the upper and lower surface of the dead tissue of old spots.

15.8.3   CAUSAL ORGANISM

Cercospora leaf spot is caused by *Cercospora gossypina* Cke. (Perfect stage: *Mycosphaerella gossypina)*

15.8.4 DISEASE CYCLE

The fungus survives as conidia and mycelium on seed and crop residue that act as primary inoculum. The pathogen spreads in the field through wind borne conidia.

15.8.5 EPIDEMIOLOGY

The disease spread rapidly under wet soil condition.

15.8.6 MANAGEMENT

- Prevention of drought stress through irrigation, application of proper dose of fertilizer to maintain plant vigor delay primary infections.
- Foliar application of mancozeb @ 0.2% or copper oxychloride @ 0.3% reduce the severity of disease outbreaks.

15.9 MYROTHECIUM LEAF BLIGHT

15.9.1 INTRODUCTION/ECONOMIC IMPORTANCE

Myrothecium leaf blight is also a serious disease of cotton. Most of the American cotton varieties are susceptible. The loss due to the disease varies from 15–37% based on disease intensity.

15.9.2 SYMPTOMS

The disease symptoms are observed on leaves, bracts, and bolls. Initially, minute, light brown to tan colored spots with reddish brown margin develop on the leaves. Later, the spots increase in size beyond 1 cm in diameter and coalesce together forming leaf blight symptom. Sometimes concentric rings of sporodochia (dark greenish black colored pinheads) may form on the spots. In advance, stage, irregular holes, shot holes form on the infected leaves which resembles ash weevil damage. Heavy spotting and defoliation reduce plant vigor and consequently the yield. In severe condition, spots may develop on petioles and stems leads to break of stems. The bolls may infect, lints become discolored and brittle with low market value.

15.9.3 CAUSAL ORGANISM

The disease caused by the fungal pathogen, *Myrothecium roridum* Tode ex Fr.

15.9.4 DISEASE CYCLE

The pathogen is seed borne in nature and infected seeds act as primary inoculum. Secondary infection spreads through rain water and wind. The pathogen also attack different collateral and alternate hosts such as okra, tomato, brinjal, coffee, cow pea, and cluster bean.

15.9.5 EPIDEMIOLOGY

High relative humidity and low temperature favor the development and spread of disease but dry weather condition suppress. It is found in several types of soil with wide range of pH.

15.9.6 MANAGEMENT

- Wider spacing or paired row system reduces the humidity and the disease.
- Early sown cotton plants are less infected by the disease.
- Foliar spraying of mancozeb @ 0.2% or copper oxy chloride @ 0.3% at 15 days interval reduces disease intensity.

15.10 RUST

15.10.1 INTRODUCTION/ECONOMIC IMPORTANCE

Rust of cotton occurs in India, Indonesia, Philippines, Sri Lanka, West Africa, West Indies and tropical parts of America. It is also found in all the cotton growing areas. Generally, the crop of 50–60 days old is attacked by the disease. Most of the American cotton varieties and perennial cottons are attacked. The yield loss due to disease is about 24%.

15.10.2 SYMPTOMS

The disease is mainly restricted to leaves, where uredia and uredospores are formed in form of pustules. Uredia are formed in purplish brown spots and they coalesce to form larger patches. Initially, they are oval and corky and later, become circular on maturity. Numerous yellowish uredospores are released from the ruptured uredia. Cluster of cylindrical aecia develop around the pustules and rupture the epidermis on the lower leaf. Severe infection leads to defoliation. In later stage, rust pustules may be seen on the cotton bolls.

15.10.3 CAUSAL ORGANISM

Rust of cotton is caused by *Phakospora gossypii* Arth.

15.10.4 DISEASE CYCLE

The disease is spread by air borne conidia and method of perpetuation during off season is unknown.

15.10.5 EPIDEMIOLOGY

Heavy rain and high relative humidity favor the development of disease.

15.10.6 MANAGEMENT

As the disease infect very late in the season, chemical control measure of the disease is not profitable. However, spraying of crop with wettable sulfur @ 0.2% control the disease.

15.11 GREY MILDEW

15.11.1 INTRODUCTION/ECONOMIC IMPORTANCE

It is also known aerolate mildew or false mildew or white mold. The disease has been reported in almost all the cotton growing areas as well as all the four species. About 21–68% yield loss occur due to the disease. However, a crop loss up to 90% has been recorded in diploid cotton. The pathogen is generally associated with the crop at physiological stress condition in late season.

15.11.2 SYMPTOMS

The disease usually appears when plant gets maturity. The disease symptom first appears under surface of the older leaves as whitish powdery growth. The entire surface of the infected leaves cover with grey or white powdery growth. Initially, infection on the upper surface of the leaves appears as triangular, square, or irregularly circular whitish spots usually 3–4 mm wide bounded by vein lets. As disease progresses, the smaller spots coalesce together forming larger one. On the lower surface of the lesion, profuse fungal growth occurs giving whitish appearance. Eventually, powdery growth spreads to the upper surface of the leaf. In advance stage, the affected leaves dry up from margin, turn yellowish brown, inward curling, and fall off premature. Infection may also spread to bracts and bolls causing premature opening of bolls and deterioration of fiber (Figure 15.4).

FIGURE 15.4   Grey mildew of cotton.

15.11.3 CAUSAL ORGANISM

The conidial stage of the pathogen is *Ramularia areola* Atk. while, ascomycetous sexual stage is *Mycosphaerella areolla* Ehr. and Wolf.

15.11.4 DISEASE CYCLE

The pathogen survives in the infected crop debris, perennial cotton crops, and volunteer plants during summer season. The viable conidia of the pathogen from the debris act as primary inoculum and infect the cotton crop during winter season. Infected crop debris, wind, irrigation water and rain splashes help the pathogen for further spread of the disease.

15.11.5 EPIDEMIOLOGY

The disease is prevalent in wet and humid condition during winter season. Intermittent rain, low temperature ranging from 20–30°C, dense canopy, excess dose of nitrogenous fertilizer and closer planting invite the disease for its development.

15.11.6 MANAGEMENT

- Removal and burn the infected crop residues.
- Rouging out the self-sown cotton plants during summer season.
- Avoidance of excessive application of nitrogenous fertilizers/manures minimize initiation of disease.
- Wider spacing or paired row system reduces the humidity and the disease.
- Adoption of resistant lines like EC 174092, G-135-9, 30805, 30814, 30826, 30838, and 30856 are found to be the immune to the disease (Mohan et al., 2006).
- Foliar spray of carbendazim (0.1%) at 15 days interval limit the spread of disease.

15.12 BACTERIAL BLIGHT OR BLACK ARM OR ANGULAR LEAF SPOT OF COTTON

15.12.1 INTRODUCTION/ECONOMIC IMPORTANCE

Bacterial blight of cotton is one of the important and serious diseases in cotton growing area of the world. It is more severe in Northern and Central India. Cotton yield loss due to bacteria varies from 1–27% due to the disease (Mishra and Krishna, 2001). In the USA, it is fifth most destructive disease of cotton.

15.12.2   SYMPTOMS

The causal organism, bacterium, attacks all stages from seedling to mature stage of cotton. Generally, five common phases of symptoms are noticed. In case of seedling blight, small, water-soaked, circular, or irregular lesions develop on the cotyledons that spread to stem through petiole and cause withering, death of seedlings. Initially tiny, dark green, water soaked spots develop on lower surface of leaves that enlarge gradually and become angular which is commonly known as angular leaf spot. The spots are visible on both the surfaces of leaves. The infection spreads veins and veinlets causing blackening and is otherwise known as vein blight or vein necrosis. On the lower surface of the leaves, bacterial oozes are formed as crusts or scales. The infection also spreads from veins to petiole and cause blighting which leads to defoliation. In black arm symptom, dark brown to black lesions are formed on the stem and fruiting branches. Later, the lesion may girdle the stem and branches to cause premature drooping off of the leaves, cracking of stem and gummosis. The typical symptom of the black arm is breaking of the stem and hanging of dry black twig from the mature fruiting cotton plants. In case of square rot or boll rot, water soaked lesions appear on the bolls which later turn into dark black and sunken irregular spots. The infection slowly spreads to entire boll and shedding occurs (Figures 15.5 and 15.6).

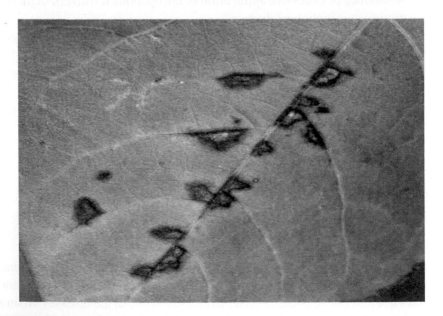

FIGURE 15.5   Angular leaf spot of cotton.

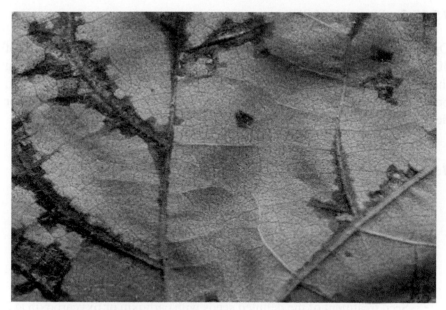

FIGURE 15.6   Vein blight of cotton.

The infection on mature bolls lead to premature bursting. The bacterium spreads inside the boll and lint gets stained yellow because of bacterial ooze and loses its appearance and market value. The pathogen also infects the seed and causes reduction in size and viability of the seeds.

15.12.3   CAUSAL ORGANISM

Bacterial blight of cotton caused by *Xanthomonas axonopodis* pv. *malvacearum* Smith (Vauterin).

15.12.4   DISEASE CYCLE

The bacterium survives on infected, dried plant debris in soil for several years. The bacterium is also seed-borne and remains in the form of slimy mass on the fuzz of seed coat. The bacterium also attacks other hosts like *Thumbergia thespesioides, Eriodendron anfructuosum,* and *Jatropha curcus.* The primary infection starts mainly from the seed. The secondary spread of the bacteria may be through rain splash, irrigation water, insects, and other implements.

15.12.5   EPIDEMIOLOGY

Optimum soil temperature of 28°C, high atmospheric temperature of 30–40°C, 85% relative humidity, rain followed by bright sunshine during the months of October and November, early sowing, delayed thinning, poor tillage, late irrigation, potassium deficiency in soil are favorable for development and further spread of bacteria as well as disease.

15.12.6   MANAGEMENT

- Removal and destruction of the infected plant debris.
- Rouging of the volunteer cotton plants and weed hosts.
- Adoption of crop rotation with non-host crops.
- Adoption of resistant varieties like Sujatha, 1412 and CRH 71.
- Early thinning and early earthing up with potash minimize the perpetuation of pathogen as well as spread of disease.
- Seed treatment of the delinted seeds with carboxin or oxycarboxin at 2 g/kg or soaking of seeds in 1000 ppm Streptomycin sulfate overnight reduce chance bacterial infection at primary growth stage of crop.
- At the initial stage of disease, spraying with Streptomycin sulfate and Tetracycline mixture 100 g along with copper oxychloride at 1.25 Kg/ha to the standing crop protect the crop from the disease.

15.13   LEAF CURL DISEASE

15.13.1   INTRODUCTION/ECONOMIC IMPORTANCE

Leaf curl disease is an important viral disease of cotton. This was first reported from Nigeria in the year 1912. In 1987 it was severe form damaging the cotton crop in Pakistan. The disease occurs in Algeria, India, Libiya, Morocco, Sudan, Tunisia, and Pakistan. The loss due to this disease is about 50.7% in India.

15.13.2   SYMPTOM

The affected young seedlings show severe curling of leaves and stunting of growth. Such plants do not bear flowers and bolls at all. In the grown up plants dark green small bead like structure on the leaf lamina which subsequently turn into thickening of veins and veinlets at the lower surface of the leaves.

Such leaves show upward or downward curling with twisted petioles. Formation of cup like structures from the nectar glands known as enation is the characteristic symptom of the disease. Severely affected plants became bushy with dark green color, short internodes, without flowers and bolls. The primary stems grow taller than healthy one. Diseased plants often found to be stunted.

15.13.3 CAUSAL ORGANISM

The leaf curl disease is caused by Gemini virus which transmitted by white fly *Bemisia tabaci.*

15.13.4 EPIDEMIOLOGY

The disease is transmitted by the insect vector white fly (*Bemisia tabaci*) which acquire the viral particles while feeding on a diseased host and transmit to the healthy host. Except cotton, virus, and the vector affect different host plants like brinjal, tomato, okra, tobacco, and chili and weed hosts like *Abutilon indicum, Tribulus terrestris, Achyranthus sp., Melilotus indica, Sida sp.*, etc. In India, the disease appears in June itself when the temperatures are high between 42–47°C (Raj et al., 2002).

15.13.5 MANAGEMENT

There is no such control measure to minimize the virus infected disease. However, application of dimethoate 30 EC @ 1.5 ml/lit. or oxydemeton-methyl 25 EC @ 2 ml/lit. or imidachloprid 200 SL @ 3 ml/10 lit., thiomethoxam 25 WG @ 2.5 g/10 lit or Acephate 95 SG @ 2 g/lit. minimize further spread of disease (Mandal and Chinara, 2014).

15.14 NEW WILT (PHYSIOLOGICAL DISORDER)

15.14.1 INTRODUCTION/ECONOMIC IMPORTANCE

Cotton new wilt is otherwise known as Para wilt, transitory wilt and Adilabad wilt. The disease was first time reported from Adilabad of Andhra Pradesh during 1978–79 on intra-*hirsutum*, JKHy 1. The loss caused by the wilt depends upon the stage of the plant when it is affected. Its incidence varies from 5–50%; however, in some areas disease covers about 80% of population (Bhale and Raj, 1984).

15.14.2 SYMPTOMS

The disease is characterized by sudden wilting of plant at flowering and boll formation stage. Young leaves turn light green and start droop down. Various degree of reddening found on leaf, petiole, stem, and branches. The top leaves wither and dry up, entire foliage dry within a week. The squares shed, bolls burst prematurely. There is no vascular discoloration found in the affected plants. Some plants revive after irrigation or rain but, such plants bear less and small bolls.

15.14.3 CAUSAL ORGANISM

No pathogen is associated with this type of wilting cotton plants. Most of the hybrid plants are affected by the wilt and is also suspected to be a genetically controlled physiological disorder.

15.14.4 EPIDEMIOLOGY

Prolong dry spell with high temperature followed by heavy rainfall or irrigation induce this wilt. The plants with large canopy and heavy boll formation are more prone to wilt. Under waterlogged condition, the higher uptake of nutrients for conversion of photosynthates into protein, and other macromolecules get reduced and similarly, the root system gets degenerated lead to new wilt. Sandy loam soil helps more severe incidence of wilting in all the susceptible plants than black cotton soil. Synthetic pyrethroids induce the development of new wilt disorder in susceptible hybrids, but there is negative role of insects in spread of disease.

15.14.5 MANAGEMENT

- *Gossypium arboreum* and *G. herbaceum* are the wilt tolerant genotypes.
- Avoidance of water logging condition as well as provision of irrigation facilities during grand growth phase to escape from prolong dry spell situation in the field.

- Making ridges a month after planting conserve the moisture and reduce the possibility of wilt.
- Excess use of farm yard manure and fertilizers may avoid the wilt in heavy soils.
- Foliar application of DAP (2%) soon after showing the symptom increase the tolerance against the wilt.

KEYWORDS

- **bacterial blight**
- *Gossypium arboreum*
- **grey mildew**
- **leaf curl disease**
- **photosynthates**
- *Tribulus terrestris*

REFERENCES

Bhale, N. L., & Raj, S., (1984). New wilt in cotton. *Proc. Group Discussion on New Wilt in Cotton* (p. 59), CICR, Nagpur.

Mandal, S. M. A., & Chinara, N., (2014). Chemical control of sucking pests in cotton. *Journal of Plant Protection and Environment, 11*(1), 60–61.

Mishra, S. P., & Ashok, K., (2001). Assessment of yield losses due to bacterial blight of cotton. *Journal of Mycology and Plant Pathology, 31,* 232–233.

Mohan, P., Mukeswar, P. M., Singh, V. V., Khadi, B. M., Amudha, J., & Deshpande, V. G., (2006). Identification of sources of resistance to grey mildew disease (*Ramularia areola*) in diploid cotton (*Gossypium arboreum*). *CICR Technical Bulletin, No. 34,* p. 10.

Monga, D., & Raj, S., (2002). *Twenty Five Years Achievements in Cotton Pathology at CICR* (pp. 42–45).

Raj, S., Monga, D., & Gupta, K. N., (2002). Leaf curl. In: Sheo, R., (ed.), *Twenty Five Years Achievements in Cotton Pathology* (pp. 16–25). At CICR (1976–2001), Central Institute for Cotton Research, Nagpur.

Singh, M., Narain, U., & Singh, M., (1984). Leaf spot of Deshi cotton caused by *Alternaria alternata. Indian Journal of Mycology and Plant Pathology, 14,* 171.

Smith, S. N., Devay, I. E., Hseih, W. H., & Lee, H. I., (2000). Soil borne populations of *Fusarium oxysporum* f.sp. *vasinfectum*, a cotton wilt fungus in California fields. *Mycologia*.

CHAPTER 16

Disease Spectrum in Jute and Their Management

RAJIB KUMAR DE

ICAR–Central Research Institute for Jute and Allied Fibers, (ICAR), Nilganj, Barrackpore, Kolkata–700120, India
E-mail: rkde@rediffmail.com

ABSTRACT

Among all diseases of jute, stem rot incited by *Macrophomina phaseolina* is economically most important. Although commonly known as stem rot, but any part of the plant may be infected by the pathogen at any stage of growth right from germination to harvest producing various symptoms, like, damping-off, seedling blight, leaf blight at seedling stage, stem rot, collar rot, stem break, root rot at adult plant stage and brown spot on pods especially in seed crop. Other important diseases of jute are anthracnose, black band, soft rot, jute mosaic, and root knot nematode. In Hooghly wilt, *Ralstonia solanacearum* is the actual pathogen, whereas stem rot fungus and root knot nematode help in facilitating the entry of the original pathogen. The disease is seed, soil, as well as air borne. Management of fungal diseases of jute and allied fiber crops involves management of soil condition for soil borne diseases, treatment of seed with fungicides for seed borne diseases and application of fungicides to check the air borne spores, or good combination of both. Rouging of infected plants and spraying of insecticides to check vectors could prevent the spread of the viral diseases.

The diseases of Jute have been discussed below with introduction/ economic importance, symptoms, casual organism, disease cycle, epidemiology, and integrated management strategies.

16.1 JUTE CROP DISEASES

16.1.1 INTRODUCTION

Jute crop (*Corchorus olitorius* L. and *C. capsularis* L.), also called as 'golden fiber' crop, is cultivated predominantly in eastern region of India in an area of 0.77 million hectares and produces of 10 million bales (one bale is equivalent to 180 kgs) with dry fibers productivity of 2329 kg per hectare. It is cultivated as pre-kharif crop mainly in the states of West Bengal, Bihar, and Assam with percentage contributions to National production are 81.29, 11.63 and 6.25%, respectively (Anonymous, 2013). In spite of declining popularity among farmers due to availability of cheap synthetic substitutes, recently jute has emerged with positive attributes like, more production of oxygen, absorption of carbon dioxide and fuel wood yield, besides its diversified products which are biodegradable.

Recently due to need for finer quality fiber and globalization of diversified products, the disease management becomes more important. Jute fiber consists mainly of cellulose, hemicellulose, and lignin. Microorganisms capable of producing enzymes that can degrade cellulose, hemicellulose, or lignin are potential pathogens resulting into significant loss in fiber production of crops like jute, flax, Mesta, sisal, ramie, andsunnhemp. In general, fiber crops except flax are grown in season having high rainfall with hot temperature and more relative humidity and that allow the crops to produce huge volume of green biomass in them and simultaneously make them more vulnerable to various diseases. Each fiber crop has its specific climatic requirements for optimum growth. Attempts made by the entomologists and pathologists to select biological control measures have been frustrated due to such wide varying meteorological conditions in various areas during normal season cycles.

Among diseases caused by fungi, stem rot incited by *Macrophomina phaseolina* is economically most important. Other major diseases affecting jute crop are anthracnose, black band, soft rot, jute mosaic, root knot nematode, and Hooghly wilt. Management of fungal jute and allied fibers diseases involves manipulation of soil condition for soil borne diseases, treatment of seed with fungicides for seed borne diseases and foliar spraying of fungicides to reduce air borne spores, or judicious combination of both. Removal and destruction of infected plants and application of insecticides to check vectors could reduce the virus diseases. Diseases of jute crop and their causal organisms are listed below:

Crop: Jute (*Corchorus olitorius* L. and *C. capsularis* L.)

Sl. No.	Name of disease	Nature	Causal organism
1	Stem rot	Fungal	*Macrophomina phaseolina*
2	Hooghly wilt	Bacterial	*Ralstonia* (= *Pseudomonas*) *solanacearum* (*Macrophomina phaseolina*, *Meloidogyne incognita* may facilitate entry of bacteria by making injury).
3	Anthracnose	Fungal	*Colletotrichum corchorum*; *C. gloeosporioides*
4	Black band	Fungal	*Botryodiplodia theobromae*
5	Soft rot	Fungal	*Sclerotium rolfsii*
6	Tip blight	Fungal	*Curvularia subulata*
7	Stem gall	Fungal	*Physoderma corchori*
8	Mildew	Fungal	*Oidium* sp.
9	Sooty mould of pods	Fungal	*Cercospora corchori*, *Corynespora cassicola*, *Alternatia* spp.
10	Root gall nematode	Nematode	*Meloidogyne incognita*, *M. javanica*
11	Jute mosaic	Viral	A Begomo virus under family Geminiviridae, vector: *Bemisiatabaci* Genn. (whitefly).
12	Jute Chlorosis	Viral	A member of Tobravirusgenus
13	Yellow vein disease	Viral	A bipartite Begomo virus, vector: whitefly (*Bemisiatabaci* Genn.)

16.2   FUNGAL DISEASES

Large number of fungal pathogens causes damage to jute crop infecting all plant parts including root, stem, leaf, and pods. Besides soil borne, fungal diseases are both seed borne and air borne in nature. Hence, it becomes complicated to manage fungal jute diseases by either management of soil condition for soil borne diseases, treatment of seed with inhibitory chemicals for seed borne diseases, and timely foliar spraying of fungicides to check the air borne spores, or good combination of both. Prophylactic measures to manage diseases are always better than curative application of toxic or hazardous chemicals because several ecological problems could be avoided by the former.

16.2.1   STEM ROT OF JUTE

16.2.1.1   ECONOMIC IMPORTANCE

Initially in 1916, K. Sawada of Japan reported this disease, also designated as die-back, for the first time from Formosa. Later, it was reported in course of time from India where it was known as Stem-Rot disease of Jute.

Among biotic constraint of raw jute production, stem rot incited by *Macrophomina phaseolina* (Tassi) Goid. is economically the most serious disease in both cultivated species affecting adversely both yield and quality of fiber. Although it is commonly known as stem rot, but any part of the jute plant may be attacked by the pathogen at any stage of growth right from germination to harvest producing various symptoms, like, damping-off, seedling blight, leaf blight at seedling stage, stem rot, collar rot, stem break, root rot at adult plant stage and brown spot on pods especially in seed crop. The disease is seed, soil, as well as air borne. It damages the crop beginning from germination to maturity in both seed and fiber crops wherever it is grown in India and other countries. Management of soil, pre-sowing seed treatment and timely application of effective fungicides or good combination of all are vital for management of jute stem rot (Roy et al., 2008).

Stem rot is the economically most threatening disease causing fiber yield reduction and deterioration of fiber quality in two commonly cultivated species of jute (*Corchorus olitorius* L. and *C. capsularis* L.). Yield of jute seed crop also declines both quantitatively and qualitatively due to this stem rot disease producing infected seeds. The crop continues to suffer from this disease during the full duration from germination to maturity. Besides soil borne, it is seed borne and air borne through spores. Hence, use of resistant/ tolerant variety is economically viable tool in present day disease management programme and is important core component of integrated disease management strategy also.

16.2.1.2 SYMPTOMS

The disease is very commonly found in all the jute growing areas not only in India but in all other countries as well wherever jute is grown. Two cultivated species, i.e., *Corchorus olitorius* and *C. capsularis* are equally damaged by this disease.

Common name is stem rot but various symptoms, like, damping-off, seedling blight, leaf blight, stem rot, collar rot, root rot, stem break and brown spot on pods especially in seed crop are observed due to the pathogen infection any part of the plant at any stage of growth right from germination to harvest. In damping off, newly emerged seedlings rot above and under the soil level. In seedling blight, cotyledons become brown to black later and seedlings die. As a result of stem rot infection during June-July, leaves become brown in color due to prevailing high humidity. Dark brown lesion on green stem may extend vertically or horizontally, 10–15 cm or higher. Plants wilt and leaves droop or stems break in high speed wind. Infected plant dries up and slowly becomes dark brown to black and stand distinctly visible in the field. Collar rot symptom consists of brown rot in the collar region. In root rot symptom, plants wilt, defoliate, become brown to black later and stand as naked stem, and finally die (Figures 16.1 and 16.2). Roots are rotten and brown, rootlets absent. In seed crop of jute, it also produces dark brown and black colored spots on pods and cause seed infection.

FIGURE 16.1   Seedling damping off and stem rot symptom of stem rot disease in jute.

FIGURE 16.2 Damping off, leaf blight, stem rot, drooping of leaves, stem break and root rot symptoms of jute stem rot disease.

16.2.1.3   CASUAL ORGANISM

Macrophomina phaseolina (Tassi) Goid

Macrophomina phaseolina (Tassi) Goid is the pycnidial stage of pathogen. The sclerotial stage is *Rhizoctonia bataticola* and perfect stage is *Orbilia obscura*. The pycnidial and sclerotial stage are mainly responsible for the disease and perfect stage is hardly seen (Ghosh, 1957; Mandal, 1990). The fungus may infect wide range of hosts. The causal fungus is seed borne, soil borne and air borne. Once introduced into the field through infected seed, the pathogen can survive for many years in soil (Mandal, 2001).

1. **Pathogenic Variation:** Wide range of variation was observed in fungus isolated from different hosts from different regions of the globe in their morphological shape and size and their range was from less than 100 to above 400 μ. Based on the size, Haige (1930) grouped them into four classes, *viz.*, A, B, C, and D. Isolates from jute collected from various jute growing places in India belonged to group C of Haige. But within such group, also there is morphological variation and pathological variation and they have been further divided into four sub-groups based on their sclerotial morphology and pathogenicity (Mandal et al., 1998). The pathogen were identified into eight races (physiologic) by workers at BJRI, Dhaka (Ahmed and Ahmed, 1969), while four strains were recognized based on virulence in India (Ghosh and Sen, 1973a). The growth rate of these four isolates in initial phase (5 days) was analyzed with their virulence. The most favorable temperature and pH were 34 ±1°C and 6.8, respectively (Ghosh and Sen, 1973b). Out of eight isolates

tested on 20 jute varieties (of both the species), six isolates showed varied degrees of disease reactions while similar resistant reaction were observed in the remaining two against all the varieties. From the cultural studies, it was observed that all the eight isolates were different from one another and eight cultural races were fitted in a dichotomous key (Ahmed and Ahmed, 2005).

2. **Detection of *M. Phaseolina*:** Detection of the pathogen from field samples by a simple method of direct PCR (dPCR) without DNA extraction involved leaf bits lysis buffer, incubation, and the lysate was used as PCR template. Lysis buffer systems were dependent on type of tissue. For leaf, the buffer system composed of tris (hydroxy-methyl aminomethane (Tris)-Cl, EDTA, sodium acetate, proteinase K, PVP, and β-mercaptoethanol. For stem, PVP was not applied with higher concentrations of other components. From both leaf and stem, *M. phaseolina* may be detected (Biswas et al., 2014).

3. **Molecular Variability:** Genetic variation in a great number of isolates of *M. phaseolina* from seeds and other infected plant parts of different crop species was studied using PCR based random amplified polymorphic DNA (RAPD) marker and many other modern advanced molecular tools. A miniprep yielding enough amount of quality DNA in comparison to other processes and a standard PCR method was reported (Biswas et al., 2012) that may amplify DNA in seed with more mucilage and secondary molecules that hinders isolation of DNA and amplification by PCR. Among the *M. phaseolina* isolates obtained from cotton, corn, sorghum, and soybean no variation was found in RFLP, RAPD, PCR analysis of ITS region, 5.8S rRNA and portion of 25S rRNA (Su et al., 2000). Presence of double-stranded RNA (dsRNA) was detected in *M. phaseolina* isolates from Mexico, Somalia, and United States and reported a close relationship between presence of dsRNA and hypovirulence in the fungus with reduction in mycelial growth as well as capacity to cause disease. Genetic diversity among *M. phaseolina* isolates from various hosts and locations did not suggest any significant difference among primer pairs and AT% in AFLPs (Pecina et al., 2000). Relationship between micro-sclerotia size and pathogenicity were not observed in diversity pattern of 30 (15 Mexican and 15 non-Mexican) isolates of *M. phaseolina* obtained from different hosts analyzed on the basis of morphological parameters and pathogenicity on seeds of bean and endoglucanase genes (Beas-Fernandez et al., 2006).

Variation in virulence and sclerotial morphology was found in the *M. phaseolina* isolates causing stem rot of jute. Among 13 isolates of from Assam, Bihar, Orissa, and West Bengal, the isolate from Sorbhog (Assam) showed maximum virulence and it was followed by Barpeta (Assam) isolate (Mandal et al., 1998). Using a simplified AFLP technique and two different methods of analysis, assessment of genetic relation among *M. phaseolina* isolates was reported (Vandemark et al., 2000). The primer OPA-13 (5'-CAGCACCCAC-3') showing profiles of fingerprint clearly distinguished the 43 *M. phaseolina* isolates and UPGMA based analysis divided these isolates into five groups (Jana et al., 2003). Among 12 universal rice primers using PCR, 5 were efficient in showing polymorphic fingerprints profiles among the 40 *M. phaseolina* isolates (Jana et al., 2005). Isolates from a single host were found to be ordinarily similar to one another as observed in RAPD and PCR-RFLPs of the ITS region of *M. phaseolina*, but those from other hosts were distinctly different. In the same host, isolates sensitive to Chlorate were distinct from chlorate-resistant ones. In restriction patterns of the ITS region, including part of 25S rDNA, a high degree of polymorphism has been reported by Purakayastha et al. (2005). *M. phaseolina* (114) isolates from Asteraceae, Euphorbiaceae, Fabaceae, and Poaceae and from cultivated and non-cultivated soils showed isolates from the Poaceae were less pathogenic and formed pycnidia less frequently compared with isolates from dicot host tissue. Chlorate-utilization phenotype was unrelated to host tissue source (Mihail and Taylor, 1995).

4. **Pathogenesis:** Pectinolytic and cellulolytic enzymes play an important part in the process of pathogenesis of blight of seedlings of jute caused by *M. phaseolina*. These two enzymes were made by *M. phaseolina* constitutively and inducibly. They were isolated from diseased and surrounding regions. Distinct infected spots were found in 14-day old jute seedlings kept in enzyme solution at 21°C for 48 hours (Chattopadhyay and Raj, 1978). Of the pectic enzymes contained in three variants of *M. phaseolina,* MP-C was observed to be more virulent in causing infection and retting jute indicating a relationship of pectic enzyme with stem rot disease of jute and retting (Ali et al., 1969; Myser Ali et al., 2005).

16.2.1.4 EPIDEMIOLOGY AND FORECASTING OF STEM ROT

Environmental factors and soil parameters, *viz.*, soil moisture, air temperature and relative humidity influence the progress of stem rot of jute. Average rainfall influenced disease development along with relative humidity and soil moisture. Host and pathogen factor had also profound effect on disease progress and development. Susceptibility of jute plants to stem rot increased with age irrespective of varieties and disease reached peak or maximum during harvesting time. Rainfall, proneness of the jute plant, virulence, and inoculum power of the isolate played crucial role in disease development (Ghosh and Purakayastha, 1985). Early sown crop during March is more susceptible to *M. phaseolina* causing stem rot and root rot. Heavy rainfall with overcast cloudy condition resulting in near field capacity soil moisture, high atmospheric humidity, air temperature below 32°C and soil temperature below 30°C favor infection (Rao, 1979).

16.2.1.5 VARIETAL RESISTANCE

True resistant variety against *M. phaseolina* causing jute stem rot is not available among either of two cultivated species, at least at farmers' level. But the level of resistance/susceptibility varies from variety to variety and also within the vagaries of monsoon from place to place, which may be attributed to the pathotypes present in a particular area, climate, and soil of that place. The most ruling variety, JRO 524 and JRO 632 of *C. olitorius* and JRC 212 and JRC 321 of *C. capsularis* showed differential reactions at different places and in the same place also with pathogen, isolates form different places (Mandal et al., 1998, 2000). When the whole set of germplasm available at CRIJAF, Barrackpore were assessed against the pathogen, quite a good number were totally free from the disease at Barrackpore and Budbud field condition (Saha et al., 1994; Mahapatra et al., 1994). But when the short listed *C. capsularis* were further exposed to more vulnerable situation at Sorbhog, Assam, most of them showed symptoms of severely affected by the disease. Only six accessions (CIM 036, CIM 064, CIN 109, CIN 360, CIN 362, CIN 386) out of 196 showed their confirmed resistant reaction (Mandal et al., 2000). Some accessions of wild species of *Corchorus*, like, *C. aestuans, C. fascicularis, C. pseudo-olitorius, C. tridens* and *C. trilocularis* have shown very high degree of tolerance against the disease under Barrackpore condition (Pulve et al., 2003, 2004). Out of 14 capsularis and 27 olitorius lines, JRO 514 and IR 1 were highly resistant to *M. phaseolina*

(Thakur Ji, 1973). De and Mandal (2007c) developed a simple inoculation technique of *M. phaseolina* by leaf tip cut and wet cotton swab on resistant CIM 036 and susceptible JRC 412.

In order to identify *C. olitorius* germplasm accessions tolerant to *M. phaseolina*, De, *and* Mandal (2012) undertook an investigation at Ramie Research Station, Sorbhogin the district of Barpeta in Assam state of India, which is also called as 'the hot spot' for the disease. In the first year of investigation, only 293 germplasm accessions, which also included previously shortlisted entries and few new collections, were tested in 'the hot spot' along with 2 check varieties, viz., JRO 632 and JRO 524 and out of those 293 only 19 were selected on the basis of PDI. These 19 and two check varieties were further evaluated during 2007 and only 8 accessions were selected. These lines, namely, OIN 107, OIN 125, OIN 154, OIN 157, OIN 221, OIN 651, OIN 853 and OIJ 084 and two check varieties were again grown for rigorous screening during 2008. The two check varieties showed higher PDI during three years consecutively. Only four accessions namely OIN 125, OIN 154, OIN 651 and OIN 853 showed moderately resistant reaction based on mean PDI 5.0 or below and were selected finally. The disease reaction of these 4 entries was also confirmed in soil culture in pot tests. Among thirteen lines of jute, six lines, namely, OEX-27, OIN-125, OIN-154, OIN- 467, OIN-651 and OIN-853 were found to show moderately resistant reaction with disease rating (1.1–5) under epiphytotic condition naturally at Barrackpore and Bahraich location (Meena et al., 2015). Interestingly, all the four accessions of De and Mandal (2012), namely, OIN 125, OIN 154, OIN 651 and OIN 853 confirmed their resistant reaction against *M. phaseolina* causing jute stem rot in separate studies of Meena et al., 2015.

16.2.1.6   JUTE STEM ROT SCORING METHODOLOGY

For precise evaluation of germplasm lines for resistance to stem rot considering all the types of symptoms produced, a rating scale was developed (Mandal and De, 2007; De and Mandal, 2012) in jute. The stem rot disease was scored observing size of lesion (1–4), position of lesion on the stem (1–4) and lesion type (1–8) with maximum value of 16 (= 4+4+8) and minimum of 3 (= 1+1+1), number of stem rot infected plants (Number × score value) and dead plants by root rot infection (Number × 16).

   i.   Lesion size, namely, minor dots or lesion size of less than 0.5 cm² was scored with 1.0, lesion size of 0.6–1.0 cm² was rated as 2.0, likewise, lesion size of 1.1–2.0 cm² with 3.0 and lesion size of greater than 2.0 cm² with 4.0.

ii. Position of lesion on the first one fourth portion from the tip of the stem of the plant carried 1.0 point, similarly, on the next second quarter it was 2.0, on the next third quarter it was 3.0 and finally on the last quarter it was 4.0.

iii. Lesion type covering below 10% diameter of the stem was given 1.0 point, in the similar fashion, lesion covering 10.1–25.0% diameter of the stem was 2.0, 26.1–40.0% was 4.0 and greater than 40.0% carried 8.0 points.

iv. Then percent disease index (PDI) was found out by sum of all numerical ratings multiplied by 100 and dividing it with number of plants observed multiplied by highest value, as given below.

v. Accessions showing less disease based on the PDI were selected and further tested in with three replications.

Sum total of numerical ratings:

$$PDI = \frac{\text{Sum total of numerical ratings}}{\text{Number of plants observed} \times \text{Highest value}} \times 100$$

Selected accessions were further evaluated under pot culture condition for confirmation using suitable method of inoculation with JRO 524 and/or JRO 8432 as susceptible checks.

16.2.1.6.1 Comparison of Three Methods of Inoculation (Leaf, Soil, and Stem)

De (2016) compared three methods of inoculation namely, by leaf, soil, and stem, in order to devise a suitable method of inoculation for large scale use. Among four leaf inoculation methods, tip cut and cotton swab method was better than inoculating floating leaf disc (8 mm), floating leaf and cotton swabbed detached leaf with typical symptom with light to dark brown spots in 100% inoculated leaves within 48 hours on susceptible line JRC 412 which enlarged on further incubation.

Among three different times of soil inoculation, 7 days before sowing was more effective in causing infection in JRO 524 and JRO 8432 with 26.6 and 23.3% stem rot than soil inoculated during sowing (15%) and 7 days after sowing (10%). Resistant reaction of four lines (OIN 125, OIN

154, OIN 651 and OIN 853) was confirmed by these three soil inoculation methods with JRO 524 as check.

Stem inoculation technique was highly efficient and caused 100% infection with typical stem rot symptoms with brown spots of different length and intensity encircling the stem in JRO 524 compared to earlier methods. The variation in virulence pattern of isolates was evident, as more virulent isolate (from Sorbhog, Assam) produced distinctly clear, longer, and darker brown stem rot lesion than less virulent ones (from Barrackpore, West Bengal) (De et al., 2014).

16.2.1.7   DISEASE CYCLE

Stem rot disease of jute spreads through seed, soil, as well as air in the field. The pathogen survives in soil and/ or in infected crop residues or root stubbles for long time in absence of host crop. It had very large host range with greater than 500 plant species belonging to greater than 50 different families. Seeds collected from infected crop produce infected seeds which upon germination produce diseased plants. Besides, abundant sporulation is often observed on infected seeds and stems with pycnidia with an ostiole. Presence of airborne conidia during crop season may be responsible for secondary infection and may cause epiphytotic outbreak of stem rot disease in susceptible variety. The gestation (incubation) period, i.e., time interval between the first appearance of the stem rot symptom on stem or leaf and the death of the plant ranges between one and twenty weeks. Disease cycle figure of stem rot of jute (Mehak, 2016) is presented below.

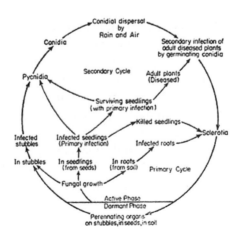

16.2.1.8 MANAGEMENT STRATEGIES

16.2.1.8.1 Cultural Management

1. **Selection of Land:** Heavy soil with low permeability is so compact when wet and hence, the crop lacks proper aeration around the root zone. Such crops remain stunted and suffer from stem rot and therefore drainage supports a good and disease free crop (Ghosh and Basak, 1965). K_2O/CaO ratio in leaf and stem of *C. capsularis* was an indicator of susceptibility or proneness to infection by *M. phaseolina* or *R. batatoticola* as assessed in an investigation on comparatively more resistant varieties JRC 918, JRC 212 and D 154 along with the highly susceptible JRC 412 (Mandal, 1976). Since the pathogen is seed, soil, and air borne, it remains in soil as either living on the dead soil-organic matter and jute stubbles as inactive sclerotia or in the weed population harboring its active phase as mycelium or pycniospores (Ghosh, 1983; Mandal, 2001). Even when the treated seeds are sown, they cannot protect the seedlings for very long time if the inoculum density of *M. phaseolina* in the soil is very high and damping off, seedling blight and collar rot symptoms are observed. The fungus from the field soil attacks root system also. When only the branch roots are attacked, the plants may grow and remain alive with reduced vigor showing rusty brown branch roots, but if the tap root is infected and damaged the plant generally dies (Som, 1977).

2. **pH of Soil and Liming:** Jute crop in acid soils with pH 6.0 and below is more readily attacked than a crop at neutral soil. With continuous cultivation of jute, soil is depleted of calcium, potassium, and other base substances whereas, the amount of iron, manganese, and aluminum increases due to acidity development. But under acidic conditions with low availability of potash, the mere addition of lime only does not help to reduce disease, and application of potash simultaneously is imperative (Mandal, 1976). Depending upon pH level, application of lime or dolomite @ 2–4 t ha^{-1} minimum 3–4 weeks before sowing once in 3–4 years is helpful in acidic soil. Selection of land, therefore, plays vital and crucial role.

   The soil where jute is grown in the state of Assam and few districts of northern West Bengal is acidic in nature and therefore the incidence of stem rot is more with high plant mortality as compared to soil of jute fields in the districts of southern West Bengal like,

Murshidabad, Nadia, Hooghly, and North 24 Parganas, where the soil is more or less neutral in nature.

3. **Selection of Variety:** Though resistant variety as such is not yet available, the cultivated varieties showed differential reaction to the pathogens of different places (Mandal et al., 2000) and accordingly, the varieties should be selected considering their status of suscepti-bility in a particular area.

4. **Adjustment of Date of Sowing:** Sowing date of crop influences the amount of damage by diseases as it determines the susceptible phases of the host crop during the period of high available inoc-ulum population of the fungal pathogen in conducive environment. In a field study with varied dates of crop sowing of jute variety JRO8432, it was conclusively evident that early sown crop showed more stem rot, caused by *M. phaseolina*, than crops sown at later dates. At 30–90 days after sowing (DAS), incidence of stem rot was reached its peak (29.03–56.55) in sowing of extra early crop. At 90 DAS, crops sown during mid-March (28.98), first April (22.48), mid-April (22.48) and end April (12.88) followed it sequentially. Stem rot disease showed declining trend in later sown crop. Crop sown very late were affected with minimum disease of 0.84%. Across all the dates of sowing, stem rot was low at 30 DAS initially but as the crop grew older it slowly progressed in crops sown later but rapidly in early crops to a reach a higher peak at 90 DAS. With the recommendation of delaying the date of sowing just by a fort-night from first part of April to middle to end, part of April signifi-cantly modified the disease picture as a whole. Dry fiber yield of 34.54 q/ha was maximum in crop sown during first part of April and it was closely followed by mid-April sown (33.9 q/ha) and mid-March sown (32.86 q/ha) jute (De, 2013).

5. **Spacing and Plant Population:** Close spacing is conducive to stem rot infection (Ghosh, 1963). Optimum spacing (line sowing in *C. capsularis* from one row to next row at 30 cm and from one plant to next plant at 5–7 cm and *C. olitorius* from one row to next row at 20–25 cm and one plant to next plant at 5–7 cm, and broadcast sowing for both *C. capsularis* and *C. olitorius* at 10–15 cm from one plant to next plant) results in low damage by disease and gave good yield. The pathogen shows a very broad host range and a large number of weeds serve as the primary source of inoculum. There-fore, weeding, and simultaneous thinning at 20–25 and 35–40 DAS and clean cultivation are essential for a good healthy crop. Since

water logging increases stem rot and root rot damage, good drainage is necessary to prevent such a situation.

De and Tripathi (2016) carried out a field investigation, to determine the effect of optimum plant population with modern sowing method on jute stem rot causing high plant death. Two different sowing methods, viz., broadcast, and line sowing and different plant population densities starting from 3–10 lakhs/ha of jute were tested. With increasing plant population density, in highest plant population density of 10 lakh per ha, the incidence of stem rot increased reaching peak of 16.3 and 27.4%, respectively, in broadcast and line sown crop. Minimum stem rot incidence of 5.4 and 7.4% was recorded in very low population density of 3 lakh per ha inline sown and broadcast crops. Stem rot incidence was at medium level of 7–9% in line sown and 10–17% in broadcast crops in optimum plant population densities of 5–6 lakh per ha. When similar level of plant population densities were compared, a low level of jute stem rot was always recorded in line sown crops than their counter part in broadcast crops. Disease dynamics of jute stem rot over the entire crop growth period in line sown crop with low plant density of 3 lakhs / ha, the disease enhanced from initial low level of 0.2% at 30 DAS to 3.7% at 90 DAS and reaching finally to 5.4% at maturity in 120 DAS. On the contrary, at 5–6 lakhs /ha plant density level, the jute stem rot disease increased from initial low level of 0.1% at 30 DAS to 3–4% at 75 DAS, 4–6% at 90 DAS and finally reached to 7–9% at 120 DAS. On the other hand, the jute stem rot appeared with 0.7% at 30 DAS, slowly increased to 5, 15, 20% at 60, 75, 90 DAS, respectively, and finally reached a peak of 27.4% at the harvest stage of the crop at 120 DAS in broadcast crop with highest plant density of 10 lakhs /ha.

6. **Fertilizer Application:** Jute needs a substantial quantity of nitrogen for required growth, vigor, and yield, but application of N beyond 80 kg ha^{-1} and in new alluvial soils even above 60 kg ha^{-1} promotes incidence of stem rot, collar rot and root rot (Ghosh, 1983). Basal application of potash in soil @ 50–100 kg K$_2$O ha^{-1} can decrease the disease to a great level (Cheng and Tu, 1970). Loam or clay loam, medium to high land with organic matter level and having pH 6.5–7.0 or 7.5 is suitable. FYM application@ 7–8 t ha^{-1} is always useful for in case of organic matter deficient soil (Ghosh and Basak, 1965). Stem rot increased with increasing level of N and it reached maximum at 80 kg ha^{-1}. Stem rot was reduced with soil application

of micronutrients (Zn, Fe, Bo) along with NPK (Thakur Ji, 1974; Thakur Ji et al., 1976).

In the era of intensive and efficient cropping system, many other crops are cultivated both before and after jute. It is vital to understand that many familiar crops are also prone to *M. phaseolina*/*Rhizoctonia bataticola* (Ghosh, 1983). Therefore, non-host crop may be wisely selected for rotation. Rice/wheat/mustard crop to follow jute but not potato or other solanaceous vegetables for greater than three years consecutively.

In a field study, response of nitrogenous, phosphatic, and potassic fertilizer schedules was studied by De (2017) on stem rot disease on a new variety JRO 8432 of jute. Application of more nitrogenous fertilizer enhanced jute stem rot. But increase in potassic and phosphatic fertilizers reduced stem rot. Among the different NPK fertilizer schedules, 120:30:30 attracted more jute stem rot with 47.9% disease at 120 DAS, followed by 120:40:40 with 40.5%, 100:30:30 with 36.1% and in 80:40:40 with 35.2% stem rot. With similar dosage of nitrogen, when phosphate and potash were increased, lower stem rot was recorded. Phosphate and potash fertilizer moderated the effect of deleterious effect of nitrogen by reducing the stem rot incidence. In check, stem rot increased from initial 0.3% to 2.2% in 45 days and finally reached to 21.6% during the harvesting stage of the crop. But in different NPK fertilizer treated plots, dynamics of jute stem rot was different showing varied interaction. The disease progress over time period was typically slowest in NPK 40:30:30. Higher level of nitrogen with phosphorus and potash not only increased susceptibility of jute plants and killed more plants due to high prevalence of stem rot disease but also increased dry fiber yield.

16.2.1.8.2   Chemical Management

Once the pathogen enters into the field soil through infected seed, the fungus can survive in the soil for number of years (Mandal, 2001). Seed treatment is a most effective and economic step to reduce the losses due to this disease.

1. **Seed Treatment:** Elimination of the pathogen from seed is possible by the treatment with organo-mercuric compounds (Mukherjee and Basak, 1973), but these compounds have been established to be

hazardous from several considerations. It is for this fact some very effective non-mercurials were searched out for seed treatment.

Fungicide may directly reduce the seed borne inoculums at the first hand and may create a protective barrier around seed and later on growing radical and plumule and further to young seedlings, if the fungicide is systemic to plant system, and provide protection from soil borne inoculums depending upon the weather and nature of the chemical used. Treatment of seeds can be visibly evident by physical presence of powder or liquid on the surface of treated seeds. However, liquid formulation of fungicides is not preferred for seed treatment, as it may enter into the embryo and hamper germination, as especially in case of tebuconazole and hexaconazole. Investigations are in progress on combined pre-sowing seed treatment with both insecticides and fungicides with good compatibility (using specialized formulation products) with more promising results against early insect pests and diseases. Economics of seed treatment shows that it results in higher benefit cost ratio than any other agricultural practice or input, known so far.

2. **Spraying:** of Blitox 50 WP or Fytolan 50 WP (copper oxychloride) @ 5.0–7.0 g l^{-1} or Bavistin 50 WP (carbendazim) @ 2.0 g l^{-1} of water is helpful to keep the disease under control. In higher infection, alternate sprays of Blitox and Bavistin at 10 days intervals (total three sprays) are recommended.

3. **Seed Transmission:** Infection of seed of jute ranged 1.4–2.1% under natural field condition. *M. phaseolina* treated seeds sown in sterilized soil reduced germination by 38% due to pre-emergence rotting (46%) and post-emergence rotting (7%). In component plating method of jute seed, *M. phaseolina* was mostly restricted to seed coat part of the seed (Sarkar, 2014). Disease-free quality jute seed production was studied by altering the sowing dates and fungicide schedule (Sarkar et al., 2016).

    Pod formation was the most susceptible stage for seed infection for the seed borne pathogens of jute and fungicide applied at this stage produced healthy seeds. Single spray of carbendazim or copper oxychloride at pod formation stage may be necessary to manage seed infection by the pathogens (Mandal and De, 2010; De, 2012b; De and Mandal, 2012b).

4. **Effect of Pre-Sowing Soil Application of Bleaching Powder:** The hyphal growth of *M. phaseolina* was completely checked in food poisoning technique *in vitro* at 5000 μg/ml of bleaching powder or

calcium hypochlorite [Ca(OCl)$_2$] both at 24 and 48 hours after incubation. At lower dose growth inhibition decreased. At 100 and 1000 µg/ml, 87.3, 93.3% and 93.6, 96.6 inhibitions were observed after 24 and 48 hours, respectively. During further incubation of 24 hrs at a concentration of 5000 µg/ml of bleaching powder, no fresh growth was observed indicating inhibition of *M. phaseolina* completely.

In the field experiment, out of all doses of bleaching powder, soil application @ 30 kg/ha 7 days ahead of sowing was found best against jute stem rot compared to check (with no soil application) and higher (50–150 kg/ha) and lower (5–20 kg/ha) doses in all the four dates (30–120 DAS) of observations. It restricted jute stem rot to 2.1 and 6% in comparison to 15.1 and 24% in untreated check at 90–120 DAS, respectively. As the dose of soil application of bleaching powder enhanced from @ 5–120 kg/ha, the stem rot of jute decreased slowly reaching minimum at 30 kg/ha indicating that this dose being most effective against jute stem rot. Atmospheric carbon dioxide and water react with bleaching powder to release hypochlorous acid which gives a characteristic smell to the bleaching powder. Hypochlorous acid decomposes readily to atomic oxygen. This atomic oxygen acts as bleaching and lethal agent through oxidation. Increasing soil pH and adding calcium to soil, besides direct detrimental effect on pathogen itself may be responsible reason for reduction of jute stem rot using bleaching powder in the field.

The soil microbial population in jute rhizosphere was studied post soil application of bleaching powder. Drastic decline in soil population of bacteria, fungi, and other microbes was observed was observed seven days after pre-sowing soil application of bleaching powder in jute. Soil population of bacteria, fungi, and other microbes was, however, increased 15 days after soil application. Finally, fifteen days after soil application, population of bacteria, fungi, and other microbes was stabilized and at par with initial level (De et al., 2014).

5. **Effect for New Fungicides:** Lowest stem rot incidence of 25% was recorded in pre-sowing seed treatment followed by foliar spraying of tebuconazole one month after sowing and it was at par with carbendazim (28.5%) and hexaconazole (28.6%) compared to 45.2% in check. They were the most effective fungicides against jute stem rot pathogen and it was also observed in- *in vitro* observation. Tricyclazole, copper oxychloride and mancozeb followed

them, respectively, with 33.4, 33.5, and 35.9% of stem rot incidence. Among the other fungicides, thiophanate methyl was least effective against stem rot. In tebuconazole and carbendazim, disease progress was slowest in all the dates of crop growth from 30 to 90 days after sowing. It is also worth to mention that propiconazole (10 µg/ml), turmeric oil (10 µg/ml), carbendazim 50 WP (25 µg/ml), copper oxychloride 50 WP (50 µg/ml), tebuconazole EC (50 µg/ml), hexaconazole EC (100 µg/ml), curcumin mixture (100 µg/ml) and tricyclazole (10000 µg/ml) were responsible for causing complete inhibition of *M. phaseolina* under *in vitro* tests (De, 2014).

Effect of thiram, carbendazim, copper oxychloride, and mancozeb was investigated on *M. phaseolina* causing jute stem rot *in vitro*. Carbendazim completely checked the radial growth of isolates of *M. phaseolina* from Barrackpore only at 25 µg/ml. Copper oxychloride followed carbendazim closely with 100% inhibition in 50 µg/ml concentration. The radial growth inhibitions of 85–87.22 and 43.05% were separately observed in 100 µg/ml of thiram and mancozeb, respectively, in comparison to untreated check. Growth inhibitions of 91–92 and 80.61% were recorded in 1000 µg/ml of thiram and 5000 µg/ml of mancozeb. Near 50% inhibition of growth as 33–85% in 25–50 µg/ml of thiram, 40–60% in 5–10 µg/ml of copper oxychloride and 51% in 200 µg/ml of mancozeb were noted (Hembram and De, 2008, 2010).

Carbendazim (Bavistin 50 WP) @ 2.0 g kg⁻¹ or Dithane M 45 (mancozeb) @ 5.0 g kg⁻¹ and Captan are very effective compounds for this purpose (Som, 1977; Mandal, 1997). In case of severe infection, alternate sprays of Blitox and Bavistin at 10 days intervals (total three sprays) are recommended (Mandal, 1997). De et al. (2005, 2007) reported that two promising herbicides, namely, trifluraline, and quizalofop ethyl completely inhibited the growth of *M. phaseolina in vitro* at 1000 mg/ml and 25 mg/ml, respectively; apart from their usual weed management ability in jute. Two organic fungicides, Dorina, and Stup were also tested against *M. phaseolina*.

6. **Ready Reckoner of Fungicides in Stem Rot Management in Jute** *in vitro:* A large number of fungicides and chemicals were evaluated *in vitro* against *M. phaseolina* inciting stem rot in jute and following few were found promising (De, 2014; De and Mandal, 2010). Turmeric oil, propiconazole, and Carbendazim caused 100% inhibition of growth at a lower dosage of active ingredients.

Diseases	Pathogen	Fungicides	Maximum Dosage (µg/ml*) Tested	Dosage (µg/ml*) Causing 100% Inhibition	Remarks	May be Recommended (Yes/No)
Stem rot	*Macrophomina phaseolina*	Thiram	5000	2000	OK	Yes
		Carbendazim 50 WP	5000	25	Excellent	Yes
		Copper oxychloride 50 WP	5000	50	Very good	Yes
		Mancozeb 75 WP	5000	-	-	No
		Hexaconazole EC	1000	100	Good	Yes
		Tebuconazole EC	1000	50	Very good	Yes
		Propiconazole 25% EC	10,000	10	Excellent	Yes
		Tricyclazole 75% WP	10,000	10,000	Poor	No
		Turmeric oil	10,000	10	Excellent	Yes
		Curcumin mixture	10,000	100	Very good	Yes
		Bleaching powder [Ca(OCl)$_2$]	5000	5000	Excellent	Yes
		Dorina-Organic fungicide	25,000	10,000	Very good	Yes
		Stup-Organic fungicide	10,000	-	-	-
Black band	*Botryodiplodia theobromae*	Carbendazim 50 WP	1500	5	Excellent	
		Copper oxychloride 50 WP	5000	-	-	No
		Mancozeb	5000	2000	OK	Yes
Soft rot	*Sclerotium rolfsii*	Carbendazim 50 WP	1500	-	Poor	No
		Mancozeb 75 WP	2000	2000	Good	Yes
		Copper oxychloride 50 WP	5000	-	Poor	No

16.2.1.8.3 Biological Control

1. **Disease Control through Biological Control Agents:** Growing concern for environment and human health, problems of resistance, resurgence, lethal to beneficial organisms, loss of biodiversity especially vulnerable species has restricted the wide spread application of fungicides in plant disease management. In fact, a large number of good pesticides usage have been banned. Comparatively, lesser toxic chemicals are being used now and at the same time, continuous search for new and more efficient biological control agents are in progress. Biological control is quintessential component of integrated plant disease management. It provides long term management of plant diseases in a sustainable manner. The pathogens are inhibited by competition (for nutrients, space, and oxygen, etc.), myco-parasitism (direct physical damage) and antibiosis (toxins or enzymes) and their populations were reduced or completely eliminated. Bio-control agents are isolated from nature, tested on pathogens in laboratory for efficacy, screened for most efficient strains with many desirable attributes, mass cultured on cheap media, tested for viability on long term storage and then become ready for application in field.

   In the past, a good number of fungal antagonists and plant growth promoting rhizobacteria (PGPR) were tested with promising results at the farmers' fields. The same are being tried also at present for disease control in jute and allied fiber crops. Among them, *Trichoderma viride, Aspergillus niger* AN 27 and some species of fluorescent *Pseudomonas* became very effective biocontrol agents against stem and root rot in jute. Seed inoculation with bacterial biofertilizer, PGPR, and biocontrol fungi alone and soil drenching with vesicular arbuscular mycorrhyza (VAM) *Glomus mossae* after sowing; and biocontrol fungi alone reduced root rot disease from 10–37%, increased plant biomass by 11–28% and enhanced fiber yield by 5–23%. When both were cultured in a plate, pathogen inhibition was maximum (55%) with *Ps. striata*, compared to other PGPR (Bandopadhyay and Bandopadhyay, 2004a). Inoculation of seed with PGPR, *viz.*, fluorescent *Pseudomonas, Azotobacter, Azospirillum, Ps. striata* and in soil furrow application of VAM *Glomus mossae* singly, or in compatible combinations significantly reduced root and stem rot disease complex of jute incited by *Macrophomina* and *Pseudomonas* (Bandopadhyay and Bandopadhyay, 2004b). Seed coating with *T. viride* and soil drenching with similar antagonist has decreased root disease in jute under different agro-ecological conditions (Anonymous, 1949–1989, 1990–2006).

Different combinations of *Trichoderma, Azotobactor, Aspergillus,* and fluorescent *Pseudomonas* effectively controlled root diseases of jute and increased plant biomass and fiber yield. The effectiveness of PGPR in decreasing soil borne pathogens and improving the plant growth and vigor may be explained by their ability to produce siderophores (iron chelating agents) or toxins or enzymes or growth hormones (Bandopadhyay, 2002; Bandopadhyay et al., 2006; De et al., 2009). Isolates of fungal antagonists, like, *Trichoderma, Gliocladium, Aspergillus, Penicillium,* PGPR isolates, fluorescent *Pseudomonas,* phosphobacter *Ps. striata, Bacillus, and Azotobacter* exhibited promising antagonistic characteristics by inhibition of highly virulent isolates (R 9) of *M. Phaseolina* to a large extent (Bandopadhyay et al., 2006, 2009). Biochemical characterization of most efficient bioagents, namely, isolates of fluorescent *Pseudomonas, Bacillus* spp., *Azotobacter* spp. with respect to production of volatile and nonvolatile antibiotics, enzymes, growth regulators, siderophores, and phosphorus solubilization was reported by Bandopadhyay et al., 2007a, b, 2008). *Aspergillus versicolor* grown in compost medium for 10 day under diffused light reduced jute root rot by *M. Phaseolina* by 56% in pot culture (Bhattacharyya et al., 1985).

Soil treatment with *Trichoderma viride* thrice, i.e., 7, 15, 30 DAS, at Bahraich in Uttar Pradesh, India, was best in checking seedling blight, collor rot, stem rot and root rot diseases of jute with lowest disease incidence (1.4, 3.0, 4.7 and 4.9) in comparison to untreated control (16.1, 9.4, 16.6 and 16.3, respectively). It was followed by *T. viride* as soil application twice at 7 and 15 DAS with 2.0, 4.4, 6.1 and 6.1% incidence. Carbendazim 50WP seed treatment @ 2 g kg^{-1} resulted in 8.5% seedling blight, 5.2% collar rot, 6.9% stem rot and 8.5% root rot. *T. viride* at 7, 15 and 30 DAS gave highest (25.7 q ha^{-1}) mean dry fiber yield; Carbendazim showed 22.2 q ha^{-1} while it was lowest in control (Srivastava et al., 2010).

In a field experiment with eco-friendly treatments in JRO-524 jute, it was observed that 50% N: P: K + seed treatments with *Azotobacter* and phosphorus solubilizing bacteria @ 5 g/Kg+ *Trichoderma viride* (seed treatment @ 5 g/Kg and soil application @ 2 kg/ha at 21DAS) + *Psuedomonas fluorescens* spray @ 0.2% at 45DAS was superior to farm yard manure @ 5 t/ha + seed treatment with *Azoto-bacter* and PSB @ 5 g/Kg+ *T. viride* (seed treatment @ 5 g/Kg and soil application @ 2 Kg/ha 21AS)+ *P. fluorescencs* spray @ 0.2% at 45AS (Meena et al., 2014).

Occurrence of stem rot of jute was reduced significantly from 15–39% by pre sowing seed dressing and soil application 15 DAS with formulations of consortia *(Trichoderma + Gliocladium + Asper-gillus + Pseudomonas)*. Plant biomass enhanced up to 46–68%. Talc and fly ash based fungal consortium with Azotobacter increased fiber yield 36 and 32% respectively. The benefit-cost ratio was maximum for talc and fly ash based seed dressing fungal consortium formulates (Bandopadhyaya and Das, 2017). Attempt was reported by Bhattacharyya et al. (2017) for combining biocontrol agents with fungicides, weedicides, and plant growth regulators for management of stem and root rot of jute.

2. **Botanicals as Alternative:** Bio-efficacy of different botanicals, namely, neem (*Azadirchtaindica*), ramie (*Boehemerianivea*), hatisoor (*Heliatroium indicum*), bilakhani (*Tephrosiaindica*), garlic *(Allium sativum), kalmegh (Andrographis paniculata)* and turmeric *(Curcuma longa)* were studied against *M. phaseolina* causing jute stem rot *in vitro* and field experiment with JRO 8432. The absolute aqueous extract from variety of plants were applied as overnight seed soaking and followed by a foliar spraying @ 0.6% after 45 days. Neem components, particularly leaf extract and seed kernel suspension exhibited high inhibition of both pathogens at 10–100%. *Tephrosiaindica* showed 10–30% inhibition of *M. phaseolina* whereas ramie leaf appeared ineffective on both pathogens. Extract of neem leaf at 100 and 50% concentration decreased hyphal growth of *M. phaseolina* and *P. parasitica*. Extract of neem seed kernel was effective at 100 and 50% inhibiting the pathogens. Neem oil suspension was less effective on *Macrophomina* while considerably effective against *Phytophthora* inhibiting them *in vitro*. Leaf extract of neem was most effective in decreasing incidence of jute stem rot from 22.7% in check to 12.0%. Neem was closely followed by extracts of turmeric, garlic, and kalmegh at maturity of the crop. The rate of growth of stem rot incidence was very less in neem extracted as compared to check. Turmeric extract also arrested the hyphal growth of stem rot to a large extent. Aqueous extract of *kalmegh* was also adversely active decreasing stem rot compared to check. Overnight seed soaking in turmeric, garlic, and kalmegh extract boosted

seedling growth and vigor initially indicating growth promotion effect (De, 2012). Extract of *Tephrosia candida* inhibited sclerotia formation and germination in *M. phaseolina* (De et al., 2009).

The pure extract of both younger and old leaves of *Abromaaugusta* (Devil's cotton), a Malvaceous plant, was evaluated against *M. phaseolina*. The inhibition of radial growth *M. phaseolina* was increasing with greater concentration of test botanical in both cases. The extract of younger leaves showed greater inhibition on test pathogen as comparison to older leaves owing to presence of more antimicrobial factors. Inhibition percentage of mycelial growth in separate concentrations of 5, 10, 15, 25 and 50% younger leaf extract was 8.8, 13.6, 17.5, 28.8 and 42.2% after 96 h. However, in older leaf extract percent growth inhibition was observed as 4.1, 8.6, 14.1, 16.9, and 19.7 compared to untreated check (De et al., 2015).

The specific active components in these botanicals may be identified, purified, produced in large scale and finally commercialization and this may open up a new avenue of opportunities for better and sustainable management of important diseases of jute in future.

16.2.2   ANTHRACNOSE

16.2.2.1   ECONOMIC IMPORTANCE

Anthracnose is economically more serious in *C. capsularis* jute. *Olitorius* jute is rarely affected. In case latter is affected, it occurs at the very late stage of plant growth when the plants reaches almost harvest stage and the lesions are also not deep enough to affect fiber portion of the plant. In *capsularis* jute numerous spots are visible on the stem and consequently the infected plants die in most cases. In case plants survive, the fiber is speky or knotty that leads to the 'cross bottom' fiber.

16.2.2.2   GEOGRAPHIC DISTRIBUTION

The anthracnose disease occurs regularly in the *capsularis* belt of India covering states of Assam, Northern part of West Bengal, Bihar, and Uttar Pradesh. It is also prevalent in Bangladesh (Ghosh, 1957; Mandal, 1990). In all probability, anthracnose incited by *Colletotrichum corchorum* entered India during thirties along with jute germplasm from Southeast Asia, particularly Taiwan unknowingly. No report on this disease was found in Indian jute literature prior to 1945, when it was first reported on a variety 'Jap-Red,' a *capsularis* introduction, from Formosa (Taiwan) at Daccca farm, now in Bangladesh (Ghosh, 1957). Then from Dhaka, it further spread to other portions of Bangladesh. It later entered India through Assam.

16.2.2.3 SYMPTOMS

The symptoms of anthracnose disease comprised of sunken spots of various colors on different part of plants, namely, leaves, stems, fruits, or flowers. These lesions often enlarge and later with favorable weather, they lead to wilting, withering of infected plants, and ultimately dying of infected plant tissues (Hiremath et al., 1993).

In anthracnose of jute, irregular spots appear on stem that may coalesce, because deep necrosis, girdle stem and cracks and expose the fiber (Figure 16.3). These slowly turn to brownish depressed spots. Depressed spots are observed on pods also. Infected seeds are lighter both in weight and color, shrunken, and germination is poor.

FIGURE 16.3 Anthracnose in white jute.

16.2.2.4   CAUSAL ORGANISM

Colletotrichum gloeosporioides (Penz.) Penz. and Sacc.

Two different species of *Colletotrichum* infect two different species of *Corchorus*. *Colletotrichum corchori* for *capsularis* jute and *Colletotrichum gloeosporioides* for *olitorius* jute. The pathogen was described by Ikata and Yoshida (1940) as *C. corchorum*.

During the interaction between host, pathogen produce specialized structures, viz., conidia, acervulli, setae, and appressoria. A number of acervuli and conidia are formed by *C. gloeosporioides* by colonization and injuring plant tissues. The fungal conidia are hyaline, one celled, ovoid to oblong, slightly curved or dumbbell shaped. Acervuli are black, superficial (erumpent) and scattered. Conidiophores are simple, straight, hyaline bearing conidia singly. Conidia are falcate and germinate forming appressoria (Ghosh, 1957). Conidia are hyaline, one-celled, and straight, cylindrical, and obtuse at apices.

Rain splash or overhead irrigation helps conidia to disseminate over relatively short distances and initiate new infection on other healthy plant tissues. For penetration into host tissues this fungal pathogen forms specialized infection structures called appressoria. These appressoria help the fungus to penetrate the host cuticle.

16.2.2.5   CLASSIFICATION

Kingdom: Fungi
Division: Ascomycota
Class: Sordariomycetes
Order: Phyllachorales
Family: Phyllachoraceae
Genus: *Colletotrichum*
Species: *gloeosporioides*
Teleomorph: *Glomerella cingulata* (Stoneman) Spauld. and H. Schrenk.

16.2.2.6   DISEASE CYCLE

Jute anthracnose is mainly soil, seed, and air borne. The pathogen is cosmopolitan in distribution and has very wide host range. Primary inoculum is disseminated by wind or rain (Farr et al., 2006; Purkayastha, and Sengupta, 1973).

16.2.2.7   EPIDEMIOLOGY

Warm and humid conditions are necessary for the pathogen to infect different plant hosts, which includes gymnosperm, angiosperms, ornamental, and fruit plants, vegetables, crops or even grasses. High relative humidity coupled with continuous rain and temperature around 35C are favorable for this anthracnose. Epidemic of anthracnose was observed in an exotic variety, Japanese Red at Chinsurah, West Bengal in 1950–51 (Ghosh, 1957).

16.2.2.8   MANAGEMENT

1. **Cultural:**
   - Removal of affected plants and other hosts from the field and clean cultivation are useful.
2. **Chemical:**
   - Seed treatment using Bavistin @ 2 g kg^{-1} of seed or Captan @ 5 g kg^{-1} of seed may be recommended to eliminate primary source of inoculum. Seed lots having 15% or more infection should not be used even after treatment and be discarded straightway without any consideration.
   - Spraying Bavistin @ 2 g l^{-1} or Captan @ 5 g l^{-1} or Dithane M 45 @ 5 g l^{-1} is advisable to reduce the spread of this disease.

16.2.3   SOFT ROT

16.2.3.1   ECONOMIC IMPORTANCE

Soft rot disease is still not so serious in jute crops but certainly, a caution should be maintained for the jute growers due to its increasing trend. This was also a minor disease, but now gradually increasing. The disease is present in all the jute growing areas but intensity is still low.

16.2.3.2   SYMPTOMS

Attack of soft rot begins when the jute crop is 80–90 days old. The fungus first grows from soil and later slowly infects fallen leaves of jute. From there it infects stem base.

Soft, brown wet patch appears on the stem base. Skin peels off and exposed fiber layers turn rusty brown in color and plants wilt. White cottony mycellial growth is first clearly visible at the stem base and later brown mustard seed like sclerotia are observed in large numbers at the point of infection. Stem becomes weak and often may break in high wind (Figure 16.4). Soft rot decreases if fallen leaves are destroyed from the field.

FIGURE 16.4   White cottony mycelial growth at stem base of jute and mustard seed like sclerotia at infection site of stem which often breaks.

16.2.3.3   CAUSAL ORGANISM

Sclerotium rolfsii

Sclerotium rolfsii is very notorious fungus causing huge damage to agricultural crops worldwide. This is mainly a soil borne pathogenic fungus. This pathogen can survive in soil for many years even without any host crop forming hard and resistant sclerotia in soil. The pathogen has very wide host range and majority of agricultural crops are damaged by this pathogen with different degree of intensity and assumes severe form in certain pockets. Both *capsularis* and *olitorius* jute are affected (Ghosh, 1983).

16.2.3.4   EPIDEMIOLOGY

The fungus initially grows in the litter of fallen jute leaves. In hot weather and wet soil, it grows and initiates infection in the collar region. High rainfall, low sun shine, high plant population favor soft rot.

16.2.3.5   MANAGEMENT

1. **Cultural:**
   - Deep plowing during summer season to expose soil to hot sun.
   - Clean cultivation by removing the leftover of all previous crops and destruction by burning or burying under the soil, etc. are essential to reduce primary inoculum.
   - Follow long term crop rotation using non-host crops.
   - Adapt green manuring with sunnhemp or *dhaincha*.
   - Use of farm yard manure or organic manure during land preparation.
   - Soil solarization using polythene sheet to increase inside soil temperature by 15–20°C.
   - Use of neem or mahua cake.
   - Biological control with application of *Trichoderma viride* in soil or through FYM are few good options for the checking of this harmful pathogen.

2. **Chemical:**
   - Spraying of copper oxychloride @ 4 g l^{-1} of water directed towards the basal region may contain the spread of soft rot disease.

16.2.4   BLACK BAND

16.2.4.1   ECONOMIC IMPORTANCE

This disease is still not very alarming and is of minor importance now. This was a minor disease in the past, but now gradually increasing with change in cropping pattern and introduction of new varieties of jute. Now incidence is quiet high. Often no fiber and seed may be obtained from infected plant.

16.2.4.2   SYMPTOMS

Stem may break at the place of infection causing plants death. As dark colored black spots appear on the infected stem, initially, it may often be confused with stem rot because of similarity in symptom. On rubbing with finger on the infected spots, profuse black sooty powdery spore mass adheres to the fingers, which is not observed in case of stem rot. Crops raised from infected seeds show seedling blight symptoms. It attacks two species of jute in all the jute cultivating areas but requires attention due to slow increase in intensity.

16.2.4.3   CAUSAL ORGANISM

Botryodiplodia theobromae.

The pathogen is seed borne and air borne. It has very large host range (Mandal, 1990).

16.2.4.4   MANAGEMENT

1. **Cultural:** Since the pathogen has large wide host range, clean cultivation is very crucial as the infection may come from any other host. Removal of stubbles, crop residues and timely weeding and thinning are important.
2. **Chemical:** Seed treatment using Bavistin 50 WP (carbendazim) @ 2.0 g kg^{-1} or Dithane M 45 (mancozeb) @ 4.0 g kg^{-1} may be advocated. Spraying of Blitox 50 WP or Fytolan 50 WP (copper oxychloride) @ 4.0 g l^{-1} or Bavistin 50 WP (carbendazim) @ 2.0 g l^{-1} of water showed effective management of the black band of jute (Mandal and De, 2010).

16.2.5   SEED BORNE FUNGI OF JUTE

Fungi recorded in commercially available jute seed samples included *M. phaseolina, C. gleosporiodes, C. corchori, B. theobromae, S. rolfsii, Alternarai,*

Curvularia, Fusarium, Aspergillus, and *Penicillum,* etc. Range of seed infection by *M. phaseolina, B. theobromae* was from 0.25–21.5 and 0.25–1.5% (Mandal, 1987, 2001).

Experimental samples were collected from various seed sources, viz., Andhra Pradesh, Maharashtra, and local markets in West Bengal showed presence of both *M. phaseolina* and *B. theobromae.* The extent of infection was however very low (0.25–7.25%). Other fungal pathogens recorded in jute seed were *Colletotrichum gloeosporiodes, C. corchori, Curvularia* sp., *Sclerotium rolfsii* (contamination), and different *Fusarium* and *Alternaria* species.

Highest seed infection with *M. Phaseolina* and *B. theobromae* took place when inoculation was done at pod formation stage and it was followed by stage of 50% flowering, before seed maturity and pre-flowering stage of inoculation. Pod formation was the maximum vulnerable stage for seed infection for the seed borne pathogens of jute and measure taken at this phase produced healthy seeds. *M. phaseolina* and *B. theobromae* were very sensitive to carbendazim and complete growth inhibition was obtained even at as low as 10 ppm concentration, whereas with copper oxychloride no inhibition was obtained up to 200 ppm. Carbendazim was most effective fungicide showing mean seed infection percentage only 0.8 (90.7% over check) followed by copper oxychloride with 1.2% infection (86.5% control). By a single spray of carbendazim or copper oxychloride at pod formation stage, management of seed infection by both the pathogens may be possible (Mandal and De, 2010; De, 2012b; De and Mandal, 2012b). In plating method of different component of jute seed, *M. phaseolina* was mostly confined to the seed coat (Sarkar, 2014).

16.3 BACTERIAL DISEASES

16.3.1 HOOGHLY WILT

16.3.1.1 ECONOMIC IMPORTANCE

During late forties and early fifties, a typical wilt disease in jute (*C. olitorius*) was observed for the first time in Tarakeswar areas of Hooghly district, West Bengal. The malady was so widespread and the loss to jute crop was so severe that a new Pest and Disease Control Center needed to be established at Tarakeswar in the heart of Hooghly district. This disease was later widely spread in other areas of nearby district and also adjoining areas of surrounding districts, where jute used to be followed by potato in winter.

The benchmark survey estimated 30–34% loss of jute crop every year between 1950 and 1954 (Anonymous, 1949–89). During late eighties and early nineties, 5–37% disease was recorded in Kamarkundu area of Hooghly district and 2–20% in some areas of Nadia and North 24 Parganas districts (Mandal, 1986; Mandal and Mishra, 2001). The Hooghly wilt was recorded in *olitorius* jute in North Bengal (Coochbehar) but *capsularis* jute was free (Mandal and Khatua, 1986). In a survey in West Bengal, this bacterial wilt was found prevalent in jute along with many other crops and weeds with 1–91% incidence (Mondal et al., 2011).

16.3.1.2   SYMPTOMS

Typical drooping and wilting of leaves starts from the base upwards. Affected stems are soft slimy fluid comes out on slight pressing. Ooze test for bacterial diseases is positive here with quick release of bacterial cells into clear water turning it into turbid. Later plants defoliate making only naked stem with root remaining healthy and intact to stand erect in the field which is visible from a long distance. Close examination of the infected plants clarifies the difference with wilting due to stem rot.

16.3.1.3   CAUSAL ORGANISM

Earlier it was believed that this Hooghly wilt was incited by a complex combination of *Macrophomina phaseolina, Pseudomonas solanacearum* and *Fusarium solani* and was widespread in Hooghly where jute is followed by potato (Mishra, 1980). Ghosh (1961) coined the name "Hooghly wilt" and isolated seven fungi and a bacterium from the infected or dead plants. It appears that new term 'Hooghly wilt' has been accepted by the jute pathologists (Chattopadhyaya and Sarma, 1979). Later on by repeated experimentation it was established that *Ralstonia solanacearum (=Pseudomonas solanacearum)* is the original pathogen, whereas *Rhizoctonia bataticola* and *Meloidogyne incognita* help in facilitating the entry of the original pathogen. Presence of these pathogen causing root rot (*R. bataticola/ M. phaseolina*) and nematode causing root knot (*M. incognita*) increases the Hooghly wilt as these create wound in the roots, and thus facilitate more primary bacterial pathogen, *R. solanacearum* to make entry (Mandal, 1986; Mandal and Mishra, 2001; Mandal and Ghosh, 2002).

16.3.1.4 MANAGEMENT

1. **Cultural:** Potato or other solanaceaous crops in the crop rotation need to be avoided at any costs. Jute: Paddy: Paddy or Jute: Paddy: Wheat is the most efficacious rotation. In case where solanaceaous crop is the base crop of the locality in the *Rabi* season, it needs to be replaced by paddy or wheat at least for two years. Removing wilt affected plants from the crop field, burning the dead solanaceaous plants and rejecting rotten potato tubers (Mandal, 1986; Ghosh and Mukherjee, 1967) are important agronomic practices to keep the Hooghly wilt under control. By adopting cultural practices especially the appropriate rotation of crops in Hooghly district the disease came down to 1–2% compared to above 40% during the late eighties (Mandal, 1986; Mandal and Ghosh, 2002). The most efficacious crop rotations against this Hooghly wilt disease were jute-rice-rice or jute-rice-wheat (Mahapatra et al., 2009).

2. **Chemical:** Seed treatment with Bavistin (carbendazim) @ 2 g kg^{-1} seed and spraying the same fungicide @ 2 g l^{-1} of water helps to reduce root rot damage, which favors the entry of the bacteria. Mishra (1980) obtained best control by 0.1% Bavistin [carbendazim].

16.4 VIRAL DISEASES

Jute crop or its wild relatives are not very much prone to the viral diseases. Occurrence of some viral diseases like jute mosaic and jute chlorosis was known since long. Recently, increase in severity of these established diseases along with appearance of some newer viral diseases like jute yellow vein has drawn the attention on this jute crop. With the invention of the new molecular diagnostics recently, in depth work have been initiated for proper identification of such virus diseases on jute and a summary on these virus diseases along with their recent advances are presented below.

16.5 JUTE MOSAIC (SYNONYMS: JUTE LEAF MOSAIC, JUTE YELLOW MOSAIC, JUTE GOLDEN MOSAIC)

16.5.1 ECONOMIC IMPORTANCE

A mosaic disease on leaf of *capsularis* jute was first observed in 1917 from undivided Bengal province of India by Finlow (1917). Different researchers reported the similar type of symptoms under different names like jute yellow mosaic, jute leaf mosaic and jute golden mosaic (Ahmed, 1978; Ahmed et al., 1980). Recently, the occurrence of the disease has jumped from 20% in 2004 to 40% in 2007 (Ghosh et al., 2007a).

The disease was observed in *capsularis* jute from different jute growing belts of India, Bangladesh, and recently from Vietnam. In India, the jute mosaic disease was first reported in *capsularis* jute from West Bengal (Roy et al., 2006; Ghosh et al., 2007b) and Assam (Pun and Pathak, 2005). With the *capsularis* jute, similar disease symptom was also reported in *C. trilocularis*, a wild relative of cultivated jute, from Puna region of Maharashtra (Verma et al., 1966). They have found that the affected plants tend to flower early with occasional sterility in flowers.

16.5.2 SYMPTOMS

The disease symptoms consist of appearance of small yellow irregular dots (flakes) on leaf lamina in the early stage, which then gradually increase, intermingled with green patches and produces yellow mosaic in appearance. Infected plants appear pale; leaves size is reduced, initial chlorotic tissues become yellow (Roy et al., 2008). Owing to infection sometimes the complete leaf may turn yellow (Figure 16.5). Leaves in few cases produce small enations along the mid-vein. Plants infected with leaf mosaic have lower content of cellulose, lignin, and pectin, thus the fiber strength become weak (Biswas et al., 1989). A detailed study on the response of the disease on various varieties of jute at different agroecological situations in Bangladesh revealed considerable reduction of plant height, base diameter, green weight, and leaf weight in diseased plants (Das et al., 2001).

FIGURE 16.5   Jute mosaic in white jute.

16.5.3 CAUSAL ORGANISM

Chorchorus golden mosaic virus comes under genera Begomovirus.

16.5.4 PARTICLE MORPHOLOGY

As jute produces lots of mucilaginous substances, finding the virus particle is very difficult. So long no published attempt on the purification of this virus is available and possibly that became a great hindrance to identification of particle from purified virus material. Presence of mucilage also became a great determining factor in viewing the particle morphology through "dip method."

16.5.5 DETECTION, IDENTIFICATION, AND PHYLOGENETIC RELATIONSHIP OF THE VIRUS

Based on symptoms and transmission, the virus was observed to be a member of *Begomovirus* under family Geminiviridae. PCR based detection with *Begomovirus* group specific primers amplified a 1.2 kb DNA A fragmented portion of the virus (Ghosh et al., 2007a). Cloning and sequencing of this amplicon revealed that it consisted of 1,263 nucleotides (Accession No. EU047706) and shared the maximum nucleotide identity (91.2%) with *Corchorus golden mosaic Vietnam virus* (DQ641688). The nonanucleotide sequence at the origin of replication was noticed to be CATTATTAC in lieu of the conventional TAATATTAC. Such unique feature was also noticed in *Corchorus golden mosaic Vietnam virus.* Phylogenetic analysis with other begomoviruses revealed that the begomovirus from jute classified with other begomoviruses reported to be related with *Corchorus* sp. and clustered with the new world begomoviruses. Two other primers have been designed to amplify the complete DNA A and DNA B component of the viral genome which gave expected ~2.7 kb amplicon in each case (Roy et al., 2008).

16.5.6 TRANSMISSION

The causal virus was transmitted by whitefly (*Bemisiatabaci* Gen.) vector (Verma et al., 1966; Ahmed, 1978; Ahmed et al., 1980; Pun, and Pathak, 2005; Roy et al., 2006). Some workers also reported seed transmission of the virus (Das et al., 2001). Adult whitefly can take up (acquire) the virus in 30 minutes of its access to an infected plant and the acquisition becomes

maximum after 8 hours of access. A viruliferous whitefly can transmit (inoc-
ulate) the virus to the test plants within 30 minutes of its inoculation access;
the degree of infection is maximum with an inoculation access of 4 hours or
more. A viruliferous whitefly can hold the virus up to 10 days. The typical
symptom of the jute mosaic appeared on plants of cv. JRC 7447 and JRC 212
10 days after whitefly (*Bemisiatabaci* Genn.) transmission with 60% transmission efficiency when acquisition and inoculation access periods were 24
h and 12 h, respectively. A longer incubation period of 18–24 days for development of symptom was also noticed (Pun and Pathak, 2005).

16.5.7 HOST RANGE

Experimentally, the disease was observed to be transmitted by whitefly to
different *Corchorus* sp. but no hosts other than jute were reported so far
(Verma et al., 1966; Pun, and Pathak, 2005).

16.5.8 MANAGEMENT

- No detail study was carried out. In general, roguing of infected plant
  and spraying of systemic insecticide like imidachloprid, thiameth-
  oxam, and acetameprid could check the spread of the disease.
- Studies in Bangladesh revealed that a combination of use of seeds
  collected from healthy plants, spray of one insecticide around 30
  days after emergence (DAE), along with field sanitation with roguing
  several times during the vegetative growth period and an extra booster
  dose of nitrogen application at around 45 DAE reduced the disease
  spread (Hoque et al., 2003).
- No resistant varieties against the disease are available, but in Bangladesh,
  strains resistant to jute leaf mosaic disease have been reported from inter
  specific crosses between *C. trilocularis* (wild species) and *C. capsularis*.

16.6 CHLOROSIS OF JUTE

16.6.1 ECONOMIC IMPORTANCE

Chlorosis of jute was observed to be limited to *C. olitorius* and occur in all
the jute producing regions of West Bengal. There is no report available on
the prevalence of this disease from other countries.

16.6.2 SYMPTOMS

Mild mosaics like symptoms develop initially from the basal portion of the lamina. Then distinct yellow chlorotic spots with sharp margins irregularly formed on the lamina, which gradually coalesce giving rise to yellowish lamina (Figure 16.6). Finlow (1917) first reported the light yellow and yellow patches on the surface of leaf that led to appearance of variegation and applied the term "chlorosis" to explain the phenomenon. Chlorotic plants had less surface roots. In 1939, Finlow while reviewing the work on jute compared chlorosis as a "Morphological imperfection." It has been observed that plants can be chlorotic in the very early phase or may turn chlorotic later at any period during the vegetative growth of the plant. In a mild form, the imperfection is not directly harmful but in severe infection, the leaves was crinkled and brittle with checked growth (Roy et al., 2008).

FIGURE 16.6 Jute chlorosis.

16.6.3 CAUSAL ORGANISM

Virus.

16.6.4 TRANSMISSION

No detailed investigation was carried out so far to generate information on the transmission of this virus. Experimental inoculation of non-chlorotic plants using the juice of chlorotic plants was made in 1925 and it was observed that chlorosis was not sap transmissible. Non-chlorotic plants cross pollinated with the pollen from chlorotic plants produced progenies showing a huge jump in the proportion of chlorotic plants. Ghosh and Basak (1961) claimed that virus may be responsible for it since the virus is transmitted through grafting especially inarching.

16.6.5   HOST RANGE

Not worked out.

16.6.6   PARTICLE MORPHOLOGY

Short rod-shaped particles with a conspicuous central canal, 50–200 × 22 nm.

16.6.7   DETECTION, IDENTIFICATION, AND PHYLOGENETIC RELATIONSHIP OF THE VIRUS

Based on electron microscopic study the virus was presumed to be a member of *Tobravirus* genus, but no molecular evidence has been given to confirm its taxonomic position.

16.6.8   DISEASE MANAGEMENT

Not worked out.

16.7   YELLOW VEIN DISEASE OF JUTE

16.7.1   ECONOMIC IMPORTANCE

The disease is observed in *C. olitorius* from all the jute cultivating areas of India, but the incidence is very low. Besides India, the disease was also noticed in *C. capsularis* from Vietnam and in *C. siliquosus* from Yucatan Peninsula, Mexico.

16.7.2 SYMPTOMS

Yellow vein disease of jute is very recently occurring in *Corchorus* sp. prominent bright yellowing of veins was noticed on top leaves. Influence of this disease on fiber yield is not known.

16.7.3 CAUSAL ORGANISM: VIRUS

16.7.4 TRANSMISSION

Transmitted by whitefly (*Bemisiatabaci* Genn.).

16.7.5 HOST RANGE

Not worked out.

16.7.6 PARTICLE MORPHOLOGY

Not known.

16.7.7 DETECTION, IDENTIFICATION, AND PHYLOGENETIC RELATIONSHIP OF THE VIRUS

A begomovirus with bipartite genome from Vietnam was identified to be linked with the disease (Cuong Ha et al., 2006). Analysis of the DNA B and DNA A genomic component of this virus revealed that it was more matching similar to New World begomoviruses than to viruses from the Old World and was named as *Corchorus yellow vein virus* (CoYVV) (Accession No. AY727903, AY727904). Evidence indicated that CoYVV is indigenous to the region probably and may be the remnant of a previous population of New World begomoviruses in the Old World. The entire nucleotide sequence of DNA A and DNA B molecule of another *Begomovirus* (*Corchorus yellow vein Yucatan virus* (CoYVYuV)) has also been deposited from Mexico (Accession No. DQ875868, DQ875869) which indicated that present virus differ totally from that reported from Vietnam and showed its close association with *Sida yellow vein virus* (Roy et al., 2008; Hernández-Zepeda et al., 2007).

16.7.8   DISEASE MANAGEMENT

Not worked out.

16.8   OKRA MOSAIC TYMOVIRUS ON JUTE

Sometimes *C. olitorius* plants act as alternate host for *Okra mosaic tymovirus* (Brunt et al., 1996). Details of this virus including epidemiology and disease management in *olitorius* jute is yet to be studied.

16.9   NEMATODE DISEASES

16.9.1   ROOT KNOT NEMATODE (MELOIDOGYNE IGCONITA, M. JAVANICA)

Root knot nematodes have large range of hosts. They produce white globular swellings in roots called galls due to penetration and feeding of their larvae. As a result of this, translocation of nutrients and water is blocked. Infested plants appear yellowing show stunting of growth. Sometimes they predispose the infested plant to root rot and wilt infection. Average degree of damage is 15–20%.

This was reported first in jute by Bessey (1911). Both *olitorius* and *capsularis* were prone to its infestation. *Hibiscus sabdariffa* is resistant, while *H. cannabinus* is prone to this nematode (Laha and Bhattacharya, 1984; Laha and Pradhan, 1987).

16.9.2   MANAGEMENT

1. Several insecticides and nematicides were tried for reducing the damage of root knot nematodes in jute, of which, thiometon, Nematox, Nemagon are important. Granular insecticides reduced nematode population and increased fiber yield compared to control.
2. Various organic amendments, namely, cakes of *karanj, mahua*, groundnut, sawdust, cow dung, castor, chicken manure, etc., were tried for checking the nematode infestation.
3. Root knot nematode facilitates entry of other bacterial pathogen *R. solanacearum* and fungal pathogen *M. phaseolina* by creating

injury on roots of plant and cause heavy damage upon combined inoculation.

4. Cultural practices, like, removal of stubbles, weeding, thinning, crop rotation with paddy and wheat for a period of two years reduced population of root knot nematode in jute field.

5. Screening for resistance against root knot nematode resulted in few tolerant lines in both species of jute (Laha et al., 1995a, b). JRO 524 and JRO 632 plants showed reduced height and green weight as a result of inoculation of J-2 of root knot nematodes.

6. Investigation on the seasonal variation of population indicated that population of *Meloidogyne, Helicotylenchus,* and *Hoplolaimus* gradually increased with growth of jute plants, but decreased during winter in without suitable host plant, because they are endoparasitic in nature (Laha et al., 1988).

16.10 FUTURE RESEARCH STRATEGY

Following research strategies deserve immediate attention for effective management of economically important diseases of jute and allied fiber crops in future:

- Quantification of losses caused by diseases either singly or collectively for assigning them major or minor status.
- Development of weather-based disease forecasting and prediction models.
- Ecofriendly and sustainable options for disease management utilizing biological agents botanicals (specific biomolecules) and PGPR, etc.
- Sources and inheritance pattern of resistance and susceptibility for single and multiple diseases.
- Integrated crop management strategy including insects pests, diseases, weeds, soils, water, marketing, etc.
- Disease diagnosis based on serological and DNA probes at field level.
- Molecular characterization of pathogenic variability and identification of races of different pathogens across the different host crops.
- Cross infectivity of pathogens with large host range and role of other host crops in survival and evolution of virulence of pathogens.
- Newer alternative chemicals for disease management.

KEYWORDS

- *Corchorus yellow vein virus*
- *Corchorus yellow vein Yucatan virus*
- days after emergence
- percent disease index
- plant growth promoting rhizobacteria
- vesicular arbuscular mycorrhyza

REFERENCES

Ahmed, M., (1978). A white fly vectored yellow mosaic of jute. *FAO, Plant Protect. Bull., 26,* 169–171.

Ahmed, N., & Ahmed, Q. A., (1969). Physiologic specialization in *Macrophomina phaseoli* causing stem rot of jute, *Corchorus* species. *Mycopathologia ET Mycologia Applicata., 39,* 129–138.

Ahmed, N., & Ahmed, Q. A., (2005). Physiologic specialization in *Macrophomina phaseoli* (Maubl.) Ashby., causing stem rot of jute, *Corchorus species*, Jute Research Institute, Dacca, East Pakistan. *Mycopathologia, 39*(2), 129–138.

Ahmed, Q. A., (1968). Diseases of jute in East Pakistan. *Jute and Jute Fabrics Pakistan, 7*(8), 147–151.

Ahmed, Q. A., Biswas, A. C., Farukuzzaman, A. K. M., Kabir, M. Q., & Ahmed, N., (1980). Leaf mosaic disease of jute. *Jute and Jute Fabrics Pakistan, 6,* 9–13.

Ali, M. M., Sayem, A. Z. M., Alam, S., & Ishaque, M., (2005). Relationship of pectic enzyme of *Macrophomina phaseoli* with stem-rot disease and retting of jute. *Mycopathologia, 38*(3), 289–298.

Ali, M. M., Sayem, A. Z. M., Alam, S., & Ishaque, M., (1969). Relationship of pectic enzyme of *Macrophomina phaseoli* with stem-rot disease and retting of jute. *Mycopathologia Et Mycologia Applicata, 38*(3), 289–298.

Anonymous, (1949–1989). *Annual Report, Jute Agricultural Research Institute (JARI).* Barrackpore.

Anonymous, (1990–2006). *Annual Report, Central Research Institute for Jute and Allied Fibers.* Barrackpore.

Anonymous, (1999). "*50 Years of Research (1948–1997)*". Central Research Institute for Jute and Allied Fibers (CRIJAF), Barrackpore.

Anonymous, (2013). *Statistics from Website of Directorate of Jute Development.* government of India, ministry of agriculture, department of agriculture and cooperation, ministry of agriculture, co-operation and farmers' welfare, Nizam palace, Kolkata-700020, www. http://djd.dacnet.nic.in/ (accessed on 11 January 2020).

Bandopadhyay, A., & Bandopadhyay, A. K., (2004a). Beneficial traits of plant growth promoting rhiozobacteria and fungal antagonists consortium for biological disease management in bastfiber crops. *Indian Phytopathol., 57*(3), 356, 357.

Bandopadhyay, A. K., (1983). Tip blight of jute by *Curvularia subulata*. Hand book of jute by T. Ghosh, former Director, JARI, (personal communication). *FAO Plant Production and Protection Paper, 51,* 129.

Bandopadhyay, A. K., (2002). A current approach to the management of root diseases in bastfiber plants with conservation of natural and microbial agents. *J. Mycopathol. Res., 40*(1), 57–62.

Bandopadhyay, A. K., & Bandopadhyay, A., (2004b). Biological disease management in jute with plant growth promoting rhiozobacteria. *Indian Phytopathol., 57*(3), 357(P28).

Bandopadhyay, A. K., Bandopadhyay, A., & Majumder, A., (2006). Screening and characterization of antagonistic potential of some rhizosphere fungi and PGPR against *Macrophomina phaseolina* in jute. *J. Mycopathol. Res., 44*(2), 323–330.

Bandopadhyay, A. K., De, R. K., & Bhattacharya, S. K., (2007a). Beneficial traits of PGPR mediated strategy for disease management and enhanced plant growth in bastfiber crops. Paper presented. In: *"National Symposium on Microbial Diversity And Plant Health"* (p. 60). Held between November 28 and 29, BCKV, Kalyani.

Bandopadhyay, A. K., De, R. K., & Bhattacharya, S. K., (2008). Beneficial traits of PGPR mediated strategy for disease management and enhanced plant growth in bastfiber crops. *J. Mycopathological. Res., 46*(1), 143.

Bandopadhyay, A., Bandopadhyay, A. K., & Samajpati, N., (2007b). Biochemical characterization of some plant growth promoting rhiozobacteria for antagonistic property against *Macrophomina phaseoli* inciting stem and root rot of jute. Paper presented. In: *"National Symposium on Microbial Diversity and Plant Health"* (p. 80). Held between November 28 and 29, BCKV, Kalyani.

Bandopadhyay, A., Bhattacharya, S. K., Majumdar, P., & Bandopadhyay, A. K., (2009). Characterization of some plant growth promoting rhiozobacteria in relation to biotic stress management with enhanced growth and production of jute and allied fiber crops. *Paper Presented in the First Asian PGPR Congress for Sustainable Agriculture* (pp. 96, 97). Held at ANGRA University, Hyderabad.

Bandopadhyaya, A., & Das, N., (2017). Plant growth promoting microbial consortial formulations mediated biological control of stem and root rot disease of jute caused by *Macrophomina phaseolina* ((Tassi.) Goid. *International Journal of Current Science, 20*(1), 1–15. ISSN: 2250–1770.

Beas-Fernandez, R., De Santiago-De Santiago, A., Hernanzdez-Delgado, S., & Mayek-Perez, N., (2006). Characterization of Mexican and non-Mexican isolates of *Macrophomina phaseolina* based on morphological characteristics, pathogenicity on bean seeds and endoglucanase genes. *Journal of Plant Pathology, 88*(1), 53–60.

Bhattacharya, B., Basu, S., Chattapadhyay, J. P., & Bose, S. K., (1985). Biocontrol of macrophomina root-rot of jute by an antagonistic organism, *Aspergillus versicolor*. *Plant Soil, 87,* 435–438.

Bhattacharya, S. K., Sen, K., De, R. K., Bandopadhyaya, A., Sengupta, C., & Adhikary, N. K., (2017). Integration of biocontrol agents with fungicides, weedicides, and plant growth regulators for management of stem and root rot of jute. *Journal of Applied and Natural Science Jans Foundation.org/, 9*(2), 899–904.

Biswas, A. C., Asaduzzaman, M., Sultana, K., & Taher, M. A., (1989). Effect of leaf mosaic disease on loss of yield and quality of jute fiber. *Bangladesh J. Jute Fiber Res., 14*, 43–46.

Biswas, C., Dey, P., Mandal, K., Satpathy, S., & Karmakar, P. G., (2014). In planta detection of *Macrophomina phaseolina* from jute (*Corchorus olitorius*) by sodium acetate-based direct PCR method. *Phytoparasitica, 42*(5), 673–676. doi: 10.1007/s12600-014-0407-4.

Biswas, C., Dey, P., Satpathy, S., Sarkar, S. K., Bera, A., & Mahapatra, B. S., (2012). A simple method of DNA isolation from jute (*Corchorus olitorius*) seed suitable for PCR-based detection of the pathogen *Macrophomina phaseolina* (Tassi) Goid. *Letters in Applied Microbiology, 56,* 105–110. © 2012 ISSN: 0266–8254.

Brunt, A. A., Crabtree, K., Dallwitz, M. J., Gibbs, A. J., & Watson, L., (1996). *Viruses of Plants. Descriptions and Lists from the VIDE Database.* Wallingford, UK, CAB International.

Chattopadhyay, S. B., & Raj, S. K., (1978). Role of hydrolytic enzymes in seedling blight of jute incited by *Macrophomina phaseolina. Madras Agric. J., 65*(5), 320–324.

Chattopadhyay, S. B., & Sharma, B. D., (1979). Investigation on wilt of jute *Corchorus capsularis* L. and *Corchorus olitorius* 1. Disease syndrome and incitants. *Indian J. Mycol. Res., 17,* 55–64.

Cheng, Y. H., & Tu, C. C., (1970). Effect of nitrogen and potash fertilizer on the occurrence on stem rot of jute. *Res. Rep. Taiwan Fiber Exp. Station, TARI, N.S. No. 6*(4).

Das, S., Khokon, A. R., Haque, M. M., & Ashrafuzzaman, M., (2001). Jute leaf mosaic and its effects on jute production. *Pakistan J. Biol. Sci., 4*(12), 1500–1502.

De, R. K., & Mandal, R. K., (2012a). Identification of resistant sources of jute (*Corchorus olitorius* L.*)* against stem rot caused by *Macrophomina phaseolina* (Tassi) Goid. *Journal of Mycopathological Research, 50*(2), 217–222.

De, R. K., (2013). Effect of date of sowing on the incidence of stem rot of jute (*Corchorus olitorius* L.) caused by *Macrophomina phaseolina* (Tassi) Goid. *Journal of Mycopathological Research, 51*(2), 251–257. ISSN: 0971–3719.

De, R. K., (2012a). *Botanicals as Good Alternative to Manage Stem rot of Jute (Corchorus olitorius L. C. capsularis L.) caused by Macrophomina phaseolina (Tassi) Goid* (pp. 10, 130). Paper presented at National symposium on blending conventional and modern plant pathology for sustainable agriculture held at Indian Institute of Horticultural Research, Hesaraghatta Lake Post, Bangalore 560089 on 4–6 December, 2012.

De, R. K., (2012b). Seed infection in jute (*Corchorus olitorius* L. and *C. capsularis* L.) and its management. *Paper Presented at National Symposium on Biotic and Abiotic Stresses in Plants Under Changing Climate Scenario* (pp. 6, 7). Held at Uttar Banga Krishi Viswavidyalaya, Pundibari, Coochbehqar 736165, West Bengal, BS 4.

De, R. K., (2014). Search for new fungicides against stem rot of jute (*Corchorus olitorius* L. and *C. capsularis* L.*)* caused by *Macrophomina phaseolina* (Tassi) Goid. *Journal of Mycopathological Research, 52*(2), 217–225. ISSN: 0971-3719.

De, R. K., (2016). Comparison of three methods (leaf, soil and stem) of inoculation for stem rot jute (*Corchorus olitorius* L. and *C. capsularis* L.) caused by *Macrophomina phaseolina* (Tassi) Goid. *Journal of Mycopathological Research, 54*(1), 67–69.

De, R. K., (2017). Effect of different NPK fertilizer schedules on the stem rot of jute (*Corchorus olitorius* L.*)* caused by *Macrophomina phaseolina* (Tassi) Goid. *Journal of Mycopathological Research, 55*(1).

De, R. K., & Mandal, R. K., (2010). Effect of different fungicides on *Phytophthora parasitica* var. *sabdariffae* and *Sclerotium rolfsii* infecting mesta. *Pestology, XXXIV*(9), 23–27.

De, R. K., & Tripathi, A. N., (2016). Effect of plant population density and sowing methods on stem rot of jute (*Corchorus olitorius, C. capsularis*) Caused by *Macrophomina phaseolina*

(Tassi) Goid. *International Journal of Tropical Agriculture* (Vol. 33, No. 4, pp. 1–4). © Serials Publications, ISSN: 0254-8755.

De, R. K., & Mandal, R. K., (2007c). Simple method of inoculation of stem rot pathogen *Macrophomina phaseolina* on jute (*Corchorus* spp.) *Paper Presented in "National Symposium on Microbial Diversity and Plant Health"* (p. 38). Held between November 28 and 29, BCKV, Kalyani.

De, R. K., & Mandal, R. K., (2012b). Seed infection of jute (*Corchorus olitorius, C. capsularis*) and its management. *J. Mycol. Plant Patholo.*, *42*(4), 541.

De, R. K., Bandopadhyay, A., Bhattacharya, S. K., Biswas, B., Mitra, R. P., & Bandopadhyay, A. K., (2009). Investigation on some botanicals and safer chemicals for the management of jute and Mesta pathogens. *Paper Presented in Fifth International Conference on Plant Pathology in the Globalized Era* (pp. S-15, 369, 693). Held at New Delhi.

De, R. K., Chowdhury, H., & Chakraborty, L., (2014). New stem inoculation technique of stem rot of jute (*Corchorus olitorius* L. and *C. capsularis* L.) caused by *Macrophomina phaseolina* (Tassi) Goid. *Paper Presented in International Symposium on "Role of Fungi and Microbes in the 21st Century: A Global Scenario"* (p. 52). Held on 20-22 February, 2014 at Science City, Kolkata by Department of Botany, University of Calcutta, Kolkata (Ref.: www.imskolkata.org (accessed on 25 January 2020)).

De, R. K., Ghorai, A. K., Kumar, V., & Abantika, M., (2014). Bleaching powder [$Ca(OCl)_2$], an ecofriendly alternative for management of stem rot of jute (*Corchorus olitorius* L. and *C. capsularis* L.) caused by *Macrophomina phaseolina* (Tassi) Goid. *Paper Presented in International Symposium on "Role of Fungi and Microbes in the 21st Century: A Global Scenario"* (p. 51). To be held on 20–22 February, 2014 at Department of Botany, University of Calcutta, Kolkata (Reference: www. imskolkata.org).

De, R. K., Mandal, R. K., Sarkar, S., & Ghorai, A. K., (2005). Inhibitory effect of herbicides on *Macrophomina phaseolina* causing stem rot of jute. *Jaf News: A CRIJAF Newsl.*, *3*(1), 16.

De, R. K., Mandal, R. K., Sarkar, S., & Ghorai, A. K., (2007). Non-target effect of herbicides on *Macrophomina phaseolina* causing stem rot of jute. *Environ. Ecol.*, *25*(2), 475–478.

De, R. K., Tripathi, A. N., & Chowdhury, S. B., (2015). *Abroma* as potential botanical for organic management of stem rot of jute, *Macrophomina phaseolina* (Tassi) Goid. *Oral Paper Presented at National Symposium on "Advances in Phytopathological Research in Globalized Era With Reference to Eastern Region"* (p. 21). Held on 29–30 January, 2015 at University Department of Botany, Ranchi University, Ranchi, Jharkhand, India organized by Indian Phytopathological Society zonal Symposium-(EZ).

Farr, D. F., Aime, M. C., Rossman, A. Y., & Palm, M. E., (2006). Species of *Colletotrichum* on agavaceae. *Mycol. Res.*, *110,* 1395–1408.

Finlow, R. S., (1917). Historical notes on experiments with jute in Bengal. *Agric. J. India.*, *12*, 3–29.

Ghosh, P. K., & Purakayastha, R. P., (1985). Epidemiological studies on stem rot of jute (*Corchorus olitorius* L.) *Indian J. Plant Pathol.*, *3*(2), 174–184.

Ghosh, R., Paul, S., Das, S., Palit, P., Acharyya, S., Mir, J. I., Roy A., & Ghosh, S. K., (2007a). Association of a begomovirus complex with yellow mosaic disease of jute in India. In: *"Proceedings of 10th Plant Virus Epidemiology Symposium"* (p. 115). Held between October 15–19, ICRISAT, India.

Ghosh, S. K., & Sen, C., (1973a). Nitrogen requirement of four isolates of *Macrophomina phaseolina* (Maubl) Ashby. *Proc. Indian Natl. Sci. Acad.*, *39B*(2), 221–227.

Ghosh, S. K., & Sen, C., (1973b). Comparative physiological studies on four isolates of *Macrophomina phaseolina. Indian Phytopathol., 26,* 615–621.

Ghosh, T., & Basak, M., (1961). *Transmission of Jute Mosaic (Chlorosis).* Jute Bull.

Ghosh, T., (1957). Anthracnose of jute. *Indian Phytopathol., 10,* 63–70.

Ghosh, T., (1961). Studies on the diseases of jute caused by Macrophomina Phaseolina (Maulb) Ashby. Calcutta University. *D.Phil. Thesis.*

Ghosh, T., (1963). Importance of spacing in relation of incidence of disease in jute and its yield. *Jute Bull., 26*(6).

Ghosh, T., & Basak, M., (1965). Possibility of controlling stem rot in jute. *Indian J. Agric. Sci., 35*(3), 90–100.

Ghosh, T., & Mukherjee, N., (1967). *Macrophomina phaseolina* (Maulb) Ashby on jute: Plant disease problems. In: Raychoudhury, S. P., (ed.), *"Proceedings of the First International Symposium on Plant Pathology"* (pp. 363–369). Indian Phytopathological Society.

Ghost, T., (1983). *"Handbook on Jute"* (Vol. 51, p. 219). FAO Plant Production and Protection Paper.

Haige, J. C., (1930). *Peradeniya Ann. Rep. Bot. Gard. II., 23,* 213–239.

Hath, T. K., & Chakraborty, A., (2004). Towards development of IPM strategy against insect pests, mites and stem rot of jute under North Bengal conditions. *J. Entomol. Res., 28*(1), 1–5.

Hembram, S., & De, R. K., (2010). Effect of thiram, carbendazim, copper oxychloride and mancozeb on *Macrophomina phaseolina* causing stem rot of jute *in vitro.* In: Palit, P., Sinha, M. K., Meshram, J. H., Mitra, S., Laha, S. K., Saha, A. R., & Mahapatra, B. S., (eds.), *Jute and Allied Fibers Production, Utilization And Marketing* (pp. 286–290). Indian fiber society, Eastern region, Central Research Institute for Jute and Allied Fibers (CRIJAF), Barrackpore, Kolkata 700120.

Hembram, S., & De, R. K., (2008). Effect of thiram, carbendazim, copper oxychloride and mancozeb on *Macrophomina phaseolina* causing stem rot of jute *in vitro. Presented in the International Symposium on Jute and Allied Fibers Production, Utilization and Marketing.* Held at CRIJAF, Barrackpore, Kolkata 700120 on 10–12, January, 2008.

Hernández-Zepeda, C., Idris, A. M., Carnevali, G., Brown, J. K., & Moreno-Valenzuela O. A., (2007). Molecular characterization and phylogenetic relationships of two new bipartite begomovirus infecting malvaceous plants in Yucatan, Mexico. *Virus Genes, 35*(2), 369–377.

Hiremath, S. V., Hiremath, P. C., & Hegde, R. K., (1993). Studies on cultural characters of *Colletotrichum gloeosporioides* a causal agent of Shisham blight. *Karnataka J. Agricul. Sc., 6,* 30–32.

Hoque, M. M., Khalequzzaman, H. K. M., Khan, M. A. M., Alauddin, M., & Ashrafuzzaman, M., (2003). Management of jute leaf mosaic through vector control and cultural practices. *Asian J. Plant Sci., 2*(11), 826–830.

Ikata, S., & Yoshida, M., (1940). A new anthracnose of jute plant. *Ann. Phytopath. Soc. Japan, 10,* 141–149.

Jana, T. K., Singh, N. K., Koundal, K. R., & Sharma, T. R., (2005). Genetic differentiation of charcoal rot pathogen, *Macrophomina phaseolina,* into specific groups using URP-PCR. *Canadian Journal of Microbiology, 51*(2), 159–164.

Jana, T., Sharma, T. R., Prasad, R. D., & Arora, D. K., (2003). Molecular characterization of *Macrophomina phaseolina* and *Fusarium* species by a single primer RAPD technique. *Microbiol. Res., 158*(3), 249–257.

Laha, S. K., & Bhattacharaya, S. P., (1984). Occurrence of root knot (*Meloidogyne incognita*) nematode in Congo jute *Urenalobata. Sci. Cult., 50,* 330–331.

Laha, S. K., & Pradahan, S. K., (1987). Susceptibility of bastfibre crops to *Meloidogyne incognita*. *Nematode. Medit.*, *15*, 163, 164.

Laha, S. K., Mandal, R. K., & Dasgupta, M. K., (1995a). Studies on the varietal susceptibility of jute against root-knot nematode and root-rot disease under field condition. *Indian Biol.*, *26*(2), 27–29.

Laha, S. K., Pradahan, S. K., Sasmal, B. C., & Dasgupta, M. K., (1995b). *Nematode. Medit.*, *23*(1), 51–52.

Laha, S. K., Singh, B., & Mishra, C. D., (1988). Plant parasitic nematodes and their population fluctuation associated with jute and allied fiber crops. *Environ. Ecol.*, *6*(1), 100–102.

Mahapatra, A. K., Mandal, R. K., Saha, A. K., Sinha, M. K., Guha, R. M. K., Kumar, D., & Gupta, D., (1994). *Information Bulletin* (pp. 1–31). CRIJAF.

Mahapatra, B. S., Mitra, S., Ramasubramanian, T., & Sinha, M. K., (2009). Research on jute (*Corchorus olitorius* and *C. capsularis*) and kenaf (*Hibiscus cannabinus* and *H. sabdariffa*): present status and future perspective. *Indian Journal of Agricultural Sciences*, *79*(12), 951–967.

Mandal, A. K., (1976). Nutrition and disease resistance in of jute in relation to potassium. *Bull. Indian Soil Sci.*, *10*, 278–284.

Mandal, R. K., (1986). *Wilt Diseases of Jute and its Management (in Bengali) Sabuj Sona March Issue*.

Mandal, R. K., (1987). "Proceedings of the training programme in seed pathology under indo-danish project for staff of seed testing laboratories." In: Siddique, S. A., (ed.), *Division of Seed Science and Technology*. IARI, New Delhi.

Mandal, R. K., (1990). Jute diseases and their control. In: *"Proceedings of the National Workshop cum Training on Jute, Mesta, Sunnhemp and Ramie."* CRIJAF, Barrackpore.

Mandal, R. K., (1997). Current approach in disease management in jute and allied fiber crops. In: Saraswat, V. N., (ed.), *"Proceedings of the Training Programme for Senior Management/ Research/Extension Officers."* CRIJAF, Barrackpore.

Mandal, R. K., (2001). Seed borne fungi of jute. *J. Interacad.*, *5*(3), 402–405.

Mandal, R. K., & De, R. K., (2010). Jute seed infection by *Botryodiplodia theobronae* and its management. In: Palit, P., Sinha, M. K., Meshram, J. H., Mitra, S., Laha, S. K., Saha, A. R., & Mahapatra, B. S., (eds.), *Jute And Allied Fibers Production* (pp. 291–295). Utilization and marketing, Indian fiber society, Eastern region, Central Research Institute for Jute and Allied Fibers (CRIJAF), Barrackpore, Kolkata 700120.

Mandal, R. K., & Mishra, C. D., (2001). Role of different organisms in inducing Hooghly wilt symptom in jute. *Environ. Ecol.*, *19*(4), 969–972.

Mandal, R. K., & Khatua, D. C., (1986). Incidence of two important diseases of jute in high rainfall areas. *Jute Dev. J.*

Mandal, R. K., Sarkar, S., & Saha, M. N., (2000). Field evaluation of white jute (*Corchorus capsularis* L.) germplasm against *Macrophomina phaseolina* (Tassi) Goid under Sorbhog condition. *Environ. Ecol.*, *18*(4), 814–818.

Mandal, R. K., Sinha, M. K., Guha, R. M. K., Mishra, C. B. P., & Chakrabarty, N. K., (1998). Variation in *Macrophomina phaseolina* (Tassi) Goid. Causing stem rot of jute. *Environ. Ecol.*, *16*(2), 424–426.

Meena, P. N., De, R. K., Roy, A., Gotyal, B. S., Satpathy, S., & Mitra, S., (2015). Evaluation of stem rot disease in jute (*Corchorus olitorius*) germplasm caused by *Macrophomina phaseolina* (Tassi) Goid. *Journal of Applied and Natural Science, 7*(2), 857–859.

Meena, P. N., Roy, A., Gotyal, B. S., Mitra, S., & Satpathy, S., (2014). Eco-friendly management of major diseases in jute (*Corchorus olitorius* L.) *Journal of Applied and Natural Science, 6*(2), 541–544.

Mehak, N., (2016). *Stem-Rot of Jute: Symptoms and Control* | *Plant Diseases.* http://www. biologydiscussion.com/plants/plant-diseases/stem-rot-of-jute-symptoms-and-control-plant-diseases/58696 (accessed on 11 January 2020).

Mihail, J. D., & Taylor, S. J., (1995). Interpreting variability among isolates of *Macrophomina phaseolina* in pathogenicity, pycnidium production, and chlorate utilization. *Canadian Journal of Botany, 73*(10), 1596–1603, 10.1139/b95–172.

Mishra, C. B. P., (1980). Evaluation of a systemic fungicide (Bavistin) in controlling Hooghly wilt of jute. *Pesticides, 14*(12), 22, 23.

Mondal, B., Bhattacharya, I., & Khatua, D. C., (2011). Crop and weed host of *Ralstonia solanacearum* in West Bengal. *Journal of Crop and Weed, 72*(2), 195–199.

Mukherjee, N., & Basak, M., (1973). Chemotherapeutic control of foot and stem rot of Roselle *(Hibiscus sabdariffa* var. *altissima*) caused by *Phytophthora parasitica* var. *Sabdariffae. Indian Phytopatho., 26*(3), 576, 577.

Palve, S. M., Sinha, M. K., & Mandal, R. K., (2003). Preliminary evaluation for wild species of jute (*Corchorus* species). *Plant Genetic Resource Newsl., 134,* 10–12.

Palve, S. M., Sinha, M. K., & Mandal, R. K., (2004). Sources of stem rot *Macrophomina phaseolina* (Tassi) Goid. Resistance in wild species of jute. *Trop. Agric. (Trinidad), 81*(1), 23–27.

Pandit, N. C., Mandal, R. K., Ghorai, A. K., Chakraborty, A. K., De, R. K., & Biswas, C. R., (2004a). Impact of integrated pest management in jute and Mesta. *Paper Presented in "National Seminar on Raw Jute"* (p. 24) held between April, 16 and 17 at CRIJAF, Barrackpore. Book of abstract

Pandit, N. C., Mandal, R. K., Ghorai, A. K., Chakraborty, A. K., De, R. K., & Biswas, C. R., (2004b). Impact of integrated pest management in jute and Mesta. *Paper Presented in "National Seminar on Raw Jute"* (pp. 51–55). Held between April, 16 and 17 at CRIJAF, Barrackpore.

Pecina, V., María De, J. A., Héctor, W. A., Rodolfo, D. L. T. A., & George, J. V., (2000). Detection of double-stranded RNA in *Macrophomina phaseolina. Mycologia, 92*(5), 900–907.

Pun, K. V., & Pathak, S., (2005). Occurrence of yellow mosaic disease of jute in Assam. *Indian J. Virol., 16,* 53.

Purkayastha, R. P., & Sen, G. M., (1973). Studies on conidial germination and appressoria formation in *Colletotrichum gloeosporioides* Penz. Causing anthracnose of jute (*Corchorus olitorius* L.). *Zeitschrift fur Pflanzenkrankheiten und Pflanzenschutz, 80,* 718–724.

Purkayastha, S., Kaur, B., Dilbaghi, N., & Chaudhury, A., (2005). Characterization of *Macrophomina phaseolina*, the charcoal rot pathogen of cluster bean, using conventional techniques and PCR-based molecular markers. *Plant Pathology, 55*(1), 106–116. doi: 10.1111/j.1365–3059.2005.01317.x.

Rao, P. V., (1979). Analysis of growth environmental relationship in jute (*Corchorus olitorius* L.). *Agric. Met., 16,* 107–112

Rao, P. V., (1980). Effect of rainfall and temperature on yield of tossa jute. *Indian J. Agric. Sci., 50*(8), 608–611.

Roy, A., De, R. K., & Ghosh, S. K., (2008). Diseases of bastfiber crops and their management. In: *Book, Jute and Allied Fibers Updates, Production and Technology* (pp. 217–241).

Central Research Institute for jute and allied fibers, (Indian Council of Agricultural Research), Barrackpore, Kolkata 700120, India.

Roy, A., Mir, J. I., Ghosh, R., Paul, S., & Ghosh, S. K., (2006). A report on the occurrence of jute leaf mosaic virus at CRIJAF farm. *Jaf. News: A CRIJAF News Letter, 4*(1), 11.

Saha, A. K., Mandal, R. K., Mahapatra, A. K., Sinha, M. K., GuhaRay, M. K., Kumar, D., & Gupta, D., (1994). *Information Bulletin* (pp. 1–54). CRIJAF.

Sarkar, S. K., (2014). Transmission and location of *Macrophomina phaseolina* in jute seed. In: Nag, D., Roy, D. P., Ganguly, P. K., Kundu, D. K., Ammayappan, L., Roy, A. N., et al., (eds.), *Jute and Allied Fibers: Issues and Strategies* (pp. 141–144). The Natural Fiber Society, 12 Regent Park, Kolkata 700040, India.

Sarkar, S. K., Roy, A., & Satpathy, S., (2014). Management of stem rot of jute under integrated management system. In: Nag, D., Roy, D. P., Ganguly, P. K., Kundu, D. K., Ammayappan, L., Roy, A. N., et al., (eds.), *Jute and Allied Fibers: Issues and Strategies* (pp. 123–127). The natural fiber society, 12 Regent Park, Kolkata 700040, India.

Sarkar, S. K., Chowdhury, H., & Satpathy, S., (2016). Disease-free jute seed production in West Bengal. *Bangladesh J. Bot., 45*(3), 561–565.

Som, D., (1977). *Recent Concept on Jute Disease and Control Measures.* Jute Bull. April-March issue 1–4.

Srivastava, R. K., Singh, R. K., Kumar, N., & Singh, S., (2010). Management of macrophomina disease complex in jute (*Corchorus olitorius*) by *Trichoderma Viride. Journal of Biological Control, 24*(1), 77–79.

Thakur, J., (1973). Resistance in some jute varieties against stem rot incited by *Macrophomina phaseoli. Indian J. Mycol. Plant Pathol., 3*(1), 104.

Thakur, J., (1974). Influence of N, P, K on stem rot of *Capsularis* jute. *Indian J. Mycol. Plant Pathol., 4,* 117–120.

Thakur, J., Hiralal, & Singh, (1976). Influence of micronutrients on incidence of stem rot of *Capsularis* jute. *Indian J. Mycol. Plant Pathol., 6,* 96.

Vandemark, G., Octavio, M., Victor, P., & Maria de Jesús, A., (2000). Assessment of genetic relationships among isolates of *Macrophomina phaseolina* using a simplified AFLP technique and two different methods of analysis. *Mycologia., 92*(4), 656–664.

Verma, P. M., Rao, G. G., & Capoor, S. P., (1966). Yellow mosaic of *Corchorus trilocalaris. Sci. Cult., 32,* 466.

CHAPTER 17

Present Scenario of Diseases in Jute Crops and Their Integrated Management

KUNAL PRATAP SINGH and J. N. SRIVASTAVA

Department of Plant Pathology, Bihar Agricultural University, Sabour, Bhagalpur, Bihar, India

17.1 INTRODUCTION

Among the economically important plants, fibers producing plants comes next to food crops. Fibers used by human being for meeting his need since start of dawn of civilization. Among the fiber crops, Jute, and Allied fibers has important economic and commercial importance. Jute is an indispensible fiber crop after cotton for human. India, Bangladesh, China, Pakistan Nepal, Thailand, and Bhutan are major growing areas of jute and allied fibers. The cultivated area in India for jute cultivation was 7.42 lakh hectares with total production of 88.42 lakh bales during 2015–16. In India, jute cultivation is mainly grown in the states of West Bengal, Bihar, Assam, Andhra Pradesh, Orissa, some parts of north eastern states. The total productivity of jute in India was 25.77 q/ha in 2014–15 which is dwindling year by year. Jute fiber is mainly used in decorative bags, textiles, and geo textiles industries, and it's by products like jute sticks are also used for fuel, door panels and for making false ceiling boards. Jute plant infested by various insect pests and diseases that significantly reduces seed quality and fiber quality. Both *Capsularis* and *Olitorius* jute and Allied fibers are infested by number of fungi, bacteria, and viruses. The lists of diseases which are common in the jute field are given in Table 17.1.

TABLE 17.1   List of Diseases Affected Jute Crop

Sl. No.	Name of Disease	Causal Organism	Economic Importance
1.	Stem rot	*Macrophomina Phaseolina*	Major disease
2.	Root rot	*Macrophomina Phaseolina*	Major disease
3.	Anthracnose	*Colletotrichum corchori* and *C. gloeosporioides*	Minor disease and found mainly in *C. capsularis*
4.	Mosaic	Virus	Major disease in capsularis var.
5.	Black band	*Botryodiplodia theobromae*	Minor disease
6.	Hoogly wilt	*Pseudomonas solanacearum, M. Phaseolina*-complex disease	Minor disease
7.	Soft rot	*Sclerotium rolfsii*	Minor disease

17.2   STEM ROT OF JUTE

17.2.1   ECONOMIC IMPORTANCE

Stem rot of jute was first reported from Formosa (Taiwan) in 1916 by K. Swada of Japan. Stem-Rot of Jute may attack both species of *Corchorus capsularis* L. and *C. olitorius* L. and is prevalent in all the jute growing regions of world. The same fungus that affect stem rot may attack jute crop at different stage of plant growth causing collar rot, seedling blight and root rot (Ghosh and Mukherjee, 1970). It is a most devastating disease of jute and can occurs in an epidemic form under favorable weather conditions. Severe infection may lead to 10–30% loss in fiber yield in almost all jute growing areas. The disease infesting almost all parts of jute seedling and fibrous jute plant that causing severe reduction in fiber yield. The jute stem rot is both seed and soil borne (Varadarajan and Patel, 1943).

17.2.2   SYMPTOMS

The pathogen may attack at any stage of growth of jute plant. The blackish brown to black streaks symptoms, called damping-off, that causing soft and completely rotting of the young seedlings, are common in jute growing areas. Lesions on the stem can occur in the form of small blackish-brown colors pot which may coalesce and gradually encircle the stem causing discoloration and rot and ultimately reduces fiber quality. Lesions spots may occur on the margins and lamina of the leaves. The stem and root rot leads to shredding and wilting. Pycnidia are usually seen as small black dots which are commonly embedded in the epidermis of the stem and rotting tissue, act as source of infection. Sclerotia are common in later part of season and found on the tissues of the infected stem. In seed crop, pathogen infects capsules that become black and seeds become discolored and small.

17.2.3   CAUSAL ORGANISM

Macrophomina phaseolina (Tassi) Goid

Apycnidial stage and whose sclorotia phase is known as *Rhizoctonia batati-cola* (Taub.) Butler. The conidia are hyaline, aseptate, thin walled, and elliptical.

17.2.4   DISEASE CYCLE AND EPIDEMIOLOGY

The pathogen survivesin seed, soil, weeds like *Cyperusdistans* and on jute stubbles in sclerotia form. Under favorable conditions, hyphae germinate from sclerotia and infect the host stem/root of jute plantby penetrating the plant cell wall through physical pressure and chemical release (Ammon et al., 1974). The seed born inoculum may serves as primary source of infection than any other source of infection. The pycnidial stage of the fungus is responsible for stem rot in mature plant (Ashby, 1927). Due to primary infection, pycnidia develop on leaves and stems of jute plants that release conidia and cause secondary infection on the nearby plant which is dispersed byboth air and rain. When stem is infected by conidia, the pathogen invades the epidermis, cortex, phloem parenchyma and phloem wedges and ultimately causing stem and root rot of plant. The disease is more prevalent in hot (34±1) and humid weather condition during cropping season (Mandal, 1990). The outbreak of stem rot may occur under favorable conditions of warm and humid conditions. The pathogen perpetuates at low pH, high nitrogen level, high temperature and rainfall, low soil moisture and high humidity favor infection (Ghosh and Bask, 1965).

17.2.5 INTEGRATED DISEASE MANAGEMENT

Integrated Disease Management is highly effective, economically feasible and ecologically sound and sustainable over a period of time. In this approach, judicious combinations of all methods for management of disease are applied starting from sowing to harvesting. It keeps pest numbers below harmful level instead of eradication, protects, and conserves the environment by providing plant protection.

1. Sanitation in the Jute field is important as pathogen survives on stubbles and weeds to cut down the source of primary inoculum.
2. Use clean and healthy seeds from standard organization.
3. Seeds should be treated with Carbendazim @2 g/kg of seed before sowing.
4. Maintenance of neutral soil pH.
5. Application of balance dose of fertilizer in soil.
6. Good drainage facility in the field.
7. Proper spacing, weeding, and thinning are required.
8. Follow jute crop rotation with paddy.
9. Spraying Carbendazim @2 g/L of water or copper oxychloride (50%Cu) @5 g/L of water or Mancozeb @ 5 g/L of water at the onset of disease.
10. Timely harvesting of fiber may reduce loss.

17.3   ROOT-ROT OF JUTE

17.3.1   ECONOMIC IMPORTANCE

Root rot is similar to stem rot except affected part in earlier one is root and in later one is stem. Rootrot of Jute may attack both *C. capsularis* L. and *C. olitorius* L. and is common in all the jute growing regions of world. It is a destructive disease of jute and occurs in an epidemic form under favorable weather conditions. Severe infection of root rot may lead to greater lossof fiber yield than stem rot in almost all jute growing areas. The disease generally attacking almost lower part of jute plants especially root.

17.3.2   SYMPTOMS

The disease may attack jute plants at any stage of growth. Root lesion usually occurs at the collar region of stem and on the root as small, blackish-brown spots which may coalesce and gradually causing rot. The root rot leads to shredding and wilting of the plant. Pycnidia, seen as small black dots are usually found on the epidermis of the root.

17.3.3 CAUSAL ORGANISM

The root rot of Jute is caused by the same fungus that caused stem rot: *Macrophomina phaseolina* (Tassi) Goid whose sclorotia phase is known as *Rhizoctonia bataticola* (Taub.) Butler.

17.3.4 DISEASE CYCLE AND EPIDEMIOLOGY

The disease cycle and epidemiology is similar to stem rot. The pathogen perennates in seed, soil, weeds like *Cyperusdistans* and on stubbles in the jute fields in its sclerotia form. The jute seed may act as a primary source of infection. The severe outbreak of the root-rot may occur under warm and humid conditions. The disease cycle and epidemiology is similar to stem rot of jute.

17.3.5 INTEGRATED DISEASE MANAGEMENT

Integrated disease management of root rot is similar to IDM of stem rot.

17.4 ANTHRACNOSE

17.4.1 ECONOMIC IMPORTANCE

The anthracnose disease on jute is caused by fungus *Colletotrichum corchorum.* It is a seed-borne parasite, the mycelium originating in the seed to which the spores are adhering to external part of seed. Parasitic on stems, pods, and leaves of *Corchorus capsularis* L. The optimum temperature for growth of the fungus is 30°C. The pathogen is seed-borne, the fungus mycelium exists in the seed, and spores are adhering on the external part of seeds.

17.4.2 SYMPTOMS

The pathogen can affect stem, leaves, and pods that can manifest it as yellowish-brown to black, water soaked, depressed spot and sharply defined lesions. The disease is presently not causing any severe damage to jute crop but percentage of disease incidence has increased significantly in last two decade. This disease is mainly occurred in capsularis jute but now it is also reported from olitorious jute varieties across different region of the country. When older plants are infected, the disease are restricted on leaf and stem and badly infected leaves fall off. Initially, spots are of smaller in size and on one side of leaf but later on spot may coalesce and may cover major portion of leaf or stem and they may girdle the stem and cause stem break. The fungus invades vascular bundles, weakening bast fibers bundles. Infected Plants which remain standing at harvest may show shredding of fiber with the cankerous lesions.

17.4.3 CAUSAL ORGANISM

The anthracnose disease on capsularis jute is caused by fungus *Colletotrichum corchorum*. Tanaka and Ikata and on *Olitorious* jute by *Colletotrichum gleosporioides*.

17.4.4 DISEASE CYCLE AND EPIDEMIOLOGY

Cloudy weather and continuous rain favor the disease that can spread quickly in the jute field. Rain splash helps in dispersal of spores. The pathogen requires high rainfall, high temperature (30–32°C) and high relative humidity favor infection. The pathogen is seed borne in nature and it also survives on infected crop debris.

17.4.5 MANAGEMENT

The control of pathogen can be done by following methods:

1. Treatment of seed with carbendazim @2 g/kg seed;
2. Foliar spray of carbendazim @ 2% to control secondary infection;
3. Resistant varieties.

17.5 MOSAIC

17.5.1 ECONOMIC IMPORTANCE

Leaf mosaic was also called 'chlorosis' and affecting both Olitorius and capsularis jute varieties but it especially affect capsularis jute. Regular or irregular Yellow mosaic spots usually appear on capsularis jute by inhibiting formation of chlorophyll at any stage of growth of plant. Disease disseminated by seed, soil, and pollen. A white fly acts as vector for carrying disease.

17.5.2 SYMPTOMS

The virus causes gradual mottling of leaf with yellow and green in various patterns, and crinkling of lamina and retards growth of jute plant. The symptoms may remain in latent stage and later on express under suitable conditions. Under severe infection, plant stunted and may finally dies. The disease is widespread in capsularis jute in almost every jute growing region of India and Bangladesh.

17.5.3 CAUSAL ORGANISM

Leaf mosaic is caused by *Chorchorus golden mosaic virus* comes under genera Begomovirus. It is transmitted by white fly (*Bemiciatabaci*).

17.5.4 *DISEASE CYCLE AND EPIDEMIOLOGY*

The disease is mainly seed borne and so infected jute seeds can be responsible for reduction of fiber yield. The virus can also be carried by whitefly as vector from infected plant to healthy plant. Under hot and humid conditions, the disease may aggravate and leads to heavy disease incidence. The capsularis jute is much more susceptible to leaf mosaic and full-field infestation of jute plant may occur under severe condition.

17.5.5 *MANAGEMENT*

The control of pathogen can be done by following methods:

1. Use of certified seeds from recognized institutions;
2. Growing of seed during period of unavailability of white fly population in region;
3. Use of resistant varieties, if available;
4. Infected plants are pulled out and destroyed;
5. Management of white fly population by insecticides;
6. Avoid jute cropping near cotton growing area.

17.6 DIEBACK

17.6.1 *CAUSAL ORGANISM*

It was caused by *Diplodiacorchori.*

17.6.2 *SYMPTOMS*

In this disease, apex of main stem or branches begins to dry or wither from the tip downwards, leaves drop off, and stem turn deep brown to black at almost mature stage of crop. Sooty mass of two-celled ellipsoid dark spores cover the stem under damp atmospheric conditions. Generally, mature crop especially seed crop affected by disease under moisture stress coupled with lack of proper nutrition.

17.6.3   DISEASE CYCLE AND EPIDEMIOLOGY

The disease is less prevalent and found in jute growing region of India and Bangladesh. The fungus is weak parasite and affects nutritionally deprived plant. The disease is seed and soil borne and spores are disseminated by air.

17.6.4   MANAGEMENT

1. Raised crop in well manure soil and not exposed to moisture stress;
2. Early sowing crop to escape the disease;
3. Follow Crop rotation for 2–3 years;
4. Seed treatment with Carbendazim@2 g/kg seed;
5. Spray copper oxychloride or Mancozeb @2.5 g/l of water.

KEYWORDS

- **anthracnose**
- **dieback**
- **jute**
- **mosaic**
- **sclerotia**
- **stem rot**

REFERENCES

Ashby, S. F., (1927). *Macrophomina phaseoli* (Maubl.) Comb. Nov. The pycnidial stage of *Rhizoctonia bataticola* Taub. Butler. *Transanction of the British Mycological Society, 12,* 141–147.

Ghosh, T., & Bask, M. N., (1965). Possibility of controlling stem rot of jute. *Indian Journal of Agricultural Sciences, 35,* 90–100.

Ghosh, T., & Mukherjee, N., (1970). *Macrophomina phaseoli* (Maubl.) Ashby of jute. Plant disease problem. *International Symposium on Plant Pathology* (pp. 369, 370). New Delhi: IARI.

Hath, T. K., & Chakraborty, A., (2004). Towards the development of IPM strategy against insect pests, mites and stem rot of jute under North Bengal conditions. *J. Entomol. Res., 28*(1), 1–5.

Mandal, R. K., (1990). Jute diseases and their control. In: *Proceeding of National Workshop Cum Training, Mesta, Sunnhemp and Ramie.* CRIJAF, Barrackpore.

Varadarajan, B. S., & Patel, J. S., (1943). Stem rot of jute. *Indian Journal of Agricultural Science, 13,* 148–156.

Verma, P. M., Rao, G. G., & Capoor, S. P., (1966). Yellow mosaic of *Corchorus trilocalaris. Sci. Cult., 32,* 466.

CHAPTER 18

Nematode Diseases of Cereal Crops and Their Management

VIRENDRA KUMAR SINGH,[1] VIVEK SINGH,[1] and DEO KUMAR[2]

[1]Department of Plant Pathology, Banda University of Agriculture, Science and Technology, Banda–210001, Uttar Pradesh, India,
E-mail: virendra_singh16@yahoo.com (V. K. Singh)

[2]Department of Soil Science, Banda University of Agriculture, Science and Technology, Banda–210001, Uttar Pradesh, India

18.1 INTRODUCTION

Plant-parasitic nematodes are serious problem for economically important crops. The plant-parasitic nematodes have been recognized an important constraints in the production of agricultural crops including wheat and rice which constitute the two most important staple food crops of the world. Estimated losses due to plant-parasitic nematodes attack on different important cultivated crops all over the world by FAO are around 400 million dollars. In India, annual crop losses have been reported Rs. 242.1 billion. A large number of plant-parasitic nematodes are known to infect cereal crops in all parts of the world including India. Many plant-parasitic nematodes are elongate, eel-like; adults range in length from about 0.5 mm to 1 cm but most of them are less than 2 mm. Some nematodes have sedentary females whose bodies swell to become fusiform, irregularly saccate, or spherical. Important among these are: *Meloidogyne* spp. (wheat, barley, maize, rice), *Heterodera* spp. (wheat, maize, rice), *Hoplolaimus* spp. (maize, rice), *Ditylenchus* spp. (oat, rye, rice), *Anguina tritici* (wheat), *Pratylenchus* spp. (oat, wheat, barley, rye, maize), *Rotylenchulus* spp. (maize), *Tylenchorhynchus* spp. (wheat, oat), *Aphelenchoides* spp. (rice), and *Hirschmanniella* spp. (rice). Most of the developing countries like India lie in tropical or subtropical regions where climate is suitable for activity and multiplication of nematodes almost throughout the year. Sandy and warm soils such as found in India other developing countries in the arid zone are very favorable for nematode infection, especially in irrigated areas, which are used for crop production continuously (Taylor, 1967).

18.2   ROOT-KNOT NEMATODES: (*MELOIDOGYNE* SPP.)

Root-knot nematodes occur in all over the world but are found more frequently and in greater numbers in areas with warm or hot climates and short or mild winters. Root knot nematodes have been reported to be a serious disease causing agents in different cereals crops which grown in India. The root-knot nematodes are reducing 5–73% avoidable yield losses depending upon the population density of nematodes and environmental factors (Rao and Biswas, 1973; Gaur et al., 1993). More than 60 species of this genus are found in the world whereas in India about 12 species have been reported to occur. Root knot nematode *Meloidogyne incognita* on wheat was the first time reported from J&K (Singh and Dwivedi, 2015). However, *M. incognita, M. javanica, M. arenaria; M. hapla,* and *M. graminicola* are found in all places of India. Bio-chemical differences in protein and enzyme content among species have also been used for taxonomic identification. The above-ground symptoms: Root-knot nematode damage results in poor growth, a decline in quality and yield of the crop and reduced resistance to other stresses (e.g., drought, and other diseases). A high level of root-knot nematode damage can lead to maximum crop loss. Nematode damaged roots do not utilize water and fertilizers as effectively leading to additional losses for the grower. The below-ground symptoms: Root knot nematode cause severe galling, stunting, and chlorosis of crops. Other root swellings must not be mistaken for root-knot galls. While root knot galls have firmer tissues and contain different stages and females of the root-knot nematodes inside the gall tissues, near the fibrous vascular tissues of the root.

The male and female root-knot nematodes are easily distinguishable morphologically. The males are wormlike and about 1.2 to 1.5 mm long by 30 to 36 μm in diameter. The females are pear shaped and about 0.40 to 1.30 mm long and 0.27 to 0.75 mm wide. The life cycle includes egg, juvenile, and adult stages. A life cycle is completed in 25 days at 27°C, but it takes longer at lower or higher temperatures. Each female lays approximately 500 eggs in a gelatinous substance produced by the female.

18.3 CYST NEMATODES: (*HETERODERA* SPP.)

Nematodes belonging to the genus *Heterodera* and called cyst nematode. Cysts may be light brown to dark reddish brown and brown cysts are typically ovate to spheroid. The cyst nematode, *Heterodera zeae* was reported to be an important nematode problem of maize in Rajasthan, Punjab, M.P, Bihar, U.P., Delhi, and Maharashtra. A disease of wheat, barley, and oat locally known as 'molya' was prevalent in Rajasthan, Punjab, and Haryana since 1952. The above-ground symptoms resemble these associated with root damage and include stunting of plant in patches, yellowing of leaves and reduced size of various shoot parts within 3–4 weeks of sowing. An experienced person can absorbs cyst nematode, *Heterodera* spp. on the roots of their hosts without help of microscope. The young adult females are visible as tiny white color. After a female cyst nematode dies, her white body wall is tanned to a tough brown capsule containing several hundred eggs. The mature female bodies are found attached to roots by their head end embedded almost in the stele. The site of feeding is modified into a syncitium similar to that found in case of *Meloidogyne*. On the surface of infected roots, white to brown bodies of females can be discerned with naked eyes. The intensity of body color depends upon the maturity stage of the young female or cyst. In case of cyst, nematode matrix with eggs may also be found attached to the posterior region of the female body. The male is wormlike, about 1.3 mm long by 30 to 40 μm in diameter. Fully developed females are lemon shaped, 0.6 to 0.8 mm in length and 0.3 to 0.5 mm in diameter. Approximately 21–30 days is required for the completion of life cycle of the cyst nematode. White females can be seen in the root system by 4–6-weeks old plants. In the absence of host crop, the nematode multiplies on several graminaceous weeds.

18.4 LESION NEMATODES: (*PRATYLENCHUS* SPP.)

The root lesion nematodes are most economically important phytonematodes, and cosmopolitan in maize fields. However, more than 350 hosts have been recorded. Assessment of accurate loss by lesion nematodes has not been made possible under field conditions due to presence of mixed population of the nematodes in the field. Lesion nematodes have wide host range which can affect the selections of crop used to control the nematodes in crop rotation sequences. Soil type tillage operations have also been reported to affect lesion nematodes population dynamics. The plants show chlorosis, stunting, and general lack of vigor resulting into wilt. The plant form patches or zones in the field. Lesion nematodes reduce or inhibit root development by forming local lesions on young roots, which may then rot because of secondary fungi and bacteria. Crop yields are reduced. Both male and female of these nematodes are wormlike, 0.4 to 0.7 mm long and 20 to 25 μm in diameter. They are migratory endo-parasitic nematodes. The life cycle of Pratylenchus is completed within 45 to 65 days.

18.5   RENIFORM NEMATODES: (*ROTYLENCHULUS RENIFORMIS*)

It is one of the most widely distributed and destructive nematode pathogen of many millets in India. Maize is also a good host of this nematode. Affected plants are stunted and poor in growth. It is considered as pest of great significance after root-knot nematode. The nematode is a semi-endoparasite and remains attached to roots. Infected plants show stunting in growth with reduced and discolored root system. Damage during pre and post-emergence of seedlings leads to reduction in germination and crop stand. The total duration of life cycle is about 4–5 weeks under optimum conditions. Soil pH is also an important factor affecting reproduction of the nematode. Soil moisture and temperature have profound influence on infectivity, penetration, and the biology of the nematode. There was significant reduction in plant height, shoot, and root weight of mash at 1000 or more juveniles/ 500 g soil. Life cycle is completed in about 25 days.

18.6   LANCE NEMATODES: (*HOPLOLAIMUS* SPP.)

Hoplolaimus species is a very common nematode which can be found on plant roots in all types of soil and climate. Near about 17 species are reported in India (Sitaramaiah, 1984). Lance nematodes parasitize wide hosts. Among nematodes, which are often too small to be accurately detected by sight, the lance nematode is one of the larger species. Sometimes they feed at a particular site for a long time with nearly half of the body inside the root system (sedentary ectoparasite). In many cases, juveniles of the lance nematode completely enter the cortical tissue (endoparasite). Damage may show up as patches of yellowing and dying. These symptoms also can be caused by drought or nutrient deficiency. Lance nematodes multiply slowly in comparison to endoparasitic nematodes but they inflict significant crop damage at a lower level of infection. The availability of feeder roots and temperature are important factors for population build up of this nematode (Haider and Nath, 1992). Life cycle is completed in 13 to 38 days.

18.7 STEM AND BULB NEMATODE: (*DITYLENCHUS* SPP.)

Stem and bulb nematodes occur worldwide but are particularly prevalent and destructive in areas with temperate climate. It is one of the most destructive plant-parasitic nematodes and attacks a large number of host plants. On most crops stem nematode cause heavy losses by killing seedlings, dwarfing plants, destroying bulbs, by causing the development of distorted, swollen, and twisted stems and foliage, and, generally, by reducing yields greatly. The nematode feeds on stem, leaves, and bulbs and is rarely found in soil. The nematode is 1.0 to 1.3 mm long and about 30 μm in diameter. The females lay 200–500 eggs, mostly after fertilization by the males. The total duration of life cycle ranges from 19–25 days. Reproduction continues throughout the year.

18.8 FOLIAR NEMATODES: (*APHELENCHOIDES* SPP.)

The genus *Aphelenchoides* is a group of nematodes commonly known as leaf, bud, and foliar nematodes. This seed transmitted phytonematode is found all over the world in rice growing areas have been reported from more than 28 countries. In India foliar nematodes was serious problem on rice crop in central province Dastur (1934).

18.9 WHEAT GALL NEMATODE OR EAR COCKLE OF WHEAT: (*ANGUINA TRITICI*)

Wheat gall nematode has the historical significance of being the plant-parasitic nematode that was seen and demonstrated to cause a plant disease. The nematode is reported from all the important wheat growing regions of the world. In India, the nematode was first reported in 1919 from Punjab (Milne, 1919). Singh et al. (1953) had reported that the nematode causes a loss of 30% in the annual production of wheat in India. Nematode infected young plants show slight enlargement at the basal portion of stem. Such infected plants are mostly stunted in appearance. The leaves are twisted and curled preventing normal emergence of ears. Seed galls, which are green in the beginning later turn hard, dark brown to black. These seed galls measure about 3 to 5 mm in length and 2–3 mm in width. Each seed gall may contain 1000 to 30,000 or more nematode juveniles in a quiescent state. Even 28 years old, seed galls have been found to produce viable juveniles. The adult female is 2.64 to 4.36 mm long and adult males are 2.04 to 2.4 mm long, straighter than females, and more active. Anterior and posterior portions are slender while the middle portion is swollen. Several workers (Koshy and Swarup, 1971; Swarup, and Gupta, 1971; Bhatti and Dalal, 1976; Paruthi, and Bhatti, 1985; Gokte, and Swarup, 1987; Singh et al., 1991) have studies on lifecycle, effect of sowing time and disease record of *A. tritici* on wheat. The wheat gall nematode or ear cockle of wheat (*Anguina tritici*) on wheat was reported first time from farmer's field of Jammu Figure 18.1 (Singh et al., 2011).

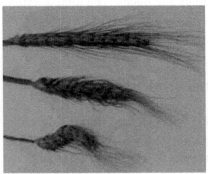 

(a) Seed gall nematode infested ear,   (b) Left: Diseased seeds and right-healthy seeds
                                                            of wheat.

FIGURE 18.1   Seed gall nematode infested plants of wheat at Jammu.

18.10   TUNDU DISEASE

Apart from ear cockle disease, the nematode is also associated in the yellow ear rot or yellow slime disease of wheat, which is commonly known as tundu disease in India. The association of *Anguina tritici* with bacterium *Corynebacterium (Clavibacter) tritici* in the development of yellow slime disease of wheat.

18.11 RICE ROOT NEMATODE: (*HIRSCHMANNIELLA* SPP.)

Rice root nematode is distributed throughout the rice growing regions of Asia and the Pacific Islands. The nematode was first reported on rice roots in Java (Indonesia) and described as *Tylenchus oryzae* by J. Von Breda de Haan in 1902, which attributed the 'mentek' disease of rice to this nematode. In India, these nematodes had been first suggested to be associated with rice since 1963 when in Shahabad district of Bihar rice crop suffered from a disease in which the foliar symptoms resembled those caused by rice root nematodes under poor moisture and nutritional conditions. Heavy clay soils supported high populations of the nematode. Mathur and Prasad (1972) was also reported 56% reduction in rice yields. Above-ground symptoms of rice, root nematode infestation can be easily confused with nutrition deficiency and imbalance of micronutrients. The symptoms that are seen on aerial parts are stunted growth, yellowing of leaves from tip downward, poor tillering, reduced number of earhead, and reduced grain weight. Adult males and females remain vermiform throughout their life. They measure 1.0 to 1.5 mm in length. From egg, stage to adult female or male stage the nematode a minimum of 30 days. Two to four generations have been completed in one crop season.

18.12 CULTURAL MANAGEMENT

The cultural practices include crop rotation, fallowing, flooding, sanitation, plowing during summer season, mulching, organic manure, spacing of plants in the field, time of sowing, resistant varieties, etc. Nematode management through intercropping and crop rotation is because some species of nematodes are able to feed and multiply only on host crop. Rotation is a very old practice for reducing soil borne problems. Rotation to non-host crops may cause many of those pests to cease reproduction and allow natural mortality factors to reduce their numbers. Flooding or fallowing may be used to help reduce numbers of nematode pests. The period of flooding appears to vary with several factors such as kind of soil, season etc. Warm conditions are said to reduce the period required for control. Nematode juveniles are more easily killed by flooding than eggs. The period of flooding needs to be worked out for each condition. Fallow periods in cropping sequences can also reduce nematode populations. The summer plowing 2–3 times during hottest period of the year, help to expose nematodes to the drying action of sun and wind and reduce the population. Hague and Prasad (1982) reported drastic reduction of plant-parasitic nematodes and significantly increased the yield with three deep plowings. The intensity of symptoms was directly related to the inoculum levels. Soil fertility and texture have a significant influence on nematode abundance and diversity. Plant-parasitic nematodes (e.g., *Meloidogyne, Rotylenchulus, Heterodera, Hoplolaimus,* etc.) generally prefer sandy loam or loam or sandy soils (Wallace, 1973; Goodell and Ferris, 1980; Singh and Singh, 2003). However, some nematodes *Hirschmaniella* and *Pratylenchus* are more prevalent in heavy soil.

Sanitation terms cover a wide range of cultural practices, including weed control, crop residue destruction and disinfestations of farm equipment before moving it from heavily infested fields to uninfected fields. In monocultures, eliminating the weed hosts can be important in reducing the populations of plant-parasitic nematodes. Soil temperature plays crucial role in the activities of plant-parasitic nematodes, the time during which crop is planted is important (Ayaub, 1980). The addition of inorganic fertilizers alone without organic manure usually increases the nematode population and disease intensity. NH_4-N reduces the disease incidence while NO_3-N may increase the same. Particular forms as well as dose and proportion of NPK may also reduce or increase the incidence of disease. Use of tolerant/ resistant varieties is most practical approach for the management of nematode diseases. Crop cultivars resistant to phytonematodes can be the most useful and cheapest means of nematode control for the small-scale farmers. Nematicidal plants with roots containing nematicidal substances have been investigated. These toxic substances reduce the population level of some nematode species. African marigolds (*Tagetes* spp.), asparagus, crotalaria, mustard, and several cruciferous plants have been reported to produce toxic substances. Various organic materials including agricultural and industrial by products, most of which are wastes, have been experimented for the control of phytonematodes infecting crops. Soil amendments with green manure, compost, oil cakes of neem, mahua, mustard, groundnut, cotton, linseed, karanj, and saw dust etc. have been found to reduce nematode populations. Neem, karanj, and groundnut cakes incorporated into soil at the rate of 500 kg/ha give good control of plant-parasitic nematodes and could be practiced wherever possible. Apart from encouraging the multiplication of natural enemies like nematode trapping fungi, the decomposition products of these organic amendments are toxic to nematodes.

Use of organic manures is of great value since the decomposition products and promotion of natural enemies decrease nematode populations (Mankau, 1963; Goswami and Swarup, 1971; Singh and Sitaramaiah, 1973; Mishra and Prasad, 1974; Haseeb et al., 1984; Singh and Singh, 1991; Singh and

Singh, 1992; Alam, 1990; Singh and Singh, 2001; Singh, 2004; Singh, 2006; Ononuja and Kpodobi, 2008; Singh, 2008). The amount of oil cakes or any organic matter to be incorporated depends on various factors like, soil type and texture, crop to be planted, the predominant nematode fauna of the soil and the amount of soil moisture present in the soil. In trap crops when grown in infested soils, the nematodes penetrate into the root system and start multiplying. Before the nematodes complete its life cycle, the plants are uprooted and destroyed. Applications of botanicals are easy, environmentally safe, and having no phytotoxic effect on crops. The inhibition of root-knot development may be due to the accumulation of toxic byproducts of decomposition (Alam et al., 1979) to increased phenolic contents resulting in host resistance (Alam et al., 1979, 1980), or to changed physical and chemical properties of soil inimical to the nematodes (Ahmad et al., 1972). One of the most economical and effective ways to control plant-parasitic nematodes is the growing of nematode resistant plant cultivars. The identification of a dominant Mi gene importing resistance to the root-knot nematodes and its linkage to an acid phosphates gene is a major breakthrough in this area of research.

18.13   BIOLOGICAL MANAGEMENT

Application of natural enemies of plant-parasitic nematodes for controlling nematode population is an essential component of eco-friendly management. Manipulation of biotic agents present in soil or introduction of such agents in soil, for reduction of nematode populations, has been one of the most important research areas engaging the attentions of the plant protection scientists all over the world. Plant-parasitic nematodes have suffered by many natural enemies,' e.g., fungi, bacteria, and predacious nematodes. Fungi have been the foremost amongst such agents. Certain fungi capture and kill nematodes in the soil. *Arthrobotrys* spp., *Dactylaria* spp., *Dactylella* spp., *Catenaria* spp., and *Trichothecium* spp., are the genera most commonly represented. The biological control of root-knot nematode on tomato under green house condition by using predaceous fungi has been reported (Singh et al., 2001; Bandyopadhya et al., 2001). Some fungi capture nematode by adhesion, but many employ specialized devices that include networks of adhesive branches, stalked adhesive knobs, nonconstricting rings and constricting rings. The surface of the nematode is penetrated and the fungus hyphae grow throughout the nematode body, digesting, and absorbing its contents. Under favorable conditions, large numbers of nematodes may be captured, and killed especially by those fungi that form adhesive networks or hyphal loops. *Trichoderma harzianum, Trichoderma virens, Aspergillus niger, Paecilomyces lilacinus, Pochonia chlamydosporia* are found promising biocontrol agents. Davide and Zorilla (1985) while investigating the biocontrol potential of *P. lilacinus* against *M. incognita* on okra found the fungus to be quite effective and economically better than nematicides. According to Shahzad and Ghaffar (1984) carbofuran @1 kg a.i /ha was less effective than *P. lilacinus* against *M. incognita*. Now mycorrhiza is not restricted to its use only as biofertilizers, its potential role in the biological control of plant-parasitic nematodes is reported by many workers. Sikora (1979) found that prior presence of VAM fungi *Glomus mosseae* has resulted into an increase in plant resistance against *Meloidogyne* spp. A bacterial parasite of nematodes *Pasteuria penetrans*, has received much attention and research effort in recent years, *P. penetrans* is probably the most specific obligate parasite of nematodes, with a life cycle remarkably well adapted to parasitism of certain phytonematodes. It directly parasitizes juvenile nematode, thus affects penetration and reproduction. *Pasteuria penetrans* can survive several years in air dried soil apparently without loss of viability (Mankau and Prasad, 1971). Seed bacterization, soil drenching and bare root dip application with *Pseudomonas fluorescens, Pasteuria penetrans, Bacillus subtilis, B. polymyxa* effectively controls plant-parasitic nematodes was also reported by many workers (Walia, 1994;Walia and Dalal, 1994; Singh et al., 1998; Rojas Miranda and Marban-Mendoza, 2000; Ravichandra and Reddy, 2008). Among the predatory nematodes, monarchs may be proved efficient predators because of stronger predatory potential, high rate of predacity and high strike rate (Bilgrami, 1998).

18.14 CHEMICAL MANAGEMENT

Two major groups of nematicides are distinguished by the manner in which they spread through the soil. Soil fumigants are gases in cylinders or liquids, which spread as gases from the point where they are infected into the soil. Non-fumigant nematicides include a variety of water-soluble compounds, which are applied to the soil as liquid or granular formulations. Most belong to the carbamate or organophosphate families of pesticides. These distributions in the soil depend on physical mixing during application and moving in solution in soil water. There are no perfect nematicides for all purposes. Nematicides vary in their effectiveness against different kinds of nematodes, ease to handling, cost effect on other classes of pests (weeds, disease organisms, and insects) behavior in different soils, toxicity to different plants and availability. The performance of the nematicides will depend on soil conditions, temperatures, and rainfall. A yield benefit is not guaranteed and nematicides are expensive. The older nematicides are mostly fumigants, which are applied in soil. These are seldom recommended because of their hazardous nature and high toxicity to nontarget organisms. Recently, large number of non-fumigant and systemic nematicides are available which are safe on plant. DBCP (nemagon) has been found very effective for standing crops against root-knot nematodes but its use has been suspended due to adverse effect on human beings. Carbofuran 3G (furadon), phorate 10G (thimet), fenamiphos (nemacur), fensulphothion (dasanit) and oxamyl have been recommended for the management of nematodes (Kalita and Phukan, 1985; Johnson, 1985; Prasad et al., 1984, 1986; Prasad, 1990; Rahman, 1991; Singh and Singh, 1991; Singh and Singh, 1991; Mishra, 1995; Singh, 2006). The development of the *H. cajani* nematode was also delayed in pot condition experiment which was treated with carbofuran, phorate, and aldicarb nematicides (Singh and Singh, 1998). The dose and method of application would vary with crops. Chemicals and Phytoalexins that have implications in the nematode control (Prasad and Swarup, 1986), synthetic pheromones to disrupt nematode reproduction (Bone, 1987), and chemicals that directly interfere with nematode chemoreceptors (Janson, 1987) should be exploited for nematode population management. Seed treatment with nematicides has been used successfully to control root-knot nematodes in various crops (Prasad and Chawla, 1991; Rahman and Das, 1994; Singh et al., 1998). *Anguina tritici* nematode: seeds are placed in 20%salt solution in suitable container and stirred vigorously for some time. The galls being lighter float on the surface and can be easily skimmed off and destroyed. The seeds are then thoroughly washed in fresh water, dried, and then can be sown or stored. Systemic nematodes like carbofuran and phorate are also very effective.

18.15 REGULATORY MANAGEMENT

Numerous attempts have been used to prevent the introduction of nematodes into countries or provinces by means of quarantine. Quarantines are established by legislative action in parliament, etc., and usually give quarantine authorities power to make and enforce regulations to accomplish the purpose. Such types of regulations usually prohibit bringing infected seeds into protected areas where similar crops might become infected.

18.16 PHYSICAL MANAGEMENT

Soil solarization with transparent polyethylene sheet has been attended as a means of raising the soil temperature to lethal levels to manage soil pathogens (Sharma and Nene, 1990). Soil solarization with double transparent sheets caused the maximum reduction of nematode population (84.11%) whereas with double black, single transparent and single black sheets reduced nematode population 80.96, 75.51, and 70.46 respectively. The scientific principle involved in thermotherapy is that the pathogens present in soil and plant materials are inactivated or eliminated at temperatures nonlethal for the host tissues. Physical means of nematode control includes heat treatment of soil, solar drying, steam sterilization, hot water treatment, and soil solarization. Soil solarization is a method of pasteurization can effectively suppress most species of nematode along with other microbes and weeds under field conditions (Katan, 1981; Kumar et al., 1993; Rao and Krishappa, 1995; Sharma, 1985a; Sharma and Nene, 1985; Singh, 2008). Generally, sheets of 50–100 µm thickness are most suitable for raising the soil temperature. The use of transparent polyethylene had yielded better results than black sheets, since transparent sheet transmit most of the incident radiation to soil (Mehrer, 1979; Mehrer and Katan, 1980). Additions of salt, sugar, charcoal, etc. create osmotic stress on nematodes and can be used for controlling nematodes. Ganguly et al. (1996) observed that mulching significantly increased rice seedling growth; the plant-parasitic nematodes were reduced by 50–87% in the solarized soil.

18.17 INTEGRATED MANAGEMENT OF NEMATODES

The development of pest management programme for economic control of any pest implies the judicious selection of those available control techniques which will reduce the effects of the pest with minimum negative impact on the environment and with overall economic soundness. Integrated nematode management is based upon the system approach, follows location specific principle, and is environment specific (Pankaj and Sharma, 1998). Utilization of the best combination of available management strategies for the pest complex at hand (nematodes, insect pests, disease organisms, weeds, etc.) constitutes an integrated crop protection system. Resistant cultivars, crop rotation, pesticides, and sanitary and cultural practices can all be employed to the best possible advantage. An integrated management strategy prevents the excessive build up any single nematode, insect, or disease population and minimizes the development of pest resistance to any single tactic. Integrated pest management systems require flexibility and depend upon the specific pest problem and locally available management options. A fixed set of recommendations may keep a pest complex in check for a limited period of time, but as the pest population shifts, recommendations will have to change also. Therefore, system development takes into account many factors including the species and race of pests present, the availability of resistant host plants, the longevity of the pest, and the crops, cropping systems, and climate of the geographical region.

18.18 APPLICATION OF MOLECULAR BIOLOGY AND BIOTECHNOLOGY

Nematode diagnostics is essential for success of any nematode management programme even more when the control method is highly specific to the control species/pathotypes/race. A number of new techniques for analyzing nucleic acids, proteins, carbohydrates, and lipids can be helpful in the identification of pests of those allozyme; monoclonal antibody and DNA based systems are most well developed for nematodes. Polyacrylamide gel electrophoresis (PAGE) has mainly focused on cyst and root-knot nematodes (Davies, 1977). PCR techniques or RFLP analysis used as supplementary tools wherever necessary. Nematode resistant transgenic plants can be designed by various approaches. The simple method is to introduce resistant genes effective against plant-parasitic nematodes from wild species to commercial cultivars.

18.19 CONCLUSION

Nematodes parasitize the roots of our carefully nurtured crops, reduce their ability to produce yields, make them weak and perhaps vulnerable to many pests. Crop resistance offer to be of very good importance in this directivity. In addition, an intensive search of plant germplasm collection for natural resistance to nematodes and their races is also required. An in-depth understanding is needed of the molecular basis of how and why plants are susceptible to nematodes. New progress is being made in studying changes in gene expression during the infestation of plants in nematode host interactions where feeding sites are formed. There is a need to reduce these avoidable yield losses by developing new environment friendly, economically acceptable and ecologically based management strategies as the current options become ineffective or unacceptable. The research on identifying the promising biocontrol agents, their mass production, application techniques and their behavior of soil under varying agro climatic conditions need to be intensified. Future nematode management must employ sustainable agricultural practices that take into account beneficial, detrimental, and other nematode species in the rhizosphere and in soil.

KEYWORDS

- **biocontrol agents**
- **nematodes**
- **pathotypes**
- **polyacrylamide gel electrophoresis**
- **rhizosphere**
- **soil solarization**

REFERENCES

Ahmad, R., Khan, A. M., & Saxena, S. K., (1972). Changes resulting from amending the soil with oil cakes and analysis of oil cakes (Abstract). *Proc. 59ᵗʰ Ses. Indian Sci. Con. Calcutta. Part III* (p. 164).

Alam, M. M., Khan, A. M., & Saxena, S. K., (1979). Mechanism of control of plant-parasitic nematodes as a result of the application of organic amendments to the soil. V. role of phenolic compounds. *Indian J. Nematol.*, *9*, 136–142.

Alam, M. M., (1990). Control of plant-parasitic nematodes with organic amendments and nematicides in nurseries of annual plants. *J. Bangladesh Acad. Sci.*, *14*, 107–113.

Ayaub, S. M., (1980). *Plant Nematology: An Agricultural Training and Nema aid Publication Sacramento, C.A.* (p. 195).

Bandyopadhyay, P., Kumar, D., Singh, V. K., & Singh, K. P., (2001). Eco-friendly management of root-knot nematode of tomato by *Arthrobotrys oligospora* and *Dactylaria brochopaga*. *Indian J. Nematol.*, *31*, 153–156.

Bhatti, D. S., & Dalal, M. R., (1976). Susceptibility of some common wheat varieties to the seed gall nematode *Anguina tritici*. *Haryana Agric. Univ. J. Res.*, *6*, 162, 163.

Bilgrami, A. L., (1981). Predatory nematodes and protozoans as biopesticides of plant-parasitic nematodes. In: Trivedi, P. C., (ed.), *Plant Nematode Management a Biological Approach* (pp 4–23).

Bone, I. W., (1987). Pheromone communications in nematodes. In: Veech, J. A., & Dickson, D. W., (eds.), *Vistas on Nematology: A Commemoration of the Twenty-Fifth Anniversary of the Society of Nematologists* (pp. 147–152). Hyattsville, Maryland, U.S.A. Society of Nematologists.

Gokte, N., & Swarup, G., (1987). Studies on morphology and biology of *Anguina tritici*. *Indian J. Nematol., 17,* 306, 307.

Goswami, B. K., & Swarup, G., (1971). Effect of oil cake amended soil on the growth of tomato and root knot nematode population. *Indian Phytopath., 24,* 491.

Haseeb, A., Alam, M. M., & Khan, A. M., (1984a). Control of plant-parasitic nematodes with chopped plant leaves. *Indian J. Plant Pathol., 2,* 180, 181.

Jansson, H. B., (1987). Receptors and recognition in nematodes. In: Veech, J. A., & Dickson, D. W., (eds.), *Vistas on Nematology: A Commemoration of the Twenty-Fifth Anniversary of the Society of Nematologists* (pp. 153–158). Hyattsville, Maryland, U.S.A. Society of Nematologists.

Kalita, M., & Phukan, P. N., (1995). *Indian J. Nematol., 25,* 110–111.

Koshy, P. K., & Swarup, G., (1971). Distribution of *H. avenae, H. zeae, H. cajani,* and *Anguina tritici* in India. *Indian J. Nematol., 1*(2), 106–111.

Kumar, B. N., Yaduraju, T., Ahuja, K. N., & Prasad, D. P., (1993). Effect of soil solarization on weeds and nematodes under tropical Indian conditions. *Weed Research, 33,* 423–429.

Katan, J., (1981). Solar heating (solarization) of soil for control of soil borne pests. *Ann. Rev. Phytopathol., 19,* 211–236.

Mankau, R., (1963). Effect of organic soil amendments on nematode population. *Phytopathology, 53,* 881–882.

Mehrer, Y., (1979). Prediction of soil temperature of a soil mulched with transparent polyethylene. *J. Appl. Meteorol., 18,* 1263–1267.

Mehrer, Y., & Katan, J., (1980). Prediction of soil temperature under polythene mulch. *Hass., 60,* 1384–1387.

Mishra, S. D., & Prasad, S. K., (1974). Effect of soil amendments on nematodes and crop yields. *Indian J. Nematol., 4,* 1–19.

Orke, E. C., Dehne, H. W., Schonbeck, F., & Weber, A., (1994). *Crop Production and Crop Protection: Estimated Losses in Major Food and Cash Crops.* Elsevier, Amsterdam.

Pankaj, & Sharma, H. K., (1998). IPM strategies for nematode management. In: Prasad, D., & Gautam, R. D., (eds.), *Potential IPM Tactics* (pp.131–145). Westville publ. House, Paschim Vihar, New Delhi.

Paruthi, T. J., & Bhatti, D. S., (1985). Estimation of loss in yield and incidence of *A.tritici* on wheat in Haryana. *Int. Nematol. Network Newsl., 2*(3) 13–16.

Prasad, D., (1990). Control of root knot nematode *Meloidogyne arenaria* on groundnut by chemical seed pelleting. *Curr. Nematol., 1,* 31–34.

Prasad, D., & Chawla, M. L., (1991). *Indian J. Nematol., 21,* 19–23.

Prasad, J. S., & Panwar, M. S., & Rao, Y. S., (1986). *Indian J. Nematol., 16,* 119–121.

Prasad, D., & Yaduraju, N. T., (1993). Solarization of filed soil for effective management of nematodes and weeds. *National Conference on Ecofriendly Approaches, 67.*

Rahman, M. L., (1991). *Curr. Nematol., 2,* 93–98.

Rao, Y. S., & Biswas, H., (1973). *Indian J. Nematol., 3,* 74.

Ravichandra, N. G., & Reddy, B. M. R., (2008). Efficacy of *Pasteuria penetrans* in the management of *Meloidogyne incognita* infecting tomato. *Indian J. Nematol., 38,* 172–175.

Rao, M. S., & Reddy, P. P., (1992). *Prospects of Management of Root-Knot Nematode on Tomato Through the Integration of Bio Control Agent and Botanicals* (p. 10). Seminar on current trends in diagnosis and management of plant diseases. I.I.H.R. Bangalore.

Rao, V. K., & Krishappa, K., (1995). Soil solarization for the control of soil borne pathogen complexes with special reference to *Meloidogyne incognita* and *Fusarium oxysporium*. *Indian Phytopathol., 48,* 300–303.

Rojas, M. T., & Marban-Mendoza, N., (2000). *Pasteuria penetrans:* Adherence and parasitism in *Meloidogyne incognita* and *Meloidogyne arachibicida. Nematropica., 29,* 233–240.

Sasser, J. N., & Carter, C. C., (1985). *An Advanced Treatise on Meloidogyne: Biology and Control* (Vol. I, pp. 1–422). North Carolina State University, Graphics.

Sasser, J. N., & Freckman, D. W., (1987). A world perspective on nematology: The role of society. *Vistas of Nematology,* 7–14.

Swarup, G., & Gupta, P., (1971). On the ear cockle and tundu diseases of wheat II. Studies on *A. tritici* and *Corynebacterium tritici. Indian Phytopath., 24,* 359–365.

Swarup, G., & Singh, N., (1962). A note on nematode bacterium complex in tundu disease of wheat. *Indian Phytopathol., 15,* 294, 295.

Sharma, S. B., (1985a). *Nematode Diseases of Chickpea and Pigeon Pea* (p. 103). Pulse pathology progress report No. 43, Patencheru, A. P., (India) International Crops Research Institute for the semi arid tropics.

Sharma, S. B., & Nene, Y. L., (1985). Effect of presowing solarization on plant-parasitic nematodes in chickpea and pigeon pea fields. *Indian J. Nematol., 15,* 277, 278

Sharma, S. B., & Nene, Y. L., (1990). Effects of soil solarization on nematodes parasitic to chickpea and pigeon pea. *J. Nematol., 22,* 658–664

Sikora, R. A., (1979). Predisposition to *Meloidogyne* infection by the endo tropic mycorrhizal fungus *Glomus mosseae.* In: Lamberti, F., & Taylor, C., (eds.), *Root-knot Nematode Meloidogyne spp. Systematic Biology and Control* (pp. 399–404). Academic Press London.

Singh, K. P., Singh, V. K., & Singh, L. P., (1991). Concomitant effect of sowing time and inoculation of *Anguina tritici* on wheat. *Curr. Nematol., 2,* 155–158.

Singh, K. P., & Singh, V. K., (1991). Nematicidal natures of Arjun bark *Terminelia Arjuna* on cyst nematode, *Heterodera cajani* of pigeon pea. *New Agric., 2,* 77, 78.

Singh, K. P., & Singh, V. K., (1992). *Terminelia Arjuna* leaf powder reduces population density of *Heterodera cajani. International Pigeon pea Newsletter, 16,* 17, 18.

Singh, K. P., Bandyopadhyay, P., & Singh, V. K., (2001). Performance of two predacious fungi for control of root-knot nematode of brinjal. *Mycopathological Research., 39,* 95–100.

Singh, M., Samar, R., & Sharma, G. L., (1998). Effect of various soil types on the development of *Pasteuria penetrans* on *Meloidogyne incognita* in brinjal crop. *Indian J. Nematol., 28,* 35–40.

Singh, R. S., & Sitaramaiah, K., (1973). *Res. Bull. No. 6* (pp. 1–289). G.B. Pant University of Agril. Sci. and Tech.

Singh, V. K., & Singh, K. P., (1998). Effect of granular nematicides on the biology of *Heterodera cajani* of pigeon pea. *Indian J. Nematol., 28,* 168–173.

Singh, V. K., & Singh, K. P., (2001). Effect of some medicinal plant leaves on the biology of *Heterodera cajani. Indian J. Nematol., 31,* 143–147.

Singh, V. K., & Singh, K. P., (2003). Penetration and development of *Heterodera cajani* on pigeon pea in relation to different types of soil. *Curr. Nematol., 14*(1/2) 17–24.

Singh, V. K., (2004). Management of *Heterodera cajani* on pigeon pea with nematicides and organic amendments. *Indian J. Nematol., 34,* 213, 214.

Singh, V. K., (2006). Management of root-knot nematode, *Meloidogyne incognita* infesting cauliflower. *Indian J. Nematol., 36,* 127, 128.

Singh, V. K., (2006). Management of root-knot nematode infesting tomato. *Indian J. Nematol., 36,* 126, 127.

Singh, V. K., (2008). Eco-friendly management of plant-parasitic nematodes in vegetable crops. In: Prasad, D., (ed.), *Insect Pest and Disease Management* (pp. 440–447). Daya Publishing House.

Singh, V. K., (2008). Effect of soil solarization for management of plant-parasitic nematodes. *Ann. Plant Prote. Sci., 16,* 541, 542.

Singh, V. K., Kalha, C. S., & Kaul, V., (2007). New record of root-knot nematode *Meloidogyne graminicola* infecting rice in Jammu. *Indian J. Nematol., 37*(1), 94.

Singh, V. K., Kalha, C. S., & Kaul, V., (2007). Occurrence of root-knot nematode *Meloidogyne incognita* on maize in Jammu. *Indian J. Nematol., 37*(1), 104.

Singh, V. K., Dwivedi, M. C., & Wali, P., (2011). Occurrence of wheat ear cockle nematode (seed gall nematode) *Anguina tritici* in Jammu. *Indian J. Nematol., 41*(1), 109, 110.

Singh, V. K., & Dwivedi, M. C., (2015). Occurrence of root-knot nematode *Meloidogyne incognita* infecting of wheat in J&K. *Indian J. Nematol., 45*(1), 119.

Singh, J., Gaur, H. S., Chahal, S. T., & Sharma, S. N., (1998). Control of Root-knot nematode in rice nursery by seed and soil treatment with nematicides and a neem production Int. *Symp. Afro Asian Soc. Nematologists.* Coimbatore.

Sitaramaiah, K., (1984). *Plant-Parasitic Nematodes of India.* Today and Tomorrow Publication, New-Delhi

Taylor, A., (1967). *Introduction to Research on Plant Nematology* (p. 133). FAO guide to the study and control of plant-parasitic nematodes.

Walia, R. K., (1994). Assessment of nursery treatment with *Pasteuria penetrans* for the control of *Meloidogyne javanica* on tomato in green house. *J. Biological Control, 8,* 68–70.

Walia, R. K., & Dalal, M. R., (1994). Efficacy of bacterial parasite, *Pasteuria penetrans* application as nursery soil treatment in controlling root knot nematode *Meloidogyne javanica* on tomato in green house. *Pest Management and Economic Zoology, 2,* 19–21.

Singh, G. (1992). Management of pod borer in chickpea. *Bhagwan Singh Award*, 38, 139, 172.

Singh, V. S. (1986). Host-density relationship and effect of plant extracts on the fecundity and egg viability in *Pseudococcus* [...]

Singh, V. S., Kapoor, K. N. [...]

Singh, Y. P. [...]

Singh, V. A., & Pawar, M. G. (1984). Occurrence of resistance in certain cultivars of chickpea [...]

Singh, Y., Saxena, S., Chahal, B. S., & Sharma, S. S. (1990). Sampling of the incidence [...] *Study* [...]

Index

C

Printed and bound by CPI Group (UK) Ltd, Croydon, CR0 4YY

23/10/2024

01777705-0009